T0100608

Self-Assembly of Nanostructures

Lecture Notes in Nanoscale Science and Technology

Volume 12

Series Editors:

For further volumes:
http://www.springer.com/series/7544

Stefano Bellucci
Editor

Self-Assembly of Nanostructures

The INFN Lectures, Vol. III

Editor
Stefano Bellucci
Istituto Nazionale di Fisica Nucleare
Laboratori Nazionali di Frascati
via E. Fermi 40
00044 Frascati RM
Italy
stefano.bellucci@lnf.infn.it

ISBN 978-1-4614-0741-6 e-ISBN 978-1-4614-0742-3
DOI 10.1007/978-1-4614-0742-3
Springer New York Dordrecht Heidelberg London

Library of Congress Control Number: 2011935363

Printed on acid-free paper

Springer is part of Springer Science+Business Media (www.springer.com)

To Gloria and the girls in my life

Preface

This is the third volume in a series of books on selected topics in *Nanoscale Science and Technology* based on lectures given at the well-known INFN schools of the same name. The aim of this collection is to provide a reference corpus of suitable, introductory material to relevant subfields, as they mature over time, by gathering the significantly expanded and edited versions of tutorial lectures, given over the years by internationally known experts.

The Nanotechnology group at INFN-LNF organizes since 2000 a series of international meetings in the area of nanotechnology. The conferences in 2008 and 2010 were devoted to recent developments in nanoscience and its manifold technological applications. They included a number of tutorial/keynote lectures, which are reflected in this volume, in addition to research talks presenting frontier nanoscience research developments and innovative nanotechnologies in the areas of biology, medicine, aerospace, optoelectronics, energy, materials and characterization, low-dimensional nanostructures and devices. Selected papers, based on conference talks and related discussions, were published in dedicated issues of international journals.

Special poster and equipment sessions were devoted to the exhibit by various firms of their institutional activities in selected areas of application where nanoscience can have a deep impact. There was also the possibility for sample testing by the participants. Tutorial lectures were delivered at the school, addressing general and basic questions about nanotechnology, such as what it is, how one goes about it, and what purposes it can serve. In tutorial sessions, the nature of nanotechnology, the instruments currently used in characterization, and the possible applications were described at an introductory level.

The conferences covered a large range of topics of current interest in nanoscience and nanotechnology, including aerospace, defense, national security, biology, medicine, and electronics. This broad focus is reflected in the decision to publish different areas of application of these technologies in different volumes. The present set of notes results, in particular, from the participation and dedication of prestigious lecturers. All lectures were subsequently carefully edited and reworked, taking into account the extensive follow-up discussions at the conferences.

A tutorial lecture by *Nunzio Motta* and collaborators (Queensland University of Technology, Brisbane, Australia) presents the analysis of the poly(3-hexylthiophene) self-assembly on carbon nanotubes and discusses how the interaction between the two materials forms a new hybrid nanostructure with potential application to future solar cells technology. After reviewing the properties of the self-assembly of polymers and biomolecules on nanotubes, the bulk characterization of polymer thin films with the inclusion of uniformly distributed carbon nanotubes is discussed. Using volume film analysis techniques (AFM, TEM, UV–Vis and Raman), the authors show how the polymer higher degree of order is a direct consequence of interaction with carbon nanotubes. The characterization of the polymer self-assembly on the nanotube surface is supported by molecular dynamic simulations, as well as a careful nanoscale analysis.

Fulvia Patella (Università di Roma Tor Vergata), building upon her tutorial lecture, together with collaborators from Università degli Studi di Modena e Reggio Emilia, as well as from Consiglio Nazionale delle Ricerche, Rome, Italy, reviewed quantum dots of III–V compounds, which are widely employed, at present, in conventional devices such as light-emitting diodes, photovoltaic cells, and quantum semiconductor lasers and, at the same time, offer appealing perspectives for more sophisticated applications in new generation devices such as single-photon emitters for nanophotonics and quantum computing. As the latter applications require the close control of both the size and position of the quantum dots on the substrate surface, the authors present atomic force and scanning tunneling microscopy, as well as reflection high-energy electron diffraction measurements, for discussing issues such as the formation and composition of the wetting layer, the evolution of the 2D–3D transition, the islands size distribution, and equilibrium shape, taking advantage also of theoretical ab initio studies of the In diffusion on the wetting layer, and simulations with the finite element method of the elastic energy relaxation of the island–substrate system.

Focusing on self-assembled quantum dots, the chapter by *Alexandr Toropov* and collaborators (Novosibirsk, Russia) provides a comprehensive review of some important aspects in the formation of quantum dots, which can play a role in fulfilling the promise of their prospective applications to new advanced devices. The use of molecular beam epitaxy for the structures growth, in conjunction with fundamental studies like the statistical approach to quantum dots formation, allows the authors to present the results of their extensive investigation of the features of droplet epitaxy, which is demonstrated as a good way to overcome growth problems, in comparison with conventional methods, such as the Stranski–Krastanov method.

The lecture by *Anna Sgarlata* and coworkers shows recent progresses toward a controlled growth of self-assembled nanostructures, dealing with the shaping, ordering, and localization in Ge/Si heteroepitaxy and reviewing recent results on the self-organization of Ge nanostructures at Si surfaces. The first part of the chapter deals with control in the positioning and the actual growth mode of epitaxial nanostructures, achieved by manipulating the intrinsic mechanisms of the SK process. In the second part, the authors analyze patterning techniques allowing harnessing of the natural self-organization dynamics of the system. A description of the use of STM nanolithography in studying the influence of elastic strain fields of defects in heterogeneous nucleation, as well as of the advantages of FIB patterning and nanoindentation patterning, wraps up this last chapter of the volume.

In concluding this effort, I wish to thank all lecturers, and especially those who contributed to this third volume in the series, as well as speakers and participants of the n&n 2008 and n&n 2010 Conferences, for having contributed to a pleasant and productive atmosphere, and fostering the creation of both a pervasive collaborative spirit and a pedagogical drive. I am confident that this third set of lectures, in turn, will provide an opportunity for those who are just now beginning to get involved with nanoscience and nanotechnology, allowing them to make contacts and obtain prime, up-to-date information from the experts. I also wish to acknowledge my wife Gloria and our wonderful daughters Costanza, Eleonora, Annalisa, Erica, and Maristella. Without their enduring dedication and caring support, I would have never been able to put this volume together.

Frascati, Italy Stefano Bellucci

References

1. Bellucci, S. (ed.): Proceedings of the School and Workshop on Nanotubes and Nanostructures 2000, Santa Margherita di Pula (Cagliari), Italy, 24 September–4 October 2000. Italian Physical Society, Bologna (2001). ISBN 88-7794-291-6
2. De Crescenzi, M., Bellucci, S. (eds.): J. Phys. Condens. Mat. **15**(34) (2003)
3. Bellucci, S. (ed.): J. Phys. Condens. Mat. **18**(33), S1967–S2238 (2006)
4. Bellucci, S. (ed.): J. Phys. Condens. Mat. **19**(39), 390301–395024 (2007)
5. Bellucci, S. (ed.): J. Phys. Condens. Mat. **20**(47), 470301, 474201–474214 (2008)
6. Bellucci, S. (ed.): Nanoparticles and Nanodevices in Biological Applications. The INFN Lectures – vol. I, vol. 4, pp. 1–198. Springer Verlag, Berlin (2009). ISBN: 978-3-540-70943-5
7. Bellucci, S. (ed.): Guest Editorial: Selected Papers from the INFN-LNF Conference on Nanoscience and Nanotechnology. Special Section of *J. Nanophoton.* **3**, 031999 (2009). doi:10.1117/1.3266501
8. Bellucci, S. (ed.): Physical Properties of Ceramic and Carbon Nanoscale Structures. The INFN Lectures – vol. II, vol. 11, pp. 1–197. Springer Verlag, Berlin (2011). ISBN: 978-3-642-15777-6

Contents

Polymer Self-assembly on Carbon Nanotubes

Michele Giulianini and Nunzio Motta

Abstract This chapter analyses the poly(3-hexylthiophene) self-assembly on carbon nanotubes and the interaction between the two materials forming a new hybrid nanostructure. The chapter starts with a review of the several studies investigating polymers and biomolecules self-assembled on nanotubes. Then conducting polymers and polythiophenes are briefly introduced. Accordingly, carbon nanotube structure and properties are reported in Sect. 3. The experimental section starts with the bulk characterisation of polymer thin films with the inclusion of uniformly distributed carbon nanotubes. By using volume film analysis techniques (AFM, TEM, UV–Vis and Raman), we show how the polymer's higher degree of order is a direct consequence of interaction with carbon nanotubes. Nevertheless, it is through the use of nanoscale analysis and molecular dynamic simulations that the self-assembly of the polymer on the nanotube surface can be clearly evidenced and characterised. In Sect. 6, the effect of the carbon templating structure on the P3HT organisation on the surface is investigated, showing the chirality-driven polymer assembly on the carbon nanotube surface. The interaction between P3HT and CNTs brings also to charge transfer, with the modification of physical properties for both species. In particular, the alteration of the polymer electronic properties and the modification of the nanotube mechanical structure are a direct consequence of the P3HT π–π stacking on the nanotube surface. Finally, some considerations based on molecular dynamics studies are reported in order to confirm and support the experimental results discussed.

M. Giulianini (✉) • N. Motta
Faculty of Built Environment and Engineering, School of Engineering Systems,
Queensland University of Technology, 2 George Street, CRICOS No. 00213J,
Brisbane, QLD 4000, Australia
e-mail: m.giulianini@qut.edu.au; n.motta@qut.edu.au

S. Bellucci (ed.), *Self-Assembly of Nanostructures: The INFN Lectures, Vol. III*,
Lecture Notes in Nanoscale Science and Technology 12,
DOI 10.1007/978-1-4614-0742-3_1,

Abbreviations

°C	Degree Celsius
AC	Alternating current
AFM	Atomic force microscopy
AR	Analytical reagent
cm	Centimetre
cm^{-1}	Wavenumber
CNT	Carbon nanotube
CuPc	Copper phthalocyanines
CVD	Chemical vapour deposition
DC	Direct current
DOS	Density of states
dTG	Differential thermal gravimetry
DWNT	Double-walled carbon nanotube
EQE	External quantum efficiency
FWHM	Full width half maximum
HOMO	Highest occupied molecular orbital
HOPG	Highly oriented pyrolytic graphite
HR-TEM	High-resolution transmission electron microscopy
I_{sc}	Short-circuit current
ITO	Indium tin oxide
LDOS	Local density of states
LUMO	Lowest unoccupied molecu lar orbital
MDMO-PPV	Poly(2-methoxy-5-(3′,7′-dimethyloctyloxy)-1,4-phenylenevinylene)
MEH-PPV	Poly[(2-methoxy-5-(2′-ethylhexoxy)-*p*-phenylene) vinylene]
mg	Milligram
MIM	Metal–insulator–metal
min	Minute
ml	Millilitre
mW	Milliwatt
MWNT	Multi-walled carbon nanotube
nm	Nanometre

NP	Nanoparticles
OPV	Organic photovoltaic
P3HT	Poly(3-hexylthiophene)
P3OT	Poly(3-octylthiophene)
PCE	Power conversion efficiency
PEDOT:PSS	Poly(3,4-ethylenedioxythiophene):poly(styrenesulfonate)
PmPV	Poly[(*m*-phenylene vinylene)-*co*-(2,5-dioctoxy-*p*-phenylenevinylene)]
PPV	Poly(phenylenevinylene)
rr-P3HT	Regioregular poly(3-hexylthiophene)
SCLC	Space charge limited current regime
SEM	Scanning electron microscopy
STM	Scanning tunnelling microscopy
STS	Scanning tunnelling spectroscopy
SWNT	Single-walled carbon nanotube
TD-SCLC	Trap-dominated space-charge-limited current regime
TEM	Transmission electron microscopy
TF-SCLC	Trap free space charge limited current regime
TG	Thermogravimetry
TGA	Thermogravimetric analysis
UHV-STM	Ultra-high vacuum scanning tunnelling microscopy
UHV-STS	Ultra-high vacuum scanning tunnelling spectroscopy
UV–Vis	Ultraviolet–visible
V	Volt
v	Volume
V_{OC}	Open circuit voltage
w	Weight

1 Introduction

Compounds of polymers and carbon nanotubes have been the object of intense research in the past years for their possible utilisation as hybrid functional systems in biology, photovoltaics [1–3], optics [4] and structural engineering [5, 6]. To fully understand the potential of such functional systems, the basic interaction between the polymer and the nanotubes must be investigated and explained.

Among all the aspects of two species interaction, the noncovalent polymer stacking (CH–π or π–π) on the nanotube surface forming highly ordered self-assembled structures is of great interest and has been investigated by several groups. In 2001, O'Connell et al. [7] proposed the polymer helical wrapping of nanotubes to explain the observed monolayer coverage of the nanotube surface. Although observed on polyvinyl-pyrrolidone (PVP), the authors considered the polymer coiling as a general phenomenon that could also be found in other polymers. Since 2001, nanotube wrapping has been proposed for enhanced solubilisation of nanotube interacting with natural biomolecules including amylose [8, 9], cyclodextrin [10–12], peptides [13], proteins [14] and DNA [15, 16]. In particular, Zheng et al. [15] proposed the helical wrapping of nanotubes through π–π interaction.

Helical wrapping of nanotubes has been studied also for conjugated polymers [17–20]. For example, by combining experimental and molecular dynamics results, Kang et al. [17] observed that the interaction of poly[*p*-[2,5-bis(3-propoxy-sulphonicacidsodiumsalt)]phenylene]ethynylene (PPES) with SWNTs gives origin to a self-assembled superstructure in which a polymer monolayer helically wraps the nanotube surface with a pitch length of approximately 13 nm. Conversely, several groups [20–23] investigated poly[(*m*-phenylenevinylene)-alt-(*p*-phenylenevinylene)] (PmPV) derivatives to study the wrapping selectivity of metallic and semiconducting nanotubes. For example, Yi et al. [19] found that PmPV selectively wraps (12,6) and (8,8) metallic SWNTs and (11,6) and (11,7) semiconducting SWNTs.

Several groups investigated the self-assembly of polymers on carbon nanotube surface with molecular dynamics (MD) simulations [17, 18, 24–27]. Naito et al. [18] explored how the polymer stiffness affects its wrapping capability. Basing their study on experimental results, they concluded that flexible and random-coiled polymers helped to debundle pristine SWNTs by helically wrapping around them. Conversely, semiflexible polymers hardly debundled the pristine SWNTs and preferentially formed thermodynamically stable liquid crystalline domains. Tallury and Pasquinelli [26] also report that flexible polymers are more likely to wrap nanotubes but helical conformations were not found. Conversely, Liu et al. [27] obtain the helical wrapping simulating the interaction between SWNTs and alginic acid in vacuum and in water. It has been found that a crucial role is played by the van der Waals attractive and hydrogen-bonding interactions.

Recently, Bernardi et al. [24] investigated the templating effect of the nanotube surface on the polymer conjugation length, whereas Caddeo et al. [25] showed how the SWNT chirality affects the final structure of the polymer.

Recent computational studies confirmed that the helical wrapping is a general phenomenon at the interface between nanotubes and chain-like polymers. It is related to the polymer stiffness and the nanotube curve surface [28, 29], so that the resultant helical pitch could be approximately determined by matching the radius of curvature of the helix with the average bending angle of the polymer, determined by its persistence length [30].

2 Conducting Polymers

2.1 Overview

Since their discovery [31, 32], conducting polymers have attracted the interest of scientists for their potential application in electronics. The main structural characteristic of conducting polymers, referred also as synthetic metals, is the π-conjugated system that extends along a number of recurrent monomers forming the polymer backbone as depicted in Fig. 1 [34]. This feature derives directly from the sp^2 hybridisation of the carbon atoms, where the σ bonds constitute the backbone of the polymer while the remaining p orbitals form the π-conjugated system that allows electrons to be transported along the polymer chain direction.

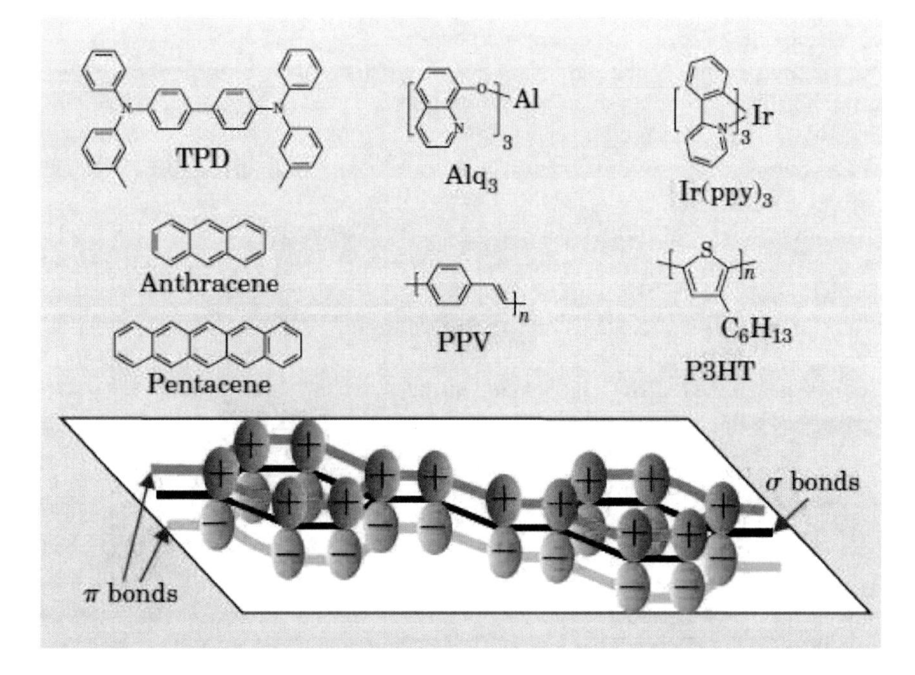

Fig. 1 Conducting polymer structures (Reproduced from [33])

The reported electrical conductivities of these systems reach 10^5 S cm^{-1} [35]. A list of the most-studied conductive polymers includes poly(acetylenes), poly (pyrroles), poly(thiophenes), poly(anilines), poly(*p*-phenylenes) and poly(*p*-phenylene-vinylenes).

Their easy processability and the possibility of tuning the electrical and optical properties for the specific application make conductive polymers ideal for electronic applications.

2.2 Electronic Structure of Conducting Polymers

All the macroscopic properties of a system can be deduced by knowing the system's wavefunction [36]. Nevertheless, it is common practice to introduce several approximations with various degrees of mathematical complexity to model the system and estimate the properties of molecular aggregates. Molecular orbital (MO) theories of wavefunction are based on MO wavefunction resulting from linear combination of atomic orbitals. If atomic orbitals are approximated by self-consistent wavefunctions, MO theories provide a reasonable agreement between theory and experiment. One of the simplest forms of MO theory is based on the division of electron population into two sets: those constituting the sigma bonds and those forming the π bonds [36]. When a periodic network of atoms is considered, the related energy levels are determined by the extent of the orbitals' overlap [37]. The creation of molecular orbitals is based on the expectation of the coupling between neighbouring bonding and anti-bonding orbitals in the molecule. In polymer chains, the interactions among molecular orbitals form quasi-continuous energy bands. The resulting energy bands are separated by energy states not allowed for electrons. At 0 K the band at lower energy, originating from the bonding orbitals (π), is fully occupied by electrons and is referred to as the highest occupied molecular orbital (HOMO). Conversely, the higher energy band, originating from the anti-bonding orbitals (π^*), is empty of electrons and is referred to as the lowest unoccupied molecular orbital (LUMO). At room temperature, according to the Fermi distribution function, a few electrons are promoted to the energy states in the LUMO band leaving the upper states of the HOMO band empty. As a matter of fact, bonding and anti-bonding σ orbitals form energy bands too, but the energy separation between π orbitals is smaller (because the overlap is weaker) and so consequently it has a predominant effect on the resulting electronic properties. When a system possesses alternating single and double bonds and p interacting orbitals, it is referred as a π-conjugated system and the energy band diagram is similar to the one shown in Fig. 2a.

As a result, electric current can flow through the conducting polymer and, as introduced in the previous paragraph, the π-conjugated system can exhibit a good electric conductivity with high charge mobilities.

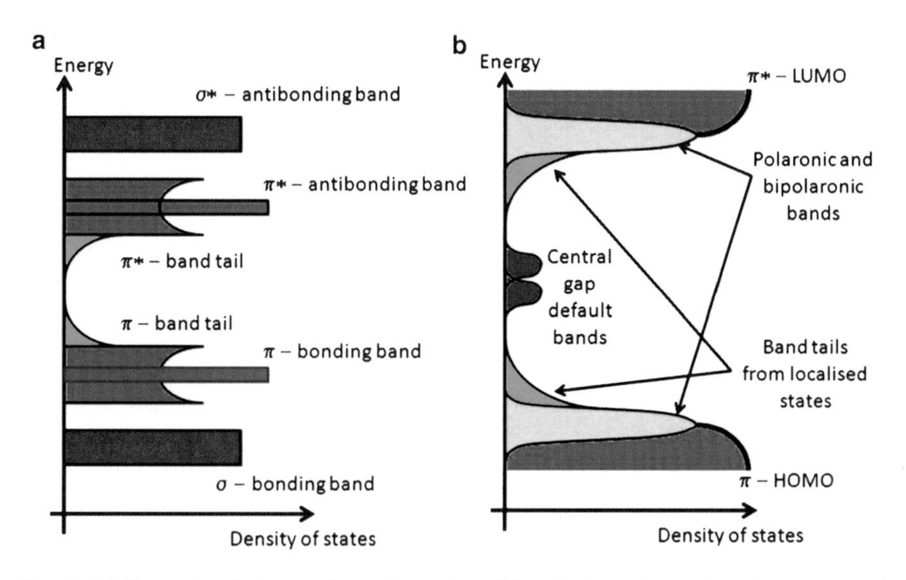

Fig. 2 (**a**) Energy bands in π-conjugated organic solids. (**b**) Energy bands in real π-conjugated organic solids (redrawn and edited from [37])

2.2.1 "Real" Organic Solids

The band diagram of Fig. 2a takes into account intrachain and interchain interactions and distortions. The overall effect is the presence of energy band tails of allowed states in the forbidden region between the HOMO and the LUMO levels that remove the abrupt discontinuity of energy level expected for isolated polymer chains. As depicted in Fig. 2b, allowed energy states and small bands can be present in the case of charge carrier chain interactions and defects of the polymer chains. These can be structural defects introduced, for example, during the chemical synthesis or caused by an excess of charge. Extra charges localised on the polymer chains can be due to photo-excitation, chemical doping (charge transfer from a donor or acceptor type atom or molecule) and charge injection. The electronic structure of the π-conjugated system is altered by the presence of defects, affecting the electric and optical properties of the conducting polymer. Referring to single polymer chains, the conjugation length represents the number of recurrent monomers with no alteration of the π-conjugated system due to a structural defect. The longer the conjugation length, the higher the grade of crystallinity and purity of the resultant polymer that approximates an ideal π-conjugated system.

2.2.2 Solitons

Concerning charge-induced defects, let us consider the polymer chain in the ground energy state. The two geometric configurations of atoms obtained by interchanging the position of double bonds can be degenerate or nondegenerate, depending on

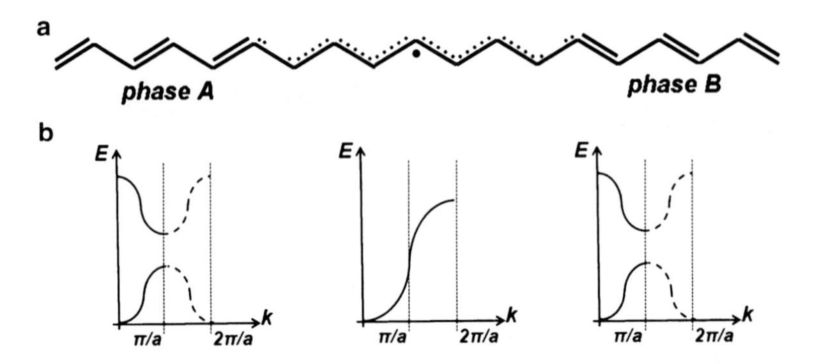

Fig. 3 (**a**) The transition region between two bond alternating phases A and B referred to as soliton. (**b**) Dispersion relation referred to the three different zones of (**a**)

whether they possess the same or different energy relative to the ground state. Figure 3a shows the backbone of polyacetylene, belonging to the class of degenerate polymers. Indeed, the reported interchange of double bonds in the chain causes no alteration of the ground state energies of the two different chain segments. The transition region between the two energetically equal segments is called "soliton." Since solitons cause no alteration to the energy state of the polymer backbone, they can travel (as a solitary wave) without distortion and/or dissipation of energy. It is estimated that this type of distortion typically forms every 2,000 atoms. As depicted in Fig. 3a, the soliton is characterised by the absence of single and double bonds and almost equal bond lengths. From an energetic point of view, it is equivalent to a dangling bond since, as shown in Fig. 3b, the energy dispersion relation is continuous at the border of the first Brillouin zone and, therefore, this energy state can accommodate either one or two electrons.

When the energy state is occupied by one electron, it is neutral (no charge carried) and carries 1/2 spin. In the other two cases, zero or two electrons present, the spin is zero and the state is charged +1 or −1, respectively.

2.2.3 Polarons and Bipolarons

In general, when a charge is introduced into an ordered ionic crystal lattice, a Coulomb interaction is expected to affect the relative position of the surrounding ions. The displacement of ion cores alters the potential of the potential wells in which the electron can be found. The lattice deformation is referred as "dielectric polaron" and is expected to alter locally the energy band structure of the affected atoms. The same phenomenon can be observed in a covalent material, where "molecular polarons" produce chain distortions and charge trapping. In degenerate ground-state polymers, polarons are responsible for the exchange of single and double bonds resulting in a higher energy ground state. As an example, the stable benzoic form of poly(*p*-phenylene) is shown in Fig. 4a. When a neutral default (or defect) occurs, the exchange of single and double bonds brings to the unstable

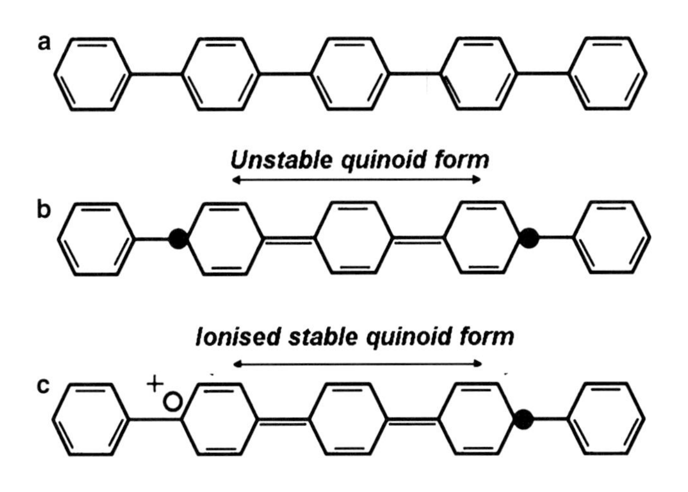

Fig. 4 Poly(*p*-phenylene): (**a**) stable benzoic form, (**b**) unstable quinoid form and (**c**) stable quinoid form due to polaron formation

quinoid form as depicted in Fig. 4b. If the default becomes charged, for example, by a charge transfer as shown in Fig. 4c, the quinoid form is ionised and stable and the polymer chain ground-state energy is higher than in the benzoic form. Polarons can be considered quasi-particles that correspond to charge–lattice interactions that self-trap within the lattice deformations they cause.

Typically, polarons extend over five benzene rings and the two defaults created interact strongly, resulting in a bonding and anti-bonding interaction. Consequently, symmetric and anti-symmetric energy states are formed and depending on the amount of charge involved, the energy state results single or double charged. The single-charged state carries a spin and is referred as polaron, whereas the double charged is called bipolaron and carries no spin.

Figures 5a–d show the possible charge/spin cases for polarons and bipolarons related to the charge–lattice distortion couplings.

2.3 *Optoelectronics in Molecules and Polymers*

As introduced before, the electronic structure of π-conjugated molecules and polymers is characterised by the bonding and anti-bonding energy bands separated by an energy gap. Following an external excitation, a molecule can pass from its fundamental energy state to an excited state. This condition is related to the transition of electrons between filled and empty states in the bonding and anti-bonding molecular orbitals ($\pi \rightarrow \pi^*$), with the consequent alteration of the charge population in the affected bands. Conversely, an excited molecule can decay to the fundamental state when electrons in higher energy states relax to lower energies ($\pi^* \rightarrow \pi$).

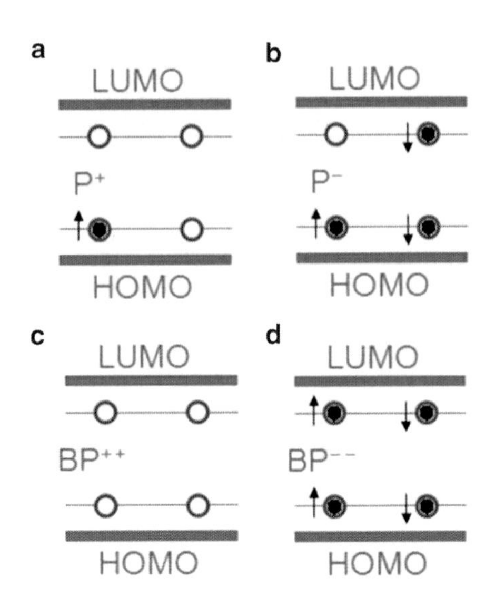

Fig. 5 Polaron and bipolaron states: (**a**) positive polaron, (**b**) negative polaron, (**c**) positive bipolaron and (**d**) negative bipolaron

In its excited state, the molecule is affected by a change of forces and equilibrium distances between neighbouring atoms [37]. Since the electronic mass is considerably lower than the nuclear mass in the molecule, transitions of electrons can be considered instantaneous (10^{-15} s) when compared to time intervals required for atoms to vibrate (10^{-13} s). Therefore, electron transitions are treated as vertical (Frank–Condon principle), with no change in the momentum and positions of atoms involved. Referring to the light–matter interaction to excite the molecule, absorption and emission spectra depend exclusively on transition selection rules related to vibronic levels due to electron–phonon coupling. The allowed dipole transitions will occur between states for which the wavefunctions exhibit opposed symmetry [36]. Moreover, the transition intensities will be proportional to the Frank–Condon factor which represents the square of the overlap integral between the vibrational wavefunctions of the start and final transition states [38, 39].

2.3.1 Excitons

In organic solids, many, but not all, optical properties of aggregates of molecules and molecular solids can be related to individual molecule excitation and decay spectra. Indeed, single molecules constituting small aggregates or crystals can be subject to collective excitations, which are a communal response of the aggregate to the exciting field. An exciton is a quasi-particle introduced to embody the collective response of the organic solid to an excitation. Molecular exciton models have provided good explanation of similarities and differences between optical properties of individual molecules and their correspondent crystalline solids [36].

The absorption of a photon can promote an electron from the lowest unoccupied molecular level (LUMO) to the highest occupied molecular level (HOMO).

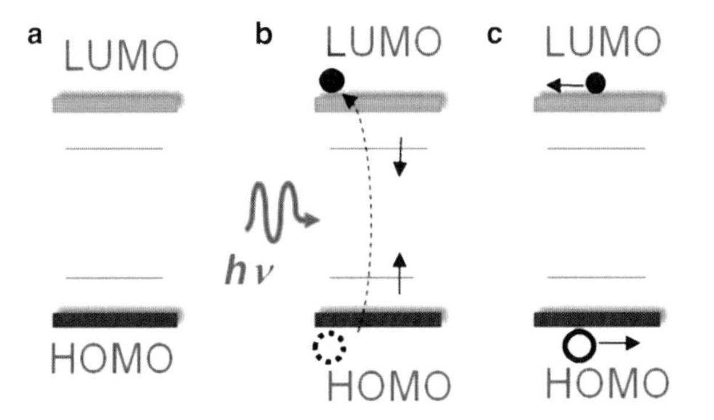

Fig. 6 (**a**) Exciton states reported with respect to the HOMO and LUMO energy levels. (**b**) Exciton formation after interaction with a photon. (**c**) Free charge carriers transported inside the energy bands

The electron–hole pair generated in this way interacts through Coulombic forces and therefore it is subject to a mutual attraction in the same physical space [37]. This pair, usually called "exciton," is viewed as a "quasi-particle" and can travel across the molecular solid if its lifetime is long enough. Depending on the type of organic solid considered, the exciton can extend across one/few molecules (Frenkel exciton) or over several molecules (Mott–Wannier exciton). Mott–Wannier excitons are characteristic of wide-energy-band semiconductors with large dielectric constants, where electrons and holes exhibit high mobilities. Generally, they are associated with inorganic semiconductors and can be easily separated. In fact, due to the large dielectric constant, the radius of the exciton is large and the interacting forces are weak. Conversely, Frenkel excitons are characteristic of insulating materials, with low mobilities and low dielectric constants that confer to the quasi-particle a small radius and strong interaction forces. Frenkel excitons are commonly associated with organic semiconductors and their lifetime is normally short (few nanoseconds) and hence the diffusion length is limited to few tens of nanometres [40]. Excitons are represented as energy states close to the HOMO and LUMO bands as shown in Fig. 6a. After the interaction with a photon of energy $h\nu$, the exciton is created as in Fig. 6b, but the charges are still interacting and the transport inside the HOMO and the LUMO is not possible. If the separation is completed, the charge carriers reach the respective energy bands and create/contribute to the current flow as shown in Fig. 6b.

Exhibiting small dimensions and strong bonds, Frenkel excitons can travel through the molecular crystal over hundreds of molecules of the same type. In fact, while the electron–hole pair is confined in the same lattice site, the exciton moves by transferring energy to adjacent molecules of the same type forming the molecular lattice [37].

If adjacent molecules of different types are considered, the mechanism of energy transfer can be through dipolar interactions (long transfer – Förster) or by the overlapping of orbitals (short transfer – Dexter). Figure 7a shows Förster and Dexter transfer mechanisms from an excited donor molecule (D*) to another

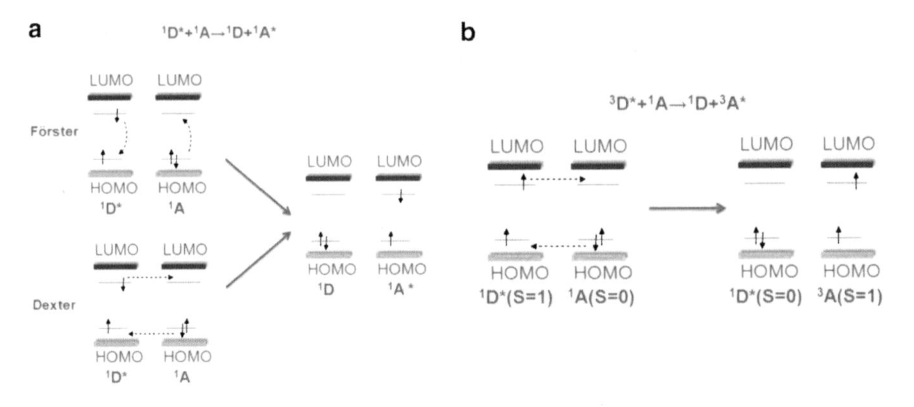

Fig. 7 (**a**) Exciton transfer between singlet states following Förster mechanism or Dexter mechanism. (**b**) Exciton transfer between singlet and triplet states following Dexter mechanism ($S = 0$ denotes a singlet state, $S = 1$ denotes as triplet state)

molecule acting as acceptor (A*). The transfer occurs only between singlet states and can be written in the form $^1D^* + {}^1A \rightarrow {}^1D + {}^1A^*$. Förster mechanism is characterised by the conservation of spin of the single molecules involved and is based on dipole–dipole interactions that are not negligible up to 10 nm (long transfer). Conversely, in the Dexter transfer, only the total spin of the system (D*A or DA*) is conserved and the mechanism does not rely upon allowed transition probabilities in donor–acceptor molecules [37]. Therefore, as depicted in Fig. 7a, the same transfer of energy between singlet states can occur leading to the same final state observed for Förster mechanism. Moreover, as illustrated in Fig. 7b, the Dexter mechanism can allow energy transfer between the donor triplet to the acceptor following the equation $^3D^* + {}^1A \rightarrow {}^1D + {}^3A^*$. In this case, the spin is conserved by the system, but is changed for single molecules. The probability of Dexter transfer is related to the surface overlap between molecular orbitals of donor and acceptor molecules. Therefore, it is a short-range transfer (up to 2 nm) and decays exponentially with distance.

2.4 *Polythiophenes*

The thiophene molecule, shown in Fig. 8a, is a hetero-cyclic, aromatic compound characterised by a flat five-membered ring identified by the formula C_4H_4S. Thiophene monomers linked through the positions 2–5 form the backbone of polythiophene polymers. In general, polythiophenes are easier to handle than other conductive polymers such as polypyrroles, since they are less sensitive to oxygen [35]. Moreover, the simple preparation of substituted monomers (for example, with alkyl groups) brings better solubility and higher electrical conductivities. Polythiophenes are synthesised with three methods: chemical oxidation, coupled with organometallic agents and electrochemical oxidation.

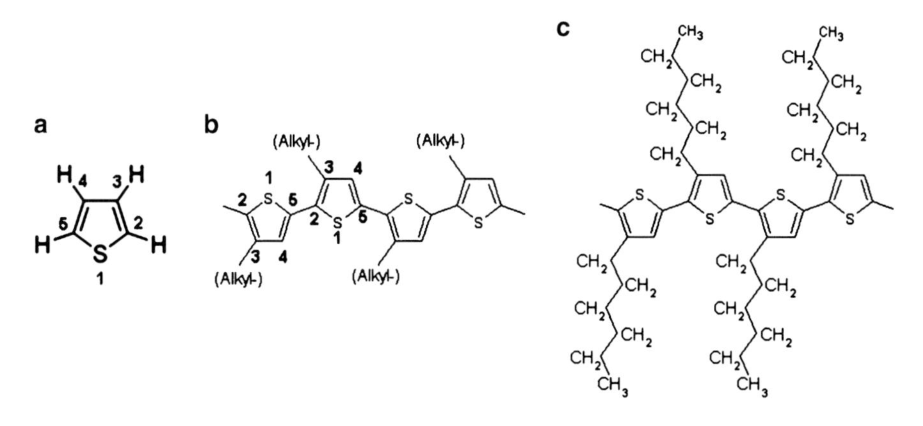

Fig. 8 (**a**) The structure of the thiophene monomer, (**b**) poly(3-alkylthiophenes), and (**c**) poly(3-hexilthiophene)

2.4.1 Poly(3-hexylthiophene)

Among polythiophenes, the class of poly(3-alkylthiophenes) or (P3ATs) is one of the most studied because of their good electrical conductivity and good processability. In particular, the instantaneous solubility in organic solvents (chloroform, chlorobenzene, dichlorobenzene, xylene, toluene, etc. [34]) has brought poly(3-alkylthiophenes) to be considered excellent candidates for electronic applications. They are characterised by the univalent radical (alkyl-) linked to the thiophene monomer in the position 3 as shown in Fig. 8b. The alkyl group can reach up to 18 atoms of carbon, and electrical conductivities up to 100 S cm^{-1} have been reported [35].

Poly-(3-hexylthiophene)'s alkyl group is characterised by six carbon atoms (hexyl-) and 13 hydrogen atoms as illustrated in Fig. 8c. Poly-(3-hexylthiophene) or P3HT is synthesised in two forms, regioregular (rr-) and regiorandom (rra-): the difference is in the alternating positions of the alkyl chains that can be respectively regular or not regular, as depicted in Fig. 9.

The grade of order (i.e., crystallinity) of P3HT directly affects its electrical and optical properties. In its crystalline form, the P3HT unit cell appears as depicted in Fig. 10 with parameters $a = 16.8$ Å, $b = 3.8$ Å and $c = 7.7$ Å [42]. Nevertheless, to explain the results of different analysis techniques, a model with two crystalline phases and a liquid crystalline mesophase has been proposed [42, 43].

The controversy on the structural properties adds further complication when considering a model for the charge transport. It has been highlighted in the previous discussion that chain distortions can be induced by charge removal, generating stable quinoid forms. Moreover, the role played by the heteroatom on the carriers transport has been questioned by some authors who consider it negligible [44], while others suggest an interaction of its p orbital with the π-conjugated system [45]. The result is that the charge transport can be related to polarons [46, 47] or conversely to bipolarons [44]. Remarkable differences have also been reported for

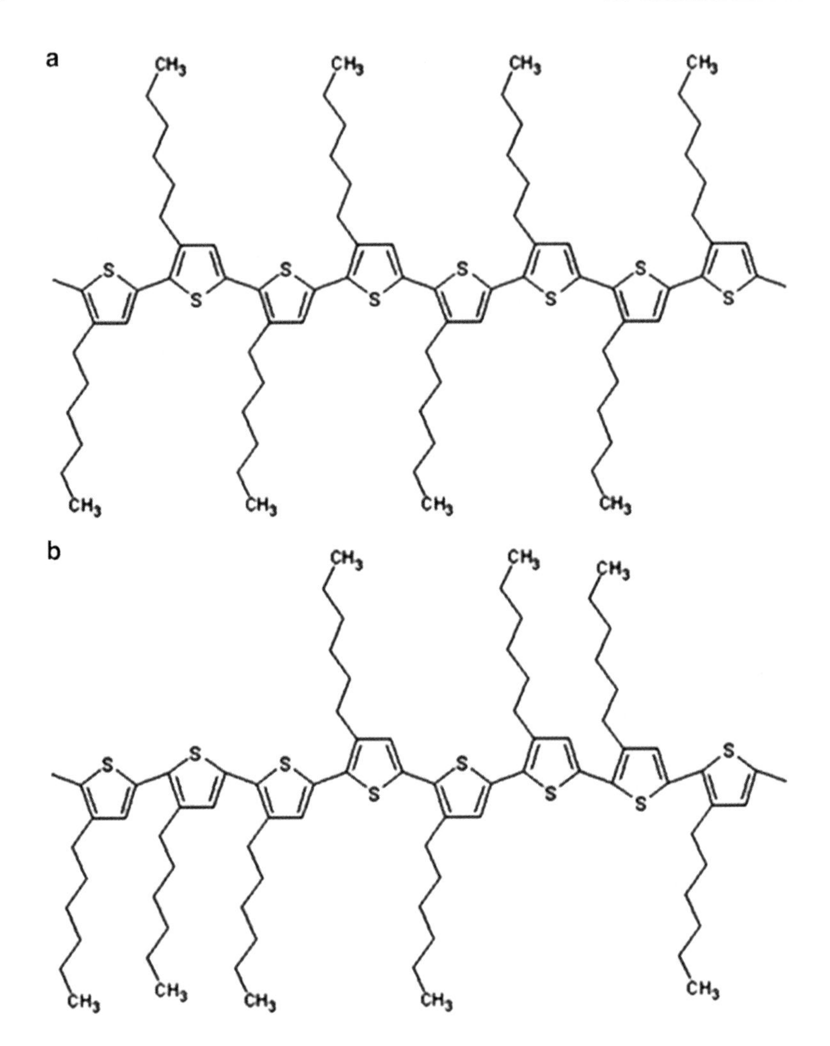

Fig. 9 Poly(3-hexilthiophene) in the form (**a**) regioregular and (**b**) regiorandom

charge carrier mobilities that can vary over several orders of magnitude passing from 0.1 $cm^2\ V^{-1}\ s^{-1}$ [48] to 10^{-4}–10^{-7} $cm^2\ V^{-1}\ s^{-1}$ [49, 50]. These differences have been attributed to the effects of different solvents [51], molecular weight [52], preparation conditions [34], etc. It has also been reported that the stretching of the deposited polymer film can result in a significant enhancement of electrical conductivity [53].

Although P3ATs are less sensitive to oxygen than other polymers, it has been reported that oxygen can diffuse rapidly into polythiophenes, altering the optical and transport properties [54]. In particular for P3HT, a charge transfer between the polymer and molecular oxygen has been reported [55]. Even if an enhancement of the electric conductivity is observed due to the increase of the charge carriers,

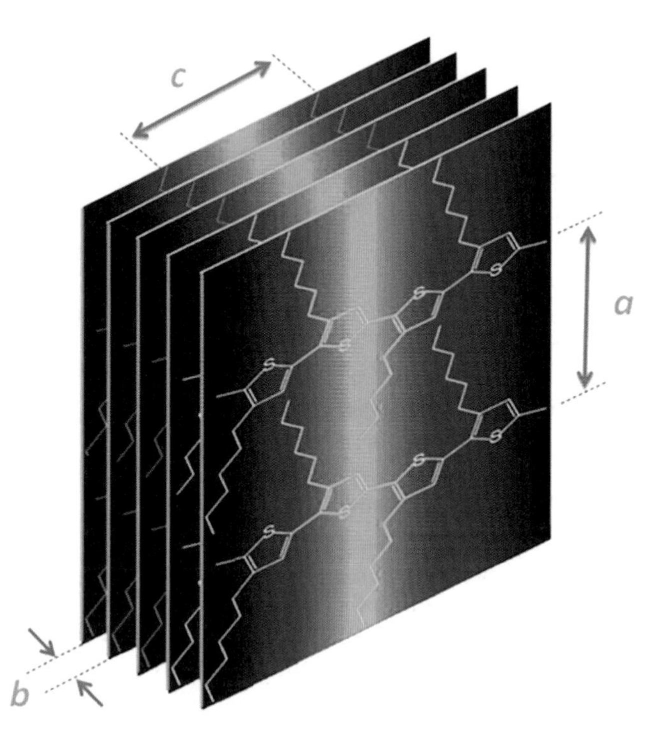

Fig. 10 P3HT in its crystalline form (adapted from [41])

the decrease of the field-effect mobility lowers the performance of the resulting device. In general, the performances of π-conjugated polymer-based devices are strongly lowered by the effects that oxygen and moisture have on the active media and on the contacts. For this reason, encapsulation techniques have been explored to reduce the contamination [56].

3 Carbon Nanotubes

3.1 Overview

Since their discovery by Iijima in 1991 [57], carbon nanotubes (CNTs) have attracted the interest of researchers because of their extraordinary properties. The small diameter (fraction of a nanometre) compared to the extended length (up to millimetres) confers to these structures a high aspect ratio identifying CNTs as de facto one-dimensional structures. The extraordinary physical and mechanical

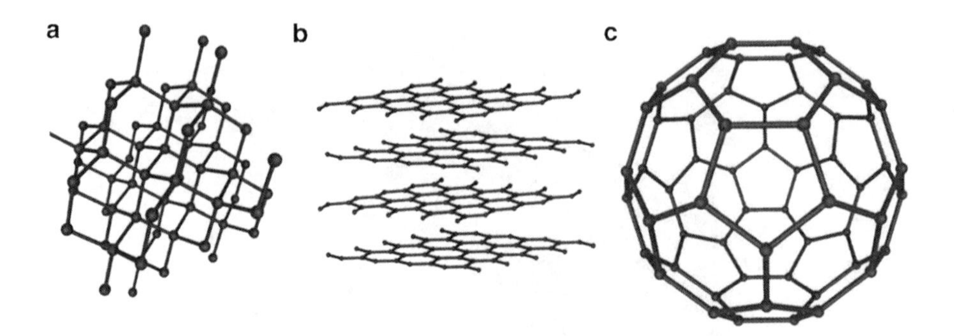

Fig. 11 Carbon allotropes: (**a**) diamond, (**b**) graphite, and (**c**) buckminsterfullerene (adapted from [63])

properties identify CNTs as superior to many other commonly used materials, making them the ideal candidate for their replacement in a large number of applications. Among the most remarkable properties, it is interesting to report the very high tensile strength and Young modulus [58, 59] that make this material tougher than any other natural or synthetic organic fibre. At the same time, carbon nanotubes exhibit high flexibility [60], high thermal conductivity [61] and low density [62].

Carbon nanotubes are one of the solid-phase carbon-different allotropic forms, which show diverse, and sometimes opposite, electrical, mechanical and physical properties. Some examples are shown in Fig. 11. In the diamond form, each sp^3 hybridised carbon atom is bonded to other four, in a perfect tetrahedral arrangement. Diamond is an extraordinarily hard material, an excellent heat conductor and an electrical insulator. Conversely, in the graphite form, the sp^2 atoms are organised in a planar layer and every atom is bonded to three others placed at 120°. The structure of each plane (referred as *graphene*) is like a honeycomb, with the electrons able to move on the plane through the unhybridised π orbital that is delocalised across the entire layer.

One of the most interesting forms of pure carbon is Buckminsterfullerenes (or simply fullerenes), with spheroidal or cylindrical molecules formed by sp^2 hybridised carbon atoms organised in hexagons and pentagons. In this category are included carbon nanotubes that can be considered as an ordered network of carbon atoms arranged in hexagons to form a seamless hollow cylinder as illustrated in Fig. 12.

In the field of electrical and electronic properties, carbon nanotubes reveal incredible features. They can exhibit a metallic behaviour with high conductivity (10^4 S cm^{-1}) and high current density (10^6 A cm^{-2}) [64]. CNTs can also be semiconducting with band gaps ranging from 0 to 1 eV [65]. This peculiar behaviour makes them suitable as replacement of semiconductors in electronic applications, especially when the need of miniaturisation brings the technology to the nanoscale.

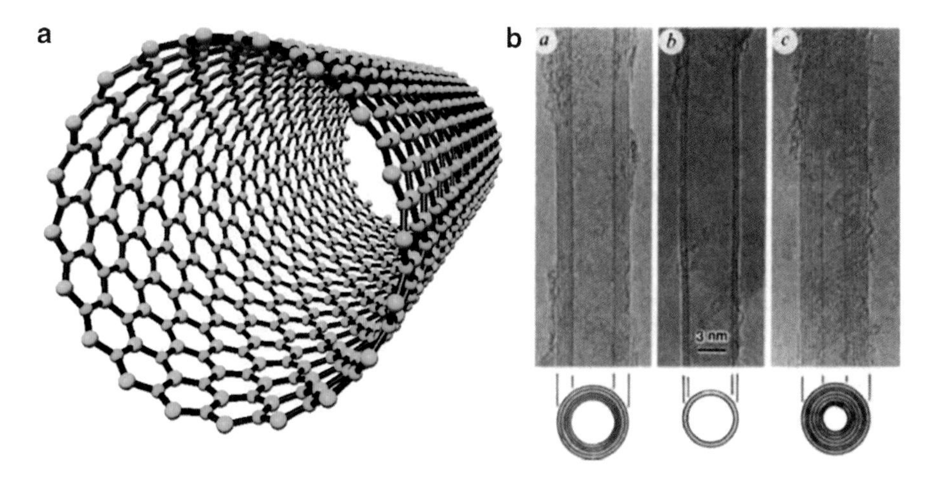

Fig. 12 (**a**) Carbon nanotube. (**b**) Carbon nanotubes as originally reported by Iijima in 1991 (from [57])

Due to their extraordinary properties, utilisation of CNTs has been proposed in several applications that include material reinforcing [5, 6], gas sensing [66–68], field emission emitters [69, 70], nanomechanics [71], atomic force microscopy tips [72], membranes [73], etc.

3.2 *Synthesis*

Carbon nanotubes are generally produced by using three main techniques: arc discharge, laser ablation and chemical vapour deposition. In arc-discharge technique, two carbon electrodes are placed at approximately 1 mm distance in controlled inert atmosphere (helium, argon) and, through a controlled power supply, a direct current of 50–120 A is driven [74]. The high-temperature discharge (3,000–4,000°C) vaporises one of the electrodes and on the other one a rod-like deposit is formed. The deposit usually contains soot, fullerenes and MWNTs. The use of transition-metallic catalysts can bring to the growth of SWNTs [75, 76]. By tailoring the arc-discharge process, the final production can be varied in terms of SWNTs' diameter and content [77, 78]. High-volume production (grams) of SWNTs can be obtained by laser ablation technique [79]. In this method, a target of cobalt/nickel and graphite is placed into a quartz tube furnace at 1,200°C with an inert atmosphere helium and argon at a pressure of 500 torr. The target is vaporised with a pulsed laser and the hot vapour plume formed expands and cools rapidly. Nanometre-size metal catalyst particles are formed in the plume and assist the growth of SWNTs. At the cold finger terminal inside the furnace, SWNTs assemble in clusters by interaction via van der Waals forces. The yields can vary from 20% to 80% of the weight that also contains graphite, fullerenes, catalysts, etc. A heat post-treatment can eliminate by-products by sublimation (fullerenes) to achieve 70% SWNTs content.

Compared to other techniques, chemical vapour deposition technique (CVD) [80] can be considered a medium temperature method since the hot zone of the growth never reaches 1,200°C. It is achieved by inserting carbon in gas phase (methane, carbon monoxide or acetylene) into a furnace containing a heated-up substrate covered with a catalyst (nickel, iron or cobalt). An energy source (plasma or resistively heated coil) transfers energy to the gaseous carbon molecules to "break" them in reactive carbon atoms that diffuse to the substrate and reach the metallic catalysts where they bind. CVD technique allows one to control growth alignment [81] and nanoscale position [82]. Typically, CVD temperature ranges from 600 to 900°C and yields to a 30% CNT content. The parameters that can be varied in the procedure (pressure, gases flow, temperatures, catalysts, etc.) affect the properties of the resulting nanotubes [83]. In the last years, many different CVD techniques have been developed such as plasma-enhanced CVD, thermal chemical CVD, alcohol catalytic CVD, vapour phase growth, aero gel-supported CVD and laser-assisted CVD.

3.3 Geometric Structure

Carbon nanotubes can be found as single individual cylinders, commonly referred as a single-wall nanotubes (SWNT), or as coaxial cylindrical structures bonded by van der Waals forces called multi-wall nanotubes (MWNTs). A special category is constituted by the double-walled nanotubes (DWNTs) which exhibit peculiar properties. A pictorial representation of single-, double- and multi-wall nanotubes is shown in Fig. 13.

The classification of the fundamental characteristics of a SWNT starts from the assumption that a carbon nanotube can be considered as a layer of *graphene* rolled up on itself to form a cylinder.

Figure 14a depicts a portion of the graphene layer; for our purpose, only the atoms included between the two dashed lines have to be taken into account. The translational vector **T** indicates the direction on which the nanotube extends. The chiral vector **C** represents the direction of the graphene layer roll and it is a linear combination of the unit vectors $\mathbf{a}_1$ and $\mathbf{a}_2$ of the hexagonal lattice:

$$\mathbf{C} = n\mathbf{a}_1 + m\mathbf{a}_2. \quad (1)$$

The indices (n,m) identify uniquely the nanotube, defining its chirality. To build an (n,m) nanotube a graphene sheet is rolled along the **C** direction, until the coincidence of the (0,0) with (n,m) atom is reached. Figure 14a shows how to build a (12,4) nanotube. The chiral angle, defined as the angle formed by the chiral vector

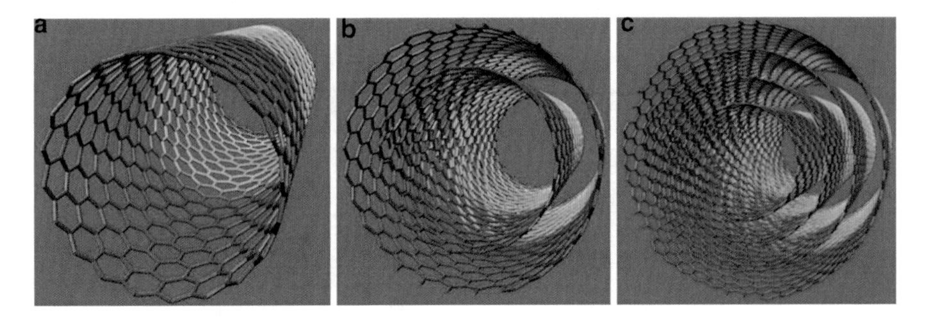

Fig. 13 (**a**) Single-wall carbon nanotube. (**b**) Double-wall carbon nanotube. (**c**) Multi-wall carbon nanotube

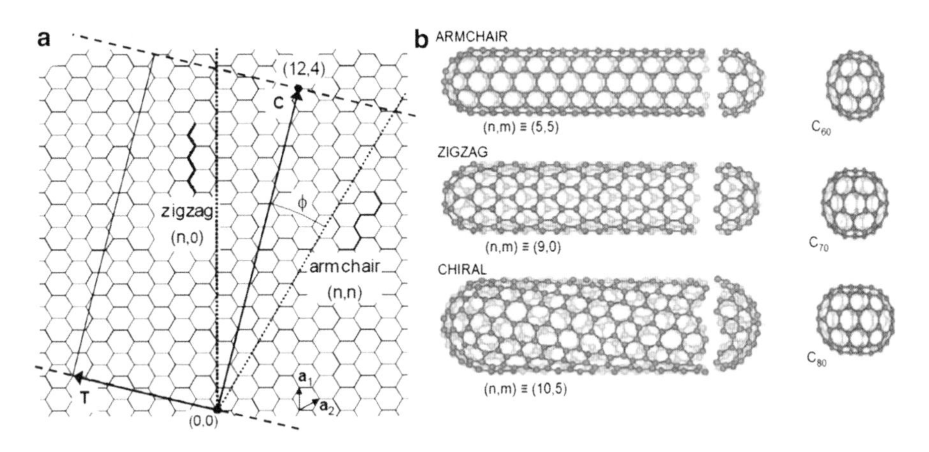

Fig. 14 (**a**) Schematic representation of graphene layer that can be folded along vector **C**. The unit vectors $\mathbf{a}_1$ and $\mathbf{a}_2$ identify the chirality indices. The vector **T** is the translational vector. (**b**) Examples of the three types of nanotube chirality armchair, zigzag and chiral and the buckminsterfullerene structures from where they can be generated (from [84])

C and the direction of $\mathbf{a}_1 + \mathbf{a}_2$, is in the range $0 \leq \varphi \leq 30$ and is defined by the following equation:

$$\varphi = \arccos\left[\sqrt{3}(n+m)\Big/\left(2\sqrt{n^2+m^2+nm}\right)\right]. \tag{2}$$

According to the direction of **C**, and hence to the indices (n,m), three different types of carbon nanotubes can be defined, as shown in Fig. 14b. When the indices (n,m) coincide, the structure is an armchair nanotube ($\varphi = 0°$); if $m = 0$ the structure is a zigzag nanotube ($\varphi = 30°$); in all the other cases, it is a chiral nanotube. It is possible to link the generating buckminsterfullerene structure to each type of carbon nanotube, as depicted in Fig. 14b. The diameter of the nanotube can be calculated from the indices (n,m) by using the following equation:

$$d = \frac{a}{\pi}\sqrt{n^2+m^2+nm}, \tag{3}$$

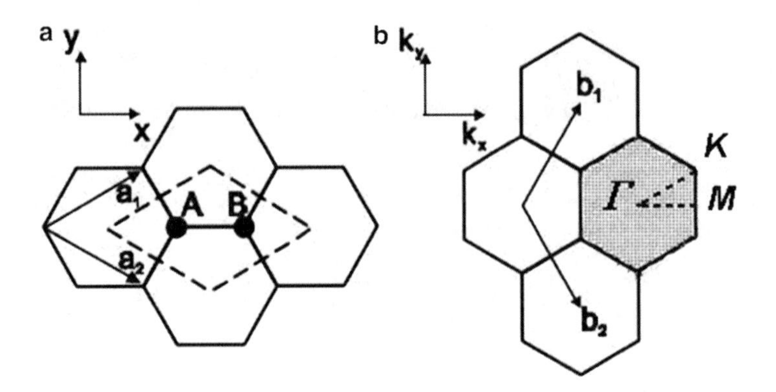

Fig. 15 (**a**) Section of graphene layer in the real space. The *dashed line* identifies the unit cell containing the two inequivalent atoms at sites A and B. (**b**) The first Brillouin zone (*dotted hexagon*) in the reciprocal space with the three high symmetry points **Γ**, **K** and **M**

where $a = 0.246$ nm is the lattice constant of graphene. The chirality is an important parameter of the carbon nanotube since it directly affects the electronic properties of the resultant structure, as described in the following section.

3.4 *Electronic Structure*

In order to easily derive the electron dispersion relations for single-wall carbon nanotubes, one can start from the electronic structure of the graphene. In this section, only the π covalent bonds perpendicular to the graphene layers are considered since the π energy bands are the most important to determine the solid-state properties of graphene and therefore of the carbon nanotubes. Figure 15 shows the unit cell for a graphene layer in the real space (a) and in the reciprocal space (b). In the real space, the dashed line marks the unit cell containing the **A** and **B** nonequivalent atoms. The first Brillouin zone is the dotted hexagon in the reciprocal space containing the high symmetry point **Γ**, **K** and **M**.

The unit vectors $\mathbf{a}_1$ and $\mathbf{a}_2$ (introduced before) for the real space and $\mathbf{b}_1$ and $\mathbf{b}_2$ for the reciprocal space can be expressed in the x, y plane and in the k_x, k_y plane, respectively, using the following relations:

$$\mathbf{a}_1 = \left(\frac{\sqrt{3}a}{2}, \frac{a}{2}\right);\ \mathbf{a}_2 = \left(\frac{\sqrt{3}a}{2}, -\frac{a}{2}\right);\ \mathbf{b}_1 = \left(\frac{2\pi}{\sqrt{3}a}, \frac{2\pi}{a}\right);\ \mathbf{b}_2 = \left(\frac{2\pi}{\sqrt{3}a}, -\frac{2\pi}{a}\right). \tag{4}$$

In the Brillouin zone, the energy dispersion relations can be calculated along the directions **Γ–K**, **Γ–M**, **K–M**. Using the tight-binding method and considering

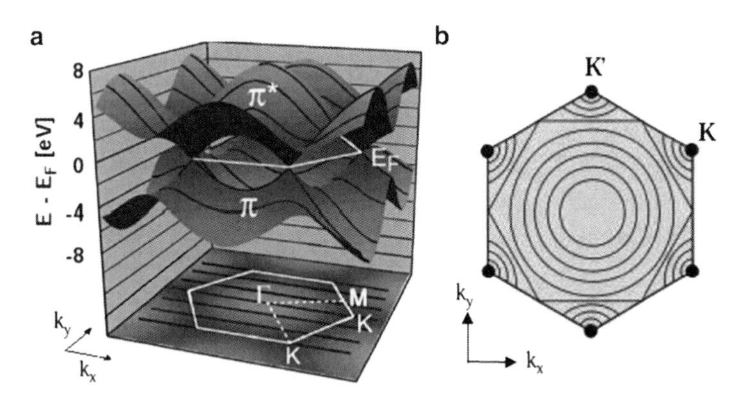

Fig. 16 (**a**) Three-dimensional representation of the energy dispersion for graphene in the first Brillouin zone. The bonding (π) and the anti-bonding (π*) orbitals meet at six points identified by **K**. (**b**) The projection in two dimension of the dispersion equation (from [85])

the overlap integral matrix diagonal [65], the energy dispersion for the electrons in the π and π* bands, referred to as bonding and anti-bonding orbitals, can be approximated by the following equation:

$$E_{\mathrm{g2D}}\left(k_x, k_y\right) = \pm\gamma_0\left[1 + 4\cos\left(\frac{\sqrt{3}k_x a}{2}\right)\cos\left(\frac{k_y a}{2}\right) + 4\cos^2\left(\frac{k_y a}{2}\right)\right]^{1/2}, \tag{5}$$

where γ_0 is the energy overlap integral between nearest neighbours. The energy dispersion relation is represented in Fig. 16 where it is possible to observe that the π and π* bands are degenerate at the **K** points which are crossed by the Fermi energy line. Since the density of energy states at the Fermi level is zero, graphene can be considered a zero-gap semiconductor. The zero gap at the **K** points results from the fact that atom sites **A** and **B** are identical.

To extend the energy dispersion relation obtained for graphene to single-wall carbon nanotubes, it is enough to apply the boundary conditions to the roll-up of the layer, which results in the quantisation of the wave vector associated with vector **C**. The wave vector k_C must satisfy the following expression:

$$k_C \cdot \mathbf{C} = 2\pi q \ \ (q = 0,\ 1,\ 2, \ldots). \tag{6}$$

For an infinite length nanotube, the wave vector associated with the translational vector **T** remains continuous. In the reciprocal space, these two conditions are represented by a discrete number of lines overlapped to the Brillouin zone, as depicted in Fig. 17. This representation is equivalent to the section of the graph of Fig. 16a along vertical planes.

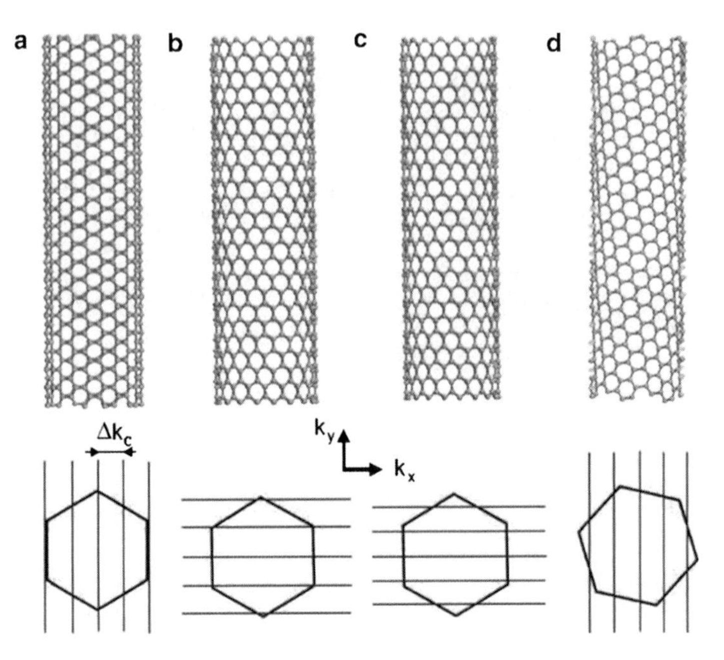

Fig. 17 Four different single-wall nanotubes reported with the respective first Brillouin zone to highlight the diverse electronic properties. In the cases (**a**) and (**b**) the parallel lines cross the **K** corners of the Brillouin zone and therefore the nanotube is metallic. When this condition is not matched, cases (**c**) and (**d**) the nanotube is semiconducting. The comparison of zigzag nanotubes (**b**) and (**c**) highlights the incidence of the diameter on the resulting electronic properties [83]

The interline spacing Δk_C is given by the following equation that relates it to the nanotube diameter d:

$$\Delta k_C = \frac{2\pi}{|\mathbf{C}|} = \frac{2}{d}. \tag{7}$$

In the case of an armchair nanotube, the quantised condition for the wave vector is

$$n\sqrt{3}k_x a = 2\pi q \ (q = 0,\ 1,\ 2, \ldots,\ n) \tag{8}$$

and the lines are vertical (along k_y). The lines always cross the **K** points of the first Brillouin zone and therefore the nanotube exhibits a zero energy gap and a metallic behaviour. For a zigzag nanotube, the lines are horizontal (along k_x) since the condition to be matched is

$$nk_y a = 2\pi q \ (q = 0,\ 1,\ 2, \ldots,\ n). \tag{9}$$

In this case, only certain values of the diameter bring to a crossing of the **K** point and it is possible to find energy states not allowed for electrons. Referring to Fig. 17, the nanotube labelled with (b) is metallic, while the one labelled with (c) is semiconducting. For chiral nanotubes, either metallic or semiconducting nanotubes can be encountered. It is possible to demonstrate that the condition for any kind nanotube to be metallic is that the difference between the indices n and m is multiple of 3 as reported in the following expression:

$$n - m = 3i \quad (i = 0, 1, 2, \ldots). \tag{10}$$

This identifies all the armchair nanotubes (n,n) as metallic and one-third of the zigzag nanotubes $(n,0)$ metallic when n is a multiple of 3.

The dispersion equation can be adapted for armchair and zigzag nanotubes as reported in the following:

$$E_{\text{armchair}}(k_y) = \pm\gamma_0 \left[1 \pm 4\cos\left(\frac{\pi q}{n}\right)\cos\left(\frac{k_y a}{2}\right) + 4\cos^2\left(\frac{k_y a}{2}\right)\right]^{1/2}, \tag{11}$$

$$E_{\text{zigzag}}(k_x) = \pm\gamma_0 \left[1 \pm 4\cos\left(\frac{\sqrt{3}k_x a}{2}\right)\cos\left(\frac{\pi q}{n}\right) + 4\cos^2\left(\frac{\pi q}{n}\right)\right]^{1/2}. \tag{12}$$

These expressions calculated for the nanotubes (5,5) and (10,0) are shown in Fig. 18 along with the respective density of states (DOS). The points where the DOS diverges are referred as Van Hove singularities. For the (5,5) nanotube, a nonzero density of state is observed at the Fermi energy, highlighting the metallic behaviour. The energy difference between the first two Van Hove singularities is referred as ΔE_{sub}. For the (10,0) nanotube, the DOS is zero at the Fermi level and no allowed states between the first two Van Hove singularities leading to the creation of the energy gap ΔE_{gap}.

The energy differences ΔE_{sub} and ΔE_{gap} can be calculated starting from the energy dispersion of a graphene sheet near the Fermi points (with $E_F = 0$) [86]:

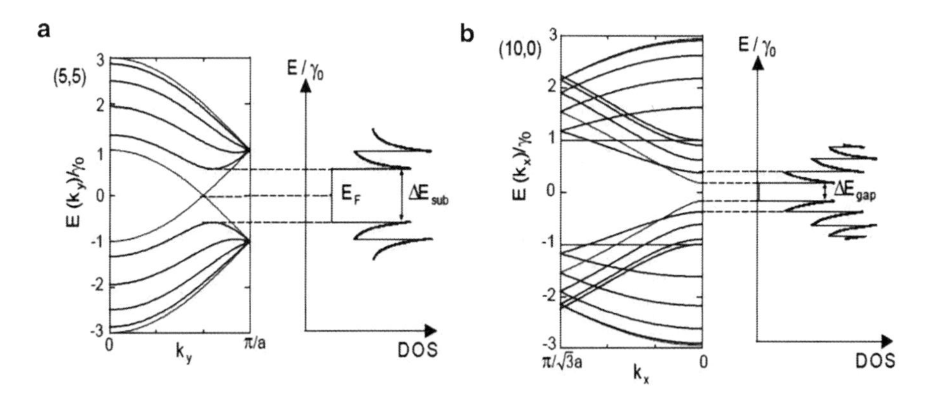

Fig. 18 The energy dispersion diagram and density of states calculated for a (5,5) and a (10,0) single-wall carbon nanotube. For the metallic nanotube (5,5), the energy difference between the first Van Hove singularities has been indicated with ΔE_{sub}. In the case of the semiconducting tube (10,0) the energy gap has been indicated with ΔE_{gap} (from [84])

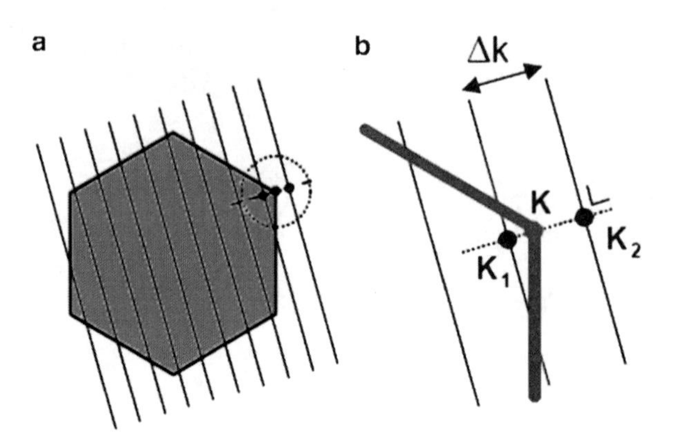

Fig. 19 (**a**) Region of the reciprocal space near the Fermi level for a generic chiral nanotube. (**b**) The detailed region with points **K** on the hexagon and $\mathbf{K_1}$ and $\mathbf{K_2}$ on the k-lines (from [85])

$$E = \frac{\sqrt{3}a}{2}\gamma_0|\mathbf{k} - \mathbf{k_F}| = \frac{3d_{\text{nn}}}{2}\gamma_0|\mathbf{k} - \mathbf{k_F}| \tag{13}$$

with $d_{\text{nn}} = 0.142$ nm being the distance between nearest neighbour carbon atoms and $|\mathbf{k} - \mathbf{k_F}|$ the distance of point **K** to $\mathbf{K_1}$ and $\mathbf{K_2}$ along the allowed k-lines as reported in Fig. 19.

For metallic nanotubes, the k-lines overlap the **K** points on the hexagon; therefore, the two closest points in neighbouring lines are exactly at a distance Δk away. Alternatively, for semiconducting nanotubes, the closest distance is $\Delta k/3$ [87]. As reported in (7), $\Delta k = \Delta k_C$ is $2/d$,with being d the diameter of the nanotube. Therefore, the energy of Van Hove singularities in each band can be calculated from (13):

$$\begin{aligned} E_1 &= \pm\frac{3\gamma_0 d_{\text{nn}}}{2}\Delta k = \pm\frac{3\gamma_0 d_{\text{nn}}}{2}\left(\frac{2}{d}\right) \quad \text{(metallic)}; \\ E_1 &= \pm\frac{3\gamma_0 d_{\text{nn}}}{2}\frac{\Delta k}{3} = \pm\frac{3\gamma_0 d_{\text{nn}}}{2}\left(\frac{2}{3d}\right) \quad \text{(semiconducting)}. \end{aligned} \tag{14}$$

Finally, the values of ΔE_{sub} and ΔE_{gap} can be calculated as the difference between the two energy levels as reported in the following:

$$\Delta E_{\text{sub}} = \frac{6\gamma_0 d_{\text{nn}}}{d}; \;\; \Delta E_{\text{gap}} = \frac{2\gamma_0 d_{\text{nn}}}{d}. \tag{15}$$

The equations reported can be verified for a large number of metallic and semiconducting nanotubes with a good agreement between theory and experimental results [88, 89]. As a consequence, CNT's behaviour is expected to become "more metallic" with the increase of the diameter since the energy gap of semiconducting tubes is reduced. This phenomenon can be seen as weakening of the one-dimensional nature and the appearance of two-dimensional effects.

The effect of the curvature of nanotubes complicates the calculations, but for both armchair and zigzag tubes no interaction or band splitting would be expected on Fermi energy level [65].

4 Polymer/Nanotube Composites

Carbon nanotubes can be considered as fullerenes extended in one direction to form a tube of several hundreds of nanometres length. For their large surface area and high electron mobility [90], the use of carbon nanotubes (CNTs) acting as electron acceptor has been proposed for bulk heterojunction solar cells. The first devices originally proposed by Kymakis and Amaratunga [3] were based on blends of P3OT and carbon nanotubes and reported efficiencies 0.04%. By improving the annealing process, PCEs of 0.22% have been reached by the same group [91]. Miller et al. made other attempts, with water-soluble polymers poly(2-(3-thienyl)-ethoxy-4-butylsulphonate) (PTEBS) [92] and PmPV [93] reaching efficiencies of 0.56% and 0.46%, respectively. The best result achieved for CNT-based devices is PCE = 1.48%, reached by Patyk et al. [94] with devices based on P3OT and polybithiophene (PBT).

Interestingly, carbon nanotube solar cells show high open circuit voltage [2, 95]. The value of V_{OC} in the range 0.65–0.75 reported by Kymakis et al. [2] was weakly related to the cathode metal work function, violating the MIM model prediction. The open circuit voltage was independent of the light intensity, being in good agreement with the energy difference between the P3OT LUMO and the CNT Fermi level. More recently, Geng et al. [95] measured V_{OC} of approximately 1.6 V and proposed the interpretation of a Fermi energy shift occurring in carbon nanotubes upon charge doping.

One of the main problems affecting CNT-based devices is the variation of properties amongst the nanotubes in the sample considered. As explained in Sect. 3.4, carbon nanotube electronic properties depend strongly on the way in which the graphene sheet is rolled (i.e., their chirality) and, as a consequence, a nanotube can be conducting or semiconducting, according to its radius and chirality. For this reason, the evaluation of the real potential of this material for photovoltaic applications has not been completed. In a recent theoretical study, Kanai and Grossman compared the charge transfer efficiency form P3HT to PCBM [96] and to CNTs [97]. This study demonstrates that while for PCBM the transfer probability and the exciton separation can be high, an efficient charge transfer has been found only for semiconducting nanotubes, whereas the metallic ones exhibit a very low charge transfer and exciton separation probability.

Nevertheless, carbon nanotubes have found their application in photovoltaic devices in part because they are easily processed and also because of their high conductivity. For instance, several authors investigated the employment of CNTs as transparent electrodes. Since highly conducting and transparent CNT films can be easily deposited on thin plastic substrates, they can be used to replace ITO deposited on glass, providing the basis for cheap, flexible and lightweight photovoltaic devices. Du Pasquier et al. [98] showed that even with sheet resistances of $R_s = 282\ \Omega/\square$ (ohm/square) and a transmittance of $T = 45\%$, it is possible to enhance the power conversion efficiency of P3HT/PCBM by replacing ITO with CNTs. Rowell et al. [99] developed films of $T = 85\%$ and $R_s = 200\ \Omega/\square$ reaching efficiencies of 2.5% close to 3% of ITO devices. Van de Lagemaat et al. [100] proved the holes' selectiveness of nanotube layer by proposing a device that can be at the same time ITO and PEDOT:PSS free, reaching a power conversion efficiency of 0.47%.

In many applications, carbon nanotubes have also been mixed with the P3HT/PCBM blend to lower the series resistance and enhance the conductivity of the active layer [101–103]. Berson et al. [101] reported efficiencies up to 2%, highlighting the interpenetration of CNTs in the active blend as an effective way to transport holes to the anode contacts. The same hole-enhancement transport has been pointed out by Pradhan et al. [102] that also involved the improvement of exciton separation due to polymer–nanotube interface. Wu et al. [103] reported the highest power conversion efficiency for a device based on a P3HT/PCBM/CNT blend reaching 3.47% from a value of 2.69% of nanotubes free compounds. The reduction of charge recombination due to a more effective hole transport was considered the reason for the better performance. Although the device architecture and the materials employed are the same, Kymakis et al. [104] attributed the improvement of performance, from 1% to 1.4%, to enhancement of electron transport related to the nanotube presence. Even if the transport mechanisms are controversial, in all cases, the device's performances were improved by the inclusion of CNTs in the active media.

5 CNT–P3HT Bulk Characterisation

5.1 Microscopic Analysis

5.1.1 Atomic Force Microscopy

Atomic force microscopy (AFM) has often been used to image carbon nanotubes interacting with molecules [16] and polymers [21, 105–107]. Figure 20a shows an example of AFM imaging of a P3HT/MWNTs compound film. High-aspect-ratio rod-like structures can be easily recognised on the film surface and buried inside the layer, corresponding to the MWNTs introduced in the polymer matrix.

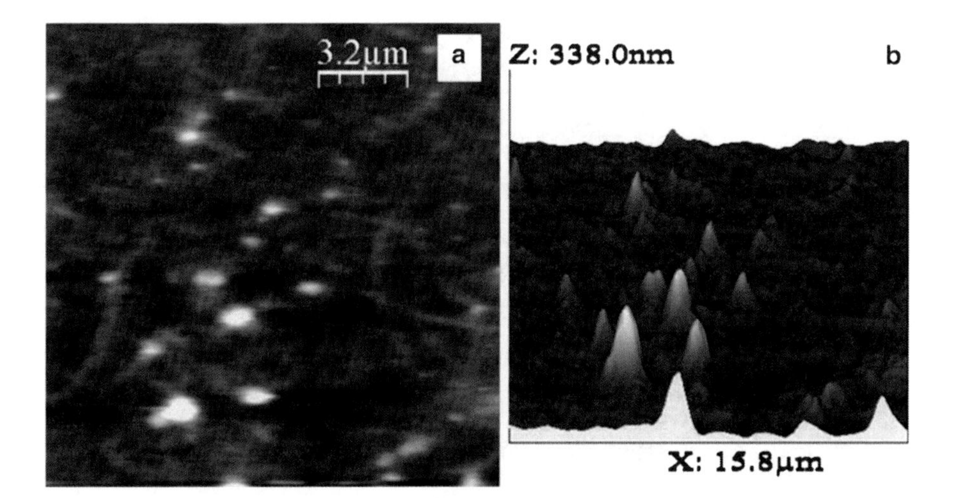

Fig. 20 AFM image of P3HT film with MWNTs included. (**a**) Two-dimensional (2D) image showing the nanotubes buried in the polymer matrix. (**b**) Three-dimensional (3D) corresponding image highlighting the nanotubes protruding from the P3HT layer (From Ref. [108])

Polymer functionalisation of MWNTs is likely to occur in the liquid phase when ultrasonically dispersed nanotubes are mixed with the polymer solution. By comparing Fig. 20a with the corresponding 3D image shown in Fig. 20b, a P3HT wrapping of the nanotube is quite evident. The 3D image allows one to identify the bright spots in the 2D image with nanotubes protruding out from the film plane.

On the other hand, the frequent occurrence of MWNTs parallel to the film's surface and emerging from the polymer layer form edge structures with a rod-like profile is a confirmation that the P3HT coverage takes place entirely along each nanotube. The diameters of these rod-like structures can be estimated in the range of 200–300 nm with lengths that range from 1 to 6 μm. Unfortunately, it is difficult to extract the diameter of the nanotube inside, as the thickness of the P3HT layer wrapping the single tube cannot be measured by this technique.

Atomic force microscopy images presented in Fig. 21a, b have been obtained on thinner films and at higher magnification. Since the film is composed only by P3HT and carbon nanotubes, the presence of tubular structures with variable diameter showing a variable height and features (i.e., peaks and valleys) of the order of hundreds of nanometres can be associated with polymer-covered nanotubes. It is easier to individuate the polymer wrapping in the corresponding 3D images of the film (Fig. 21c, d).

It can be concluded that P3HT has a strong tendency to fully cover the nanotube surface as previously reported for other polymers by O'Connell et al. [7]. Although it is clear from these images that the polymer envelops the nanotubes, smaller-scale imaging techniques are required to investigate the details of P3HT structures on the nanotube surface.

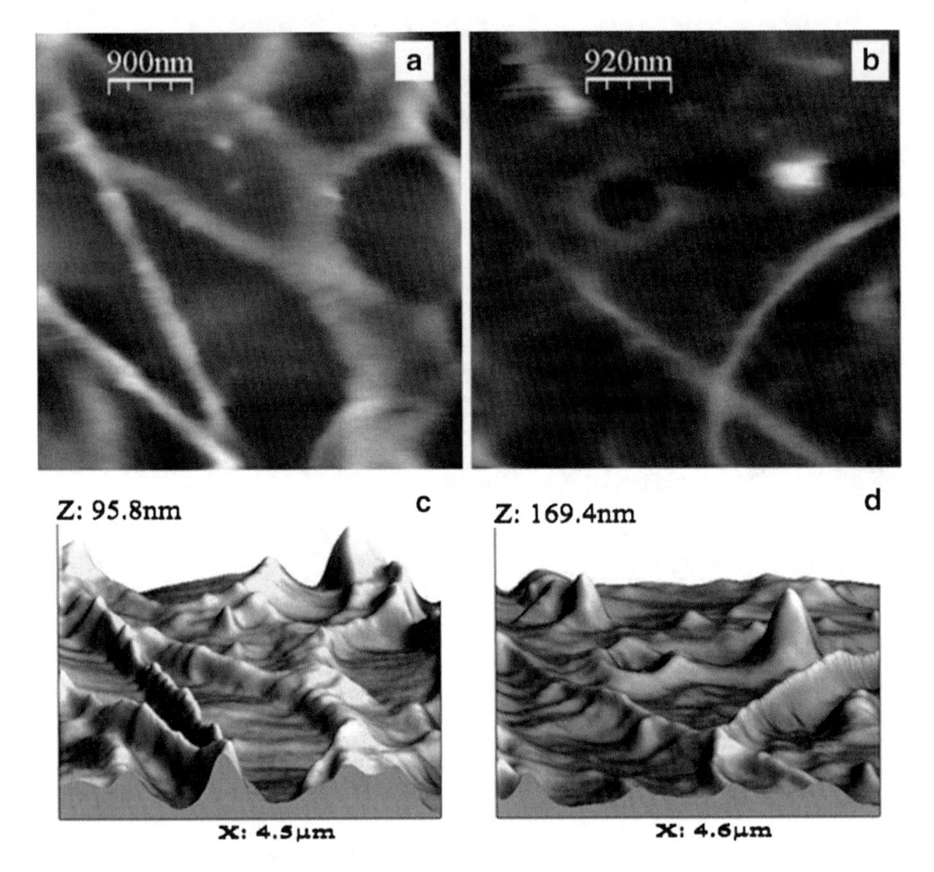

Fig. 21 (**a, b**) P3HT/MWNTs thin film AFM image of included. (**c, d**) Three-dimensional (3D) corresponding images showing polymer coverage of the carbon nanotubes along the main axis (From Ref. [108])

5.1.2 Transmission Electron Microscopy

Freestanding nanotube–polymer films were prepared in the following way: blends composed of MWNT variable loads have been deposited on glass by spin coating. Then, the samples were immersed in deionised water in order to obtain film detachment from the substrate. Successively, the film floating on the water surface was harvested with 300 mesh copper grid for TEM analysis in order to image nanotubes included in the polymer layer. Transmission electron microscopy (TEM) images reported in Fig. 22a, b show carbon nanotubes embedded in the polymer matrix. Darker zones of contrast correspond to nanotube segments near the surface or outside of the polymer film. Even if the nanotubes are completely embedded inside the polymer layer, TEM imaging allows the individuation of the entire

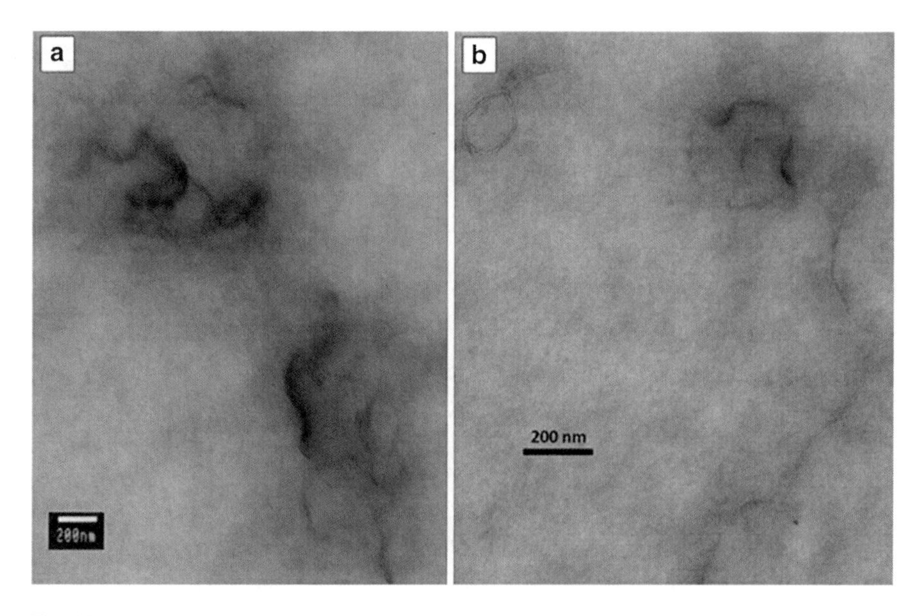

Fig. 22 TEM images of P3HT thin layer with MWNTs included. Both images show darker zones where the nanotube is coming out form the P3HT surface

nanotube structures. It is therefore possible to observe nanotubes possessing sharp bends, kinks, buckles and structural defects.

On images at higher magnification, it is possible to estimate with good precision the nanotube dimensions, as shown in Fig. 23a, b.

The diameter of the nanotube in Fig. 23a spans from 10 to 20 nm and the length is estimated to be 600 nm. On the nanotube surface (magnified in the inset), it is possible to observe darker structures, which could be tentatively assigned to adsorbed P3HT. However, the image resolution is too low to allow further considerations in this regard. The same considerations can be made for the nanotube in Fig. 23b. In this case, the diameter in the section indicted is 28 nm and the approximate length is 3.1 μm. The inset of Fig. 23b shows regular structures on the nanotube surface, which can be identified as polymer, although the resolution is insufficient to determine precisely the P3HT assembly and to clarify the role played by the nanotube structure, on the P3HT organised adsorption.

From the previous microscopy analysis (AFM and TEM), it is possible to conclude the presence of an interaction between P3HT and the MWNTs, which creates a layer of polymer wrapped around the tube. Unfortunately, the limits imposed by the resolution of the microscopes used in this experiment do not allow further conclusion on the kind of interaction and on the relationship between the nanotube surface structure and the P3HT ordering. To clarify this point, high-resolution TEM and scanning tunnelling microscopy analysis must be undertaken. The results of these studies are presented in Sect. 6.

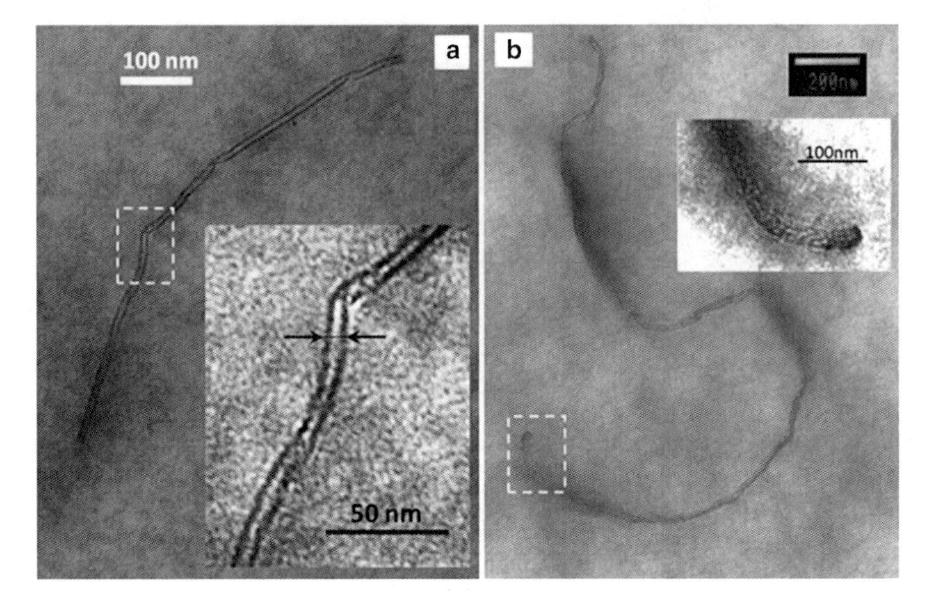

Fig. 23 Carbon nanotubes in P3HT films imaged by TEM. From the images it is possible to determine the diameter and the lengths of the nanotubes. Moreover, the *insets* show evidence of polymer structures adhered on the nanotube's surfaces (From Ref. [108])

5.2 Spectroscopy Analysis

Spectroscopy analysis can provide important information regarding the suggested polymer adhesion on carbon nanotube surface. This particular investigation can be carried out by analysing the variation of the P3HT spectrum resulting from the introduction of increasing fractions of nanotubes in the composites. The spectroscopy analysis can be usefully extended to smaller-diameter nanotubes such as double-wall carbon nanotubes (DWNTs). As the microscopy analysis reported in the previous section was restricted to large-diameter multi-wall nanotubes due to resolution limits, here the analysis is principally focussed on P3HT/DWNT composites.

5.2.1 Ultraviolet–Visible Light Absorption Spectroscopy

Ultraviolet–visible light absorption spectroscopy was used to investigate the interaction between the polymer and carbon nanotubes. Normalised UV–Vis spectra of P3HT–DWNTs compounds are shown in Fig. 24. All the relevant contributions have been determined from the analysis of the second derivative curves and have been reported in Table 1, with their position extracted.

The pristine P3HT spectrum is characterised by three contributions indicated in Fig. 24 with letter "A," "B" and "C." The maximum occurring at 512 nm (2.42 eV)

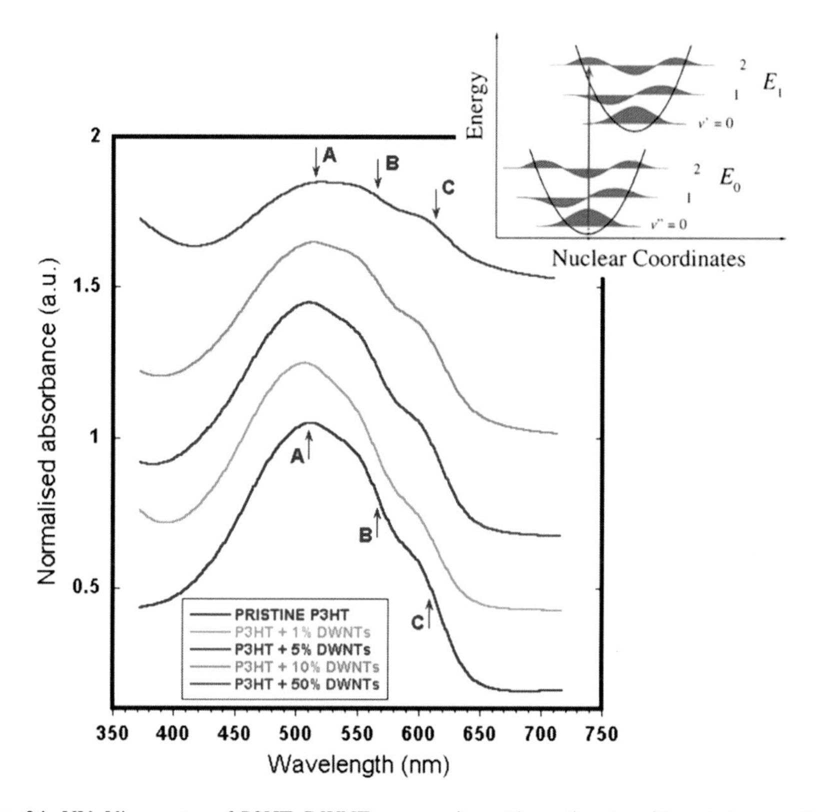

Fig. 24 UV–Vis spectra of P3HT–DWNTs composites. Absorption transitions between vibrational states (*inset*) [38]

can be identified with the transition between the vibrational levels 0 and 2, as shown in the inset [109]. Another contribution, namely B, is found for P3HT at 566 nm (2.19 eV). Trznadel et al. [109] associated this contribution to the transition 0–1 and found that its position is solvent dependent, spanning from 545 to 559 nm. It must be noticed that Brown et al. attributed this feature at 2.19 eV to the 0–0 transition [110]. Finally, the contribution labelled with C occurring at 615 nm (2.02 eV) has been attributed by Trznadel et al. to the 0–0 transition, whereas Brown et al. associated this position with P3HT interchain interaction. The value of 2.02 eV can be considered the optical band gap of the polymer.

The first consideration that can be derived from the analysis of the spectra is the increase of the relative intensities of contributions B and C with respect to A corresponding to the increase of DWNTs content. Moreover, the subgap absorption increases with the DWNT content, since carbon nanotubes show a nearly constant absorbance within the range of analysis. From the data reported in Table 1, a general red shift of the spectrum is observed as the content of nanotubes is raised. This trend is interpreted as being associated with an increase of the conjugation length of the polymer [109, 111] and hence to an overall increased order.

Table 1 UV–Vis spectra peaks and shoulders central wavelengths extracted from absorbance curves

Sample	Maximum (nm)	Contributions (nm)	
P3HT	512	566	615
P3HT + 1%DWNTs	506	563	612
P3HT + 5%DWNTs	510	564	619
P3HT + 10%DWNTs	514	567	619
P3HT + 50%DWNTs	520	569	620

The microscopic images show that the polymer covers the nanotube surface. On the other hand, the red shift in UV–Vis spectroscopy evidences a higher order in P3HT structures, or at least longer effective conjugation lengths within the composites. Therefore, it can be speculated that the nanotube surface could act as a template for P3HT, enhancing the polymer's structural order within the composites and influencing the P3HT adhesion. Nevertheless, it must also be observed that by increasing the nanotube content, the absorption below the polymer band gap (~600 nm) becomes non-negligible, suggesting a relevant degree of disorder.

Further details of polymer ordering related to nanotubes can be provided by imaging the polymer adhesion on nanotubes at atomic scale, as will be shown in Sect. 6.

5.2.2 Raman Spectroscopy

Figure 25a shows the Raman spectra of P3HT/DWNTs compounds excited with 532-nm diode laser radiation. Pristine DWNTs and P3HT spectra show peaks that are distinctive features of the material. In DWNT spectra, two bands are typically present: a so-called D-band related to nanotube defects [112] at 1,336 cm^{-1} and a G-band at 1,575 cm^{-1} associated with longitudinal and transversal vibrations of the tubes [113]. The high intensity of the G-band with respect to D-band denotes the low content of nanotube defects and carbonaceous species in the reported samples compared to other studies [112, 114].

As can be seen in Fig. 25a, the P3HT spectrum is characterised by two peaks related to C_β–$C_{\beta'}$ stretching (1,377 cm^{-1}) and $C_\beta{}^+$–H bending and C_α=C_β stretching (1,445 cm^{-1}), respectively [115]. The Raman spectra of the composites display an increase of a peak centred at 1,585 cm^{-1} as the content of nanotubes is raised. The peak does not shift with the CNT content and is related to the G-band of the DWNTs. Compared to the pristine nanotube sample spectrum, where the G-band is centred at 1,575 cm^{-1}, a considerable upshift of 10 cm^{-1} is observed. According to relevant literature, this phenomenon is likely to be related to nanotube doping [116, 117] and/or mechanical restriction of the freedom of C–C vibrations [118, 119].

Considering the possible doping effects, a downshift in the G-band of DWNT's Raman spectra may occur if carbon nanotubes act as electron acceptors. This is due to the weakening of the C–C bond strength when an additional charge is placed in

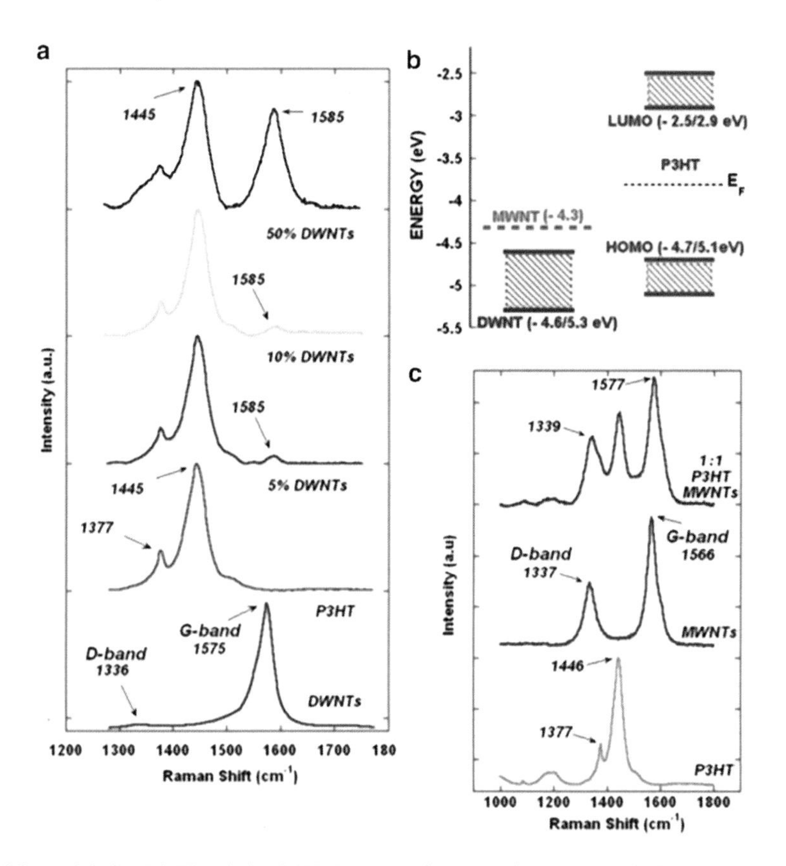

Fig. 25 (**a**) Pristine P3HT, pristine DWNTs and their composite compound's Raman spectra. (**b**) Energy band diagram of P3HT and DWNTs. (**c**) Pristine P3HT, pristine MWNTs and their 100% composite's compound (From Ref. [108])

the anti-bonding conduction band of the nanotube resulting to a downshift (or a softening) of the vibrational frequency. Therefore, the G-band can display down-shift values of up to 8 and 9 cm^{-1} in the case of interaction with Li and K, respectively [116]. Conversely, when nanotubes act as electron donors, the G-band vibrational frequency is expected to increase and an upshift in values up to 34 cm^{-1} can sometimes be observed as in the case of doping by bromine [117].

In the case of P3HT, the sulphur atom of the thiophene ring is expected to act as electron-donating functional group [95, 97]. This tendency is also supported when considering the energy band diagram for P3HT and DWNTs illustrated in Fig. 25c. From the relative position of the Fermi energy levels for DWNTs [120] and P3HT [121], the electron migration from nanotubes to the polymer can be considered unlikely. Conversely, the negative charge transfer from the polymer is energetically favoured. Thus, for significant electronic interaction between P3HT and DWNTs, a small, measurable downshift in the DWNT G-band frequency might be expected, which is clearly not the case.

Being able to discard the nanotube doping to support the observed upshift, the mechanical restriction of C–C vibration freedom is identified as the reason of the change in the Raman spectra. It can be easily related to polymer organisation on the nanotube surface, already evidenced by microscopy images, with the adhesion affected by π–π [119] or CH–π [118] noncovalent interaction. At this point, the mechanical interaction related to P3HT adhesion on the nanotube surface represents the most plausible explanation. Further analysis at nanoscale is required to confirm this assumption as discussed in Sect. 6. As a further confirmation, Fig. 25c illustrates the Raman spectra collected for pristine P3HT, pristine MWNTs and 1:1 weight/weight ratio samples. These results are in good agreement with the experimental data reported for DWNTs. The considerations made for DWNT sample upshifts are valid also for MWNT composites, where a measured G-band upshift of 11 cm^{-1} is present along with a small D-band upshift of 2 cm^{-1}.

6 Nanoscale Investigation of CNTS–P3HT Interaction

6.1 Introduction

Some of the considerations and observations made in the previous sections can be confirmed and supported by direct atomic resolution imaging. By using atomically resolved images, it is possible to investigate the nanoscale organisation of the polymer on the nanotube sidewalls. This study is highly relevant for applications since the polymer adhesion can directly affect the electrical and mechanical properties of the resulting composite [5, 103].

The adsorption of aromatic molecules on carbon nanotubes by significant π–π interaction has been the object of several theoretical studies. Some calculations [122, 123] suggest the alignment of aromatic rings along the nanotube axis. On the other hand, experimental results have also measured the selective polymer wrapping on nanotubes with large chiral angle (i.e., larger than 24.5° with respect to the CNT axis) [124]. Extended self-organised arrangements such as single- and multi-helical structures have also been proposed [7, 125] and observed with microscopic techniques [16, 17] but the role played by the nanotube radius and chirality on the final structure has not been fully evidenced yet, nor it has been supported with atomic resolved images.

Moreover, it must be observed that the nanoscale investigation of the electrical behaviour of poly(3-hexylthiophene) and carbon nanotubes as isolated materials is relevant to their application in devices. In fact, since the inclusion of CNTs in a polymer matrix has been proposed for photovoltaic application [3], it is important to evaluate the behaviour of the *p–n* junction formed. The low power conversion efficiency achieved by P3HT–CNT devices has been the object of a theoretical study by Kanai and Grossman [97] that investigated the charge separation at the polymer–nanotube interface for semiconducting and metallic nanotubes. For metallic SWNTs, inefficient exciton separation is attributed to the electrostatic potential at the interface that stops the migration of electrons on the P3HT side of the interface, causing a thermal decay of the molecule's excited state. For this purpose, scanning tunnelling spectroscopy in ultra-high vacuum (UHV-STS) has been employed. UHV-STS is recognised as a reliable investigative technique for probing the electronic properties of nanoscale metal–semiconductor interfaces [126, 127] and it has been proven extremely useful to study the local electronic structure and charge transport through carbon nanotubes [86, 128]. The local density of states (LDOS) and the transport behaviour at atomic scale of molecules [129] and polymer nanowires [130, 131] have also been investigated using UHV-STS methods.

High-resolution transmission electron microscopy (HR-TEM) image analysis has also been employed to investigate the mechanical and physical modifications that occur to nanotubes upon inclusion in the polymer matrix. In fact, since their discovery, because of their unique properties [58–61, 64], carbon nanotubes have been used as a modifier to physical properties of other materials and have been incorporated into the matrix of polymers [132], epoxies [133] and ceramics [134] to form composites showing structural reinforcement and improved mechanical properties [5, 6, 134]. After the inclusion in the polymer, the carbon nanotube's properties can change as both mechanical deformations [135, 136] and stress-induced fragmentation [137] have been reported for MWNTs in polymer matrixes. Moreover, as introduced in Sect. 5.2.2, the displacement of first-order typical vibrational modes of CNTs is often observed when carbon nanotubes are incorporated in composites [138]. As shown in Sect. 5.2, the CNT's G-band peak frequency can upshift and downshift, and this has been attributed to stress-induced compression [114, 139], molecular adsorption on nanotube surface [118] and charge transfer [117].

6.2 Scanning Tunnelling Microscopy

Figure 26a shows an STM image of an MWNT entirely wrapped by rrP3HT. The considered section shows a highly ordered, self-organised polymer structure covering the nanotube, uniformly distributed along the main nanotube axis for a length greater than 50 nm. Polymer coverage originates by the adsorption of rrP3HT chains onto the nanotube surface due to the strong π–π interactions. The conjugated thiophene rings stack against the carbon nanotube as observed on the highly oriented pyrolytic graphite (HOPG) [143].

Upon adsorption, the interdigitation of the rrP3HT alkyl chains, oriented orthogonally to the backbone, is likely to assist the polymer organisation, although there is no evidence to support this hypothesis. The rrP3HT structure in the STM image of Fig. 26a appears to be organised into a double helix around a carbon nanotube core

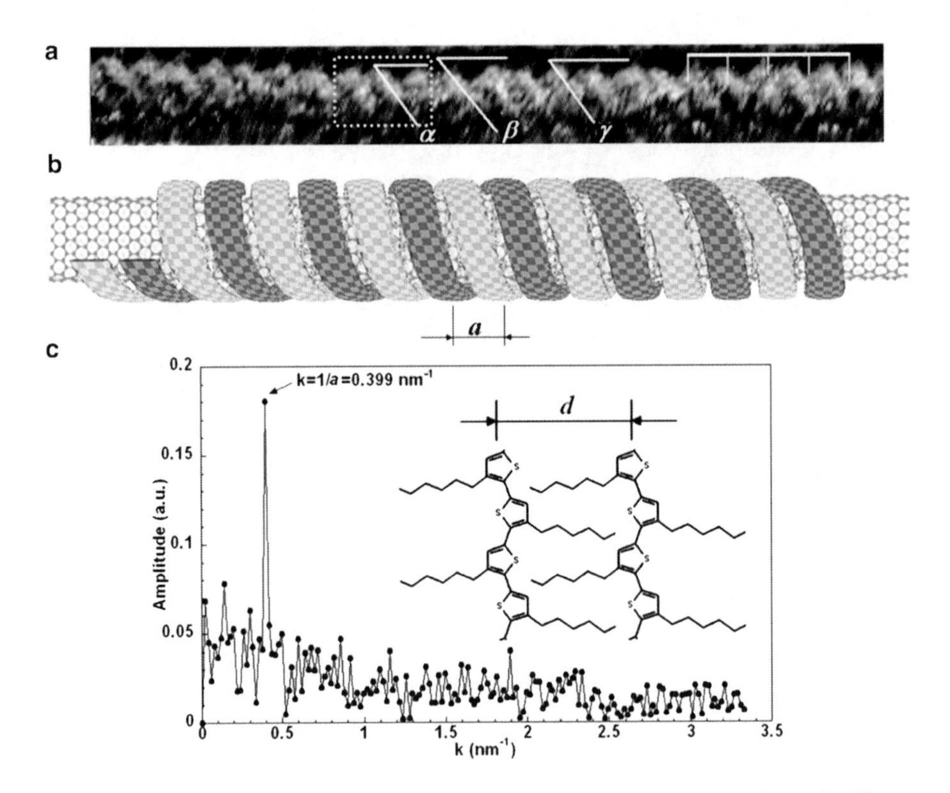

Fig. 26 (**a**) STM image of a multi-wall carbon nanotube covered by rrP3HT self-organised into a coiled structure. Unfiltered image, acquired using a bias voltage $V = -500$ mV and a current $I = 350$ pA [140]. The scale bar measures 12 nm, whereas the angles indicated with α, β and γ are 52°, 52° and 54°, respectively. The rectangle indicated is the zoom-in area used to generate Fig. 27. (**b**) Schematic representation of a double helical structure with coils equally spaced at a distance *a* [141]. (**c**) FFT power spectrum calculated from the profile line traced along the main axis of the nanotube. The peak reveals an average period of $a = 2.51$ nm. A schematic representation of rrP3HT chains interpenetrating at distance *d* (*insert*) [142]

as pictured in the schematic representation of Fig. 26b. Referring to Fig. 26a, three coiling angles between the polymer coating and central carbon nanotube main axis are identified (indicated with α, β and γ) and are used to provide an estimate of the average coiling angle $\theta = 52 \pm 2°$. Although the rrP3HT helical structure possesses some structural defects and helix deformations, the average period can be measured by extracting it from the spectrum of the profile curve as can be seen in Fig. 26c for one representative profile line. The presence of one, clearly distinguished peak, confirms the overall high degree of structural order that corresponds to a periodicity of $a = 2.51 \pm 0.15$ nm and which is equivalent to the semi-period of each helix. This value is consistent with the average calculated from several measurements and represents an estimation of the polymer-wrapping period which can be used for further analysis.

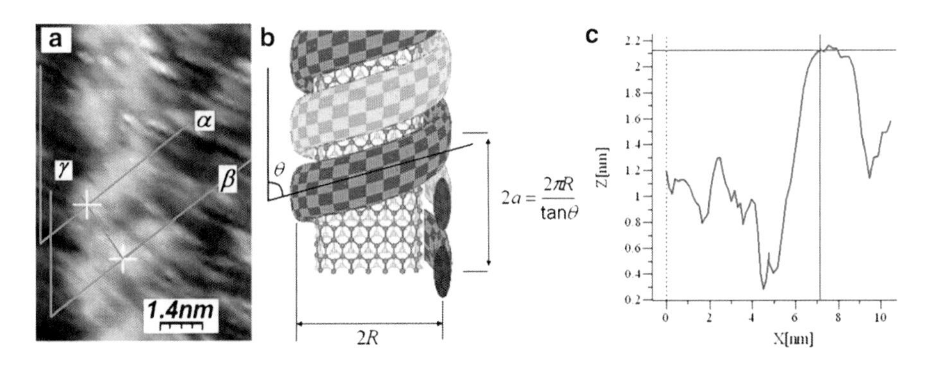

Fig. 27 (**a**) Zoom-in image of the square zone indicated in Fig. 26a. This image has been used to measure the nanotube dimension along the traced lines indicated with α and β. The diameter of the polymer coils is $2R = 2.10$ nm. (**b**) The simple geometrical model relating the coiled nanotube with radius R, the coiling angle θ and the helix period $2a$. (**c**) Height profile measured along line α of (**a**)

From the measured values of a and θ, the chain-to-chain distance d, shown in the inset Fig. 26c, can be evaluated in the direction perpendicular to the polymer backbone to give $d = 1.98 \pm 0.14$ nm. These results confirm that when the polymer is self-assembled on the nanotube, the distance d increases compared to P3HT deposited onto HOPG [144]. Indeed, previous STM studies have established that the characteristic chain-to-chain distance for the rrP3HT polymer self-organisation distance on a flat substrate is $d = 1.3$–1.4 nm [145, 146]. Considering that the measured chain-to-chain distance in a previous study on nanotubes [144] was $d = 1.68$ nm, it can be proposed that the 20–50% extension found on nanotubes compared to flat substrates could be the result of a geometrical constraint due to the regularity of the equally spaced helical structures formed. To assess this possible mechanism, the profile of the nanotube reported in Fig. 27 is considered. The maximum height averaged on two profiles (acquired on lines α and β) of the polymer-coiled section is 2.10 ± 0.10 nm as illustrated in Fig. 27c. By assuming a cylindrical symmetry, this value is assigned to $2R$, the diameter of the nanotube–polymer structure. From the same picture it is possible to extract the interchain distance $d = 1.81 \pm 0.10$ nm, measured on the line γ. Note that this value matches locally the average result obtained for the whole structure by FFT.

According to the geometrical model introduced by Coleman et al. [125] shown in Fig. 27b, the polymer helical structure lies on a cylindrical surface with a circumference of $2\pi R$, where the coils period $2a$ is related to R by the following equation:

$$2a = \frac{2\pi R}{\tan\theta}, \tag{16}$$

where R is half of the coiled nanotube diameter and θ is the coiling angle. Substituting the measured values ($2R = 2.10$ nm, $\theta = 52°$), it is found that $a = 2.57 \pm 0.35$ nm. Since this value is in good agreement with the helix semi-period of 2.51 nm that was extracted from the profile analysis, it can be asserted that the polymer organises itself into a regular double helical structure with equally spaced coils, where the chain-to-chain distance is regulated by the coiling angle and the diameter of the nanotube. The agreement between these results, obtained by averaging several measurements, also highlights the high degree of order overall and the regularity of the two equally spaced helical structures.

To further investigate these aspects and to explore the possible role played by the nanotube chirality on the coiling angle, a partially covered nanotube was imaged, as illustrated in Fig. 28a. In this image, a P3HT structure on the nanotube surface is observed, with the coiling so widely spaced that the underlying nanotube structure can easily be observed between the polymer formations. As shown in Fig. 28b, the carbon atom's alignment can clearly be discerned after appropriate zoom and high pass filtering. A self-assembled polymer structure matching the reported observation has been mathematically generated and shown in Fig. 28c. In the proposed model, although the long-range order is less evident than in the structure presented in Fig. 26, a degree of regularity and a pseudo-period is still present, as the repetition of different coiling indicated with (1) and (2) can be identified in the raw STM data of Fig. 28a.

The nanotube chirality can be determined from the high-pass-filtered zoom-in image of Fig. 28b showing the carbon atoms placed alternately and aligned at a constant distance. Referring to Fig. 29 where the profile of the bare section of the nanotube is reported, the average distance between atomic chains is 0.222 nm, matching with that of a zigzag nanotube, considering the curvature.

To determine the index of chirality n ($m = 0$ in zigzag nanotubes), the measure of the nanotube diameter is needed. Figure 30a, b shows the profile measurements performed on the unfiltered STM data along and across the wrapped nanotube. The stacking distance, namely the distance between the polymer layer and the nanotube surface, is found to be 0.36 nm, confirming the π–π interaction between the polymer backbone and the carbon atoms of the nanotube [143], in good agreement with the value of 0.34–0.38 nm encountered in literature [43, 147].

The analysis of the cross section shown in Fig. 30b provides a value of 1.19 ± 0.10 nm for the diameter of the nanotube, obtained by subtracting the stacking distance from the height of the bare section of the nanotube. From this value the nanotube index $n = 15 \pm 1$ is obtained, according to the following expression relating the diameter and the chiral angle φ:

$$n = \frac{\pi d}{a}\left(\frac{\cos\varphi}{\sqrt{3}} + \sin\varphi\right). \tag{17}$$

Here, $a = 0.246$ nm is the lattice constant of graphene and $\varphi = 30°$ for zigzag nanotubes.

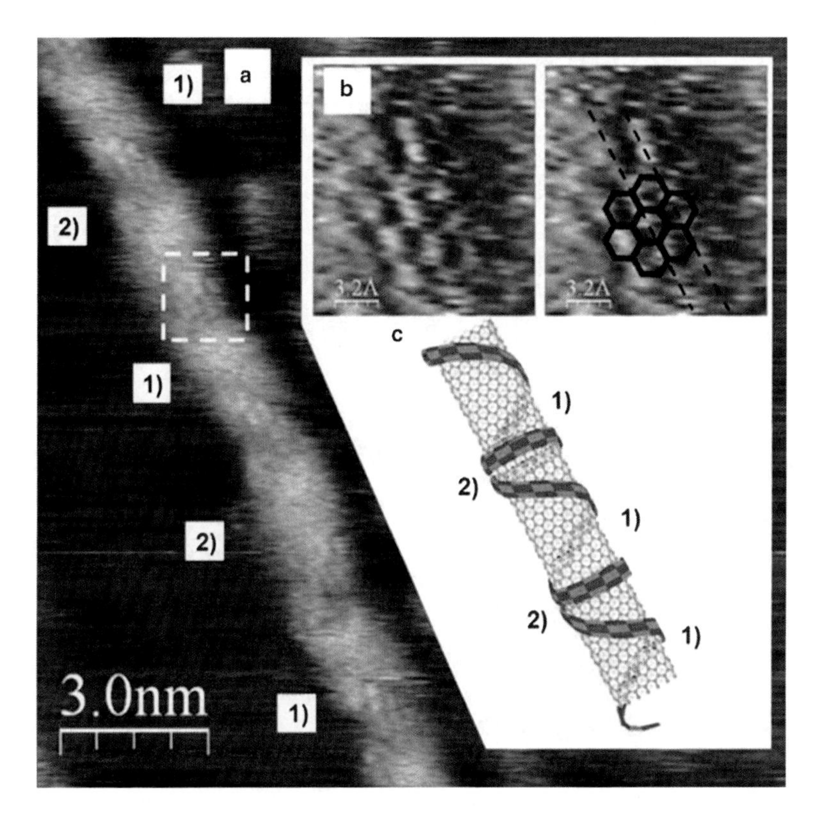

Fig. 28 (**a**) Unfiltered STM image of a partially rrP3HT-covered carbon nanotube. Image acquired using a bias voltage $V = -500$ mV and a current $I = 350$ pA. (**b**) High-pass filtered image from Fig. (a) in the zone indicated by the *white dashed line square*. The hexagonal cells are reported to highlight the nanotube structure. The distance between the lines is 0.23 nm. (**c**) Mathematical model added for comparison

In Fig. 31, the angles formed between the P3HT strands coiling the nanotube and the nanotube axis have been highlighted for different sections of the wrapped nanotube: all of them are close either to 30° or to 90°.

Considering that on a zigzag nanotube, neglecting the effect of the nanotube curvature, the contiguous carbon hexagonal cells are oriented along two main directions with respect to the nanotube axis (90° and 30°), it is possible to affirm that the polymer adsorption on the nanotube is strongly influenced by the nanotube chirality. This remarkable result provides clear evidence for a strong and preferred interaction of the rrP3HT self-organisation along established directions due to the π–π interaction with the nanotube template. The role of chirality in the polymer-wrapping process was previously suggested for poly[(*m*-phenylene-vinylene)-*co*-(2,5-dioctoxy-*p*-phenylenevinylene)] (PmPV) by McCarthy et al. [149] but the result here is supported by clearer resolution STM images that show both the nanotube structure and the polymer coils. These results also confirm that a coiling angle θ different from zero can be energetically favoured in nanotubes

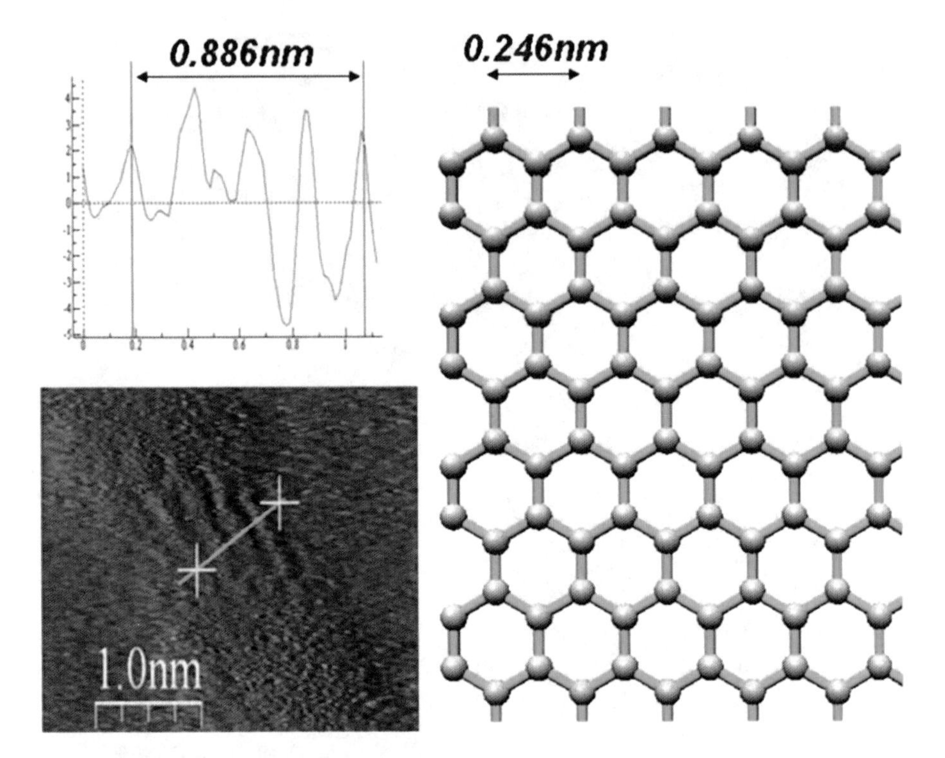

Fig. 29 High-pass-filtered image of the bare section of the wrapped nanotube. The profile analysis shows carbon atoms alignments separated by an average distance of 0.222 nm. Graphene sheet showing the distance of 0.246 nm between carbon atoms

with certain chiralities [150]. The hypothesis of the chirality-driven adhesion on carbon nanotubes is also supported considering that rrP3HT cannot be considered a rigid polymer since the persistence length values are in the range $L_p = 2.1$–2.4 nm [144, 151]. As reported before [125], polymers such as PmPV with large persistence length ($L_p = 10$ nm) are expected to show preferred coiling angles related to the minima of their coiling energy, while low L_p polymers such as polyacetylene ($L_p = 1.3$ nm) show no-preferential direction and can be more easily curved.

6.3 Scanning Tunnelling Spectroscopy

On the same partially covered nanotube, it is possible to investigate the electrical characteristic of the carbon structure and of the polymer adsorbed on nanotube. First, the electronic properties of the bare section of the coiled nanotube, identified as (15,0), have been investigated. Referring to Fig. 32, the current–voltage characteristics were collected on the point **P**. Successively, the polymer *IV* characteristics were recorded on several points of the P3HT structure evidenced in the figure by the dashed lines.

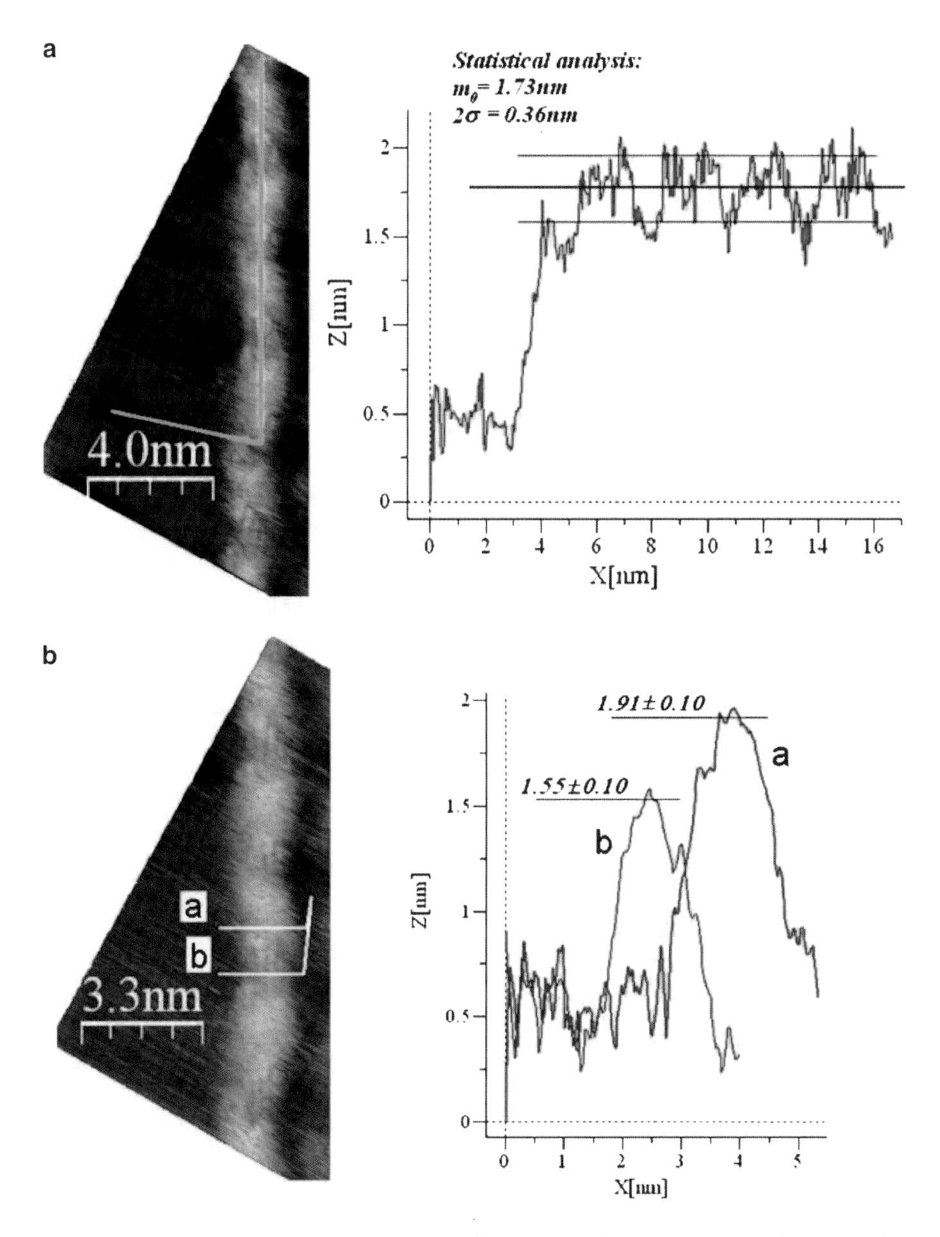

Fig. 30 Profile measurements to assess the stacking distance of the polymer (**a**) and the nanotube diameter (**b**)

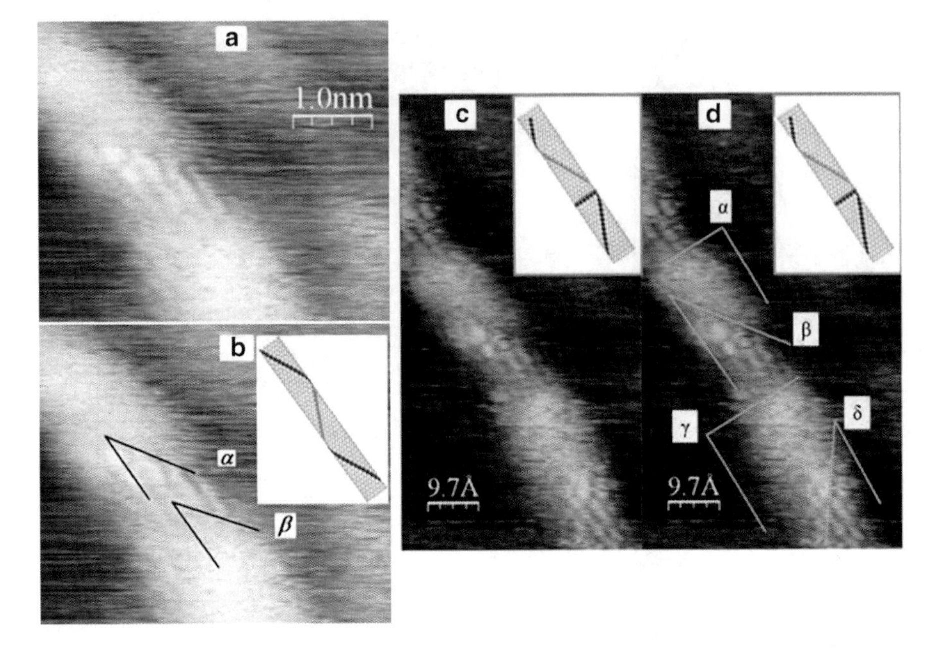

Fig. 31 Comparison between unfiltered images of the polymer-covered carbon nanotube to show the angles formed between the polymer strands and the nanotube. Image acquired using a bias voltage $V = -500$ mV and a current $I = 350$ pA. The angles indicated with α, β, γ and δ are used to evidence the preferred direction on which the polymer is aligned on the nanotube surface. In (**a, b**) the angles α, β measure 33° and 34°, respectively. In (**c, d**) the angles α, β, γ and δ are measure 90°, 33°, 90° and 34°, respectively. (15,0) Zigzag nanotube with 30° and 90° hexagonal cells alignments highlighted, reported for comparison (*insets*) [148]

The current–voltage characteristic *IV* measured on the bare nanotube section is reported in the inset of Fig. 33. The *IV* curve is symmetric with respect to the zero bias voltage, confirming that during the tunnelling process in both directions, the electrons are not affected by any charge accumulation nor do they encounter any Schottky barrier. This is expected considering that the two materials, HOPG and SWNT, have nearly the same work function (4.4 eV and 4.3 eV, respectively) [153].

The differential conductance (d*I*/d*V*) curve plotted versus the bias voltage shown in Fig. 33 was obtained by numerical differentiation of the *IV* curve. The theoretical density of states (DOS) curve near the Fermi level for a (15,0) SWNT is also reported for comparison and has been obtained independently by Akai and Saito [89] using the local density approximation in the framework of the density functional theory (DFT). The good agreement between the theoretical model and the experimental data confirms that the nanotube is in ohmic contact with the graphite substrate, as supposed previously. The experimental peaks, representing the Van Hove singularities, have been indicated with number from 1 to 6, and in Table 2 the energy positions of experimental and theoretical peaks are reported, highlighting a good degree of correspondence.

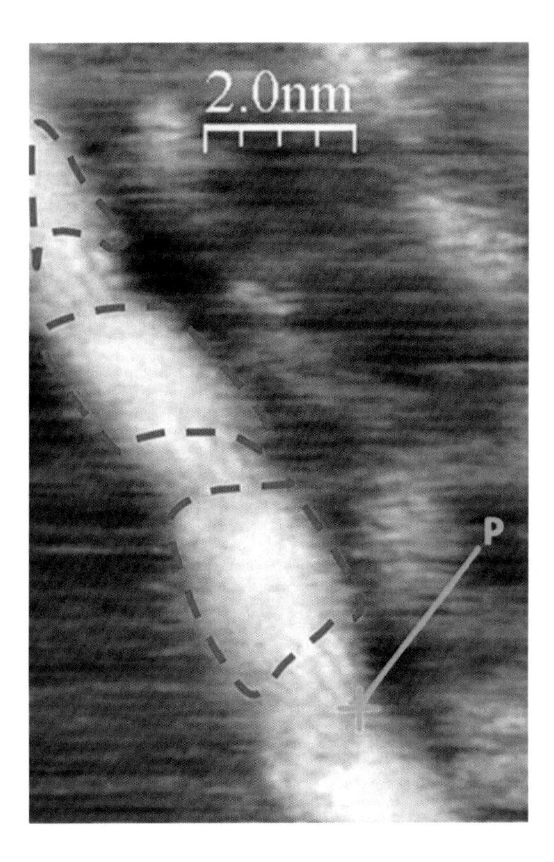

Fig. 32 One of the STM images used for the STS analysis that shows rr-P3HT self-organised into a helical self-assembly onto a carbon nanotube. Scan obtained at $V_{SAMPLE} = -500$ mV and $I = 0.350$ nA. The point **P** indicates the position of the tip where the *IV* curve reported for the bare nanotube was acquired. The areas inside the *dashed lines* highlight the P3HT covered zone where the rectifying *IV* characteristics were acquired [152]

The energy subgap ΔE_{sub}, which corresponds to the energy difference between the first Van Hove singularities, can be evaluated according to the following equation [128]:

$$\Delta E_{\text{sub}} = \frac{6 d_{\text{nn}} \gamma_0}{d}, \tag{18}$$

where $d_{\text{nn}} = 0.142$ nm is the distance between nearest-neighbour carbon atoms, d is the diameter of the nanotube and γ_0 is the energy overlap integral between nearest-neighbour carbon atoms. Substituting the measured diameter value and $\gamma_0 = 2.5$ eV which corresponds to the overlap energy of a single graphene sheet [88, 154], $\Delta E_{\text{sub}} = 1.78$ eV is obtained. The analysis confirms a very good agreement between experimental and theoretical results for the sub-band-gap value of a (15,0) nanotube. Since the experimental data of the diameter, the chirality and the LDOS are all consistent and in good agreement with the expected values for a (15,0) SWNT, there is no reason to doubt that the observed nanostructure is a single-wall carbon nanotube, excluding the presence of inner shells. Because the LDOS curve is symmetrical around the Fermi level, it can be stated that no

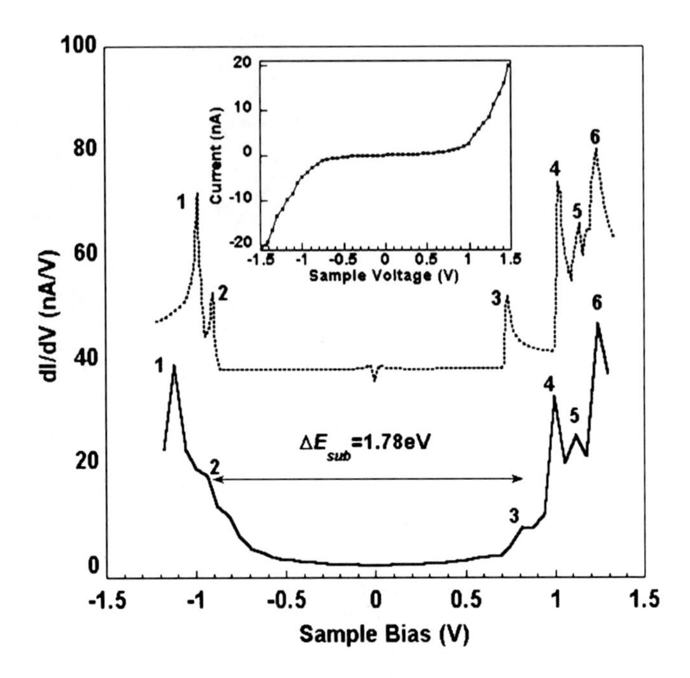

Fig. 33 Differential conductance of the bare nanotube section of Fig. 32 (continuous line) and LDOS curve from theory (*dotted line*). Van Hove singularities have been marked with numbers 1–6 with the corresponding energy positions reported in Table 2. *IV* characteristic of the bare nanotube section from which the differential conductance was obtained (*inset*)

Table 2 Comparison between experimental differential conductance peaks as marked on Fig. 33 and theoretical peaks of LDOS according to [89]

Peak number	Experimental (eV)	Theory (eV)	Deviation %
1	−1.12	−0.98	−14.30
2	−0.94	−0.90	−4.40
3	0.81	0.74	9.45
4	0.99	1.04	−4.80
5	1.12	1.14	−1.70
6	1.24	1.24	0

alteration of the nanotube electronic properties can be observed in this part of the structure and a hypothetical charge transfer from the substrate or from the polymer does not locally affect the density of states of the SWNT. Although it could be expected that the correspondence of the work functions of the nanotube and of the graphite should not lead to a charge transfer, it is remarkable that in the bare section of the polymer-wrapped structure no evidence of the surrounding polymer presence can be detected.

Under the same experimental conditions, the *IV* characteristics of the polymer-covered sections of the nanotube were investigated by collecting several STS images acquired with different tip–sample distances. On each image, a number of *IV* curves were collected at different locations on the polymer-covered parts of the nanotube and averaged.

The absence of negative differential resistance peaks in any of the acquired *IV* curves confirms the electron delocalisation on the polymer backbone and hence the presence of a continuum of electronic states in the molecular orbitals of P3HT strands. Fig. 34a shows the averaged current–voltage characteristics of P3HT strands adsorbed on the (15,0) SWNT for two different tip–sample distances.

The asymmetry of the curves denotes a rectifying behaviour that favours the electron tunnelling from the substrate to the tip, passing respectively through the nanotube and the polymer. Similar asymmetric curves have been obtained for P3HT deposited on HOPG, as reported in the inset of Fig. 34a, but minor differences have been observed between positive and negative bias currents. For poly(3-dodecylthiophene) (P3DDT) deposited on HOPG, it has been demonstrated that the rectifying characteristic is related exclusively to the polymer molecular levels [155]. However, in this case the SWNT is included between the graphite and the polymer and the evaluation is more complicated because of different phenomena involved. In particular, it must be noted that P3HT has been considered as a donor when interacting with SWNTs [95]. In order to verify the effects of the interaction between P3HT and the (15,0) SWNT on the energy band structure, the differential conductance curves are plotted in Fig. 34b at two different heights for the covered nanotube section. Here, it is possible to observe an extended plateau region associated with the energy gap of the semiconductor that is estimated as $\Delta E_{\mathrm{COND}} = 1.81$ eV and $\Delta E_{\mathrm{COND}} = 1.95$ eV at smaller and larger tip–sample distance, respectively. Both values are in accordance with previous results for P3HT on a silicon substrate, with the small difference due to the molecular orbital levels affected by the bias [131].

The important characteristic of the d*I*/d*V* curve is the shift of the Fermi level towards the highest occupied molecular orbital (HOMO) causing an asymmetry in the conductance gap. The shift is $\Delta E = 150$ meV and $\Delta E = 159$ meV at smaller and larger tip–sample distance, respectively, and is scarcely affected by the bias. This feature was not observed in the experiments for P3HT deposited on HOPG, where the midgap energy coincided with the Fermi energy E_{F}. The shift of the Fermi level toward the polymer HOMO energy band is most likely related to the transfer of electrons to the underlying nanotube from the sulphur atoms of the polymer thiophene rings (i.e., charge transfer occurs upon binding of the polymer to the nanotube). Kanai and Grossman [97] predicted a large charge transfer from P3HT to metallic single-wall nanotubes. The electron migration favours adhesion of the polymer onto the carbon nanostructure and shifts the Fermi energy level closer to the HOMO level of P3HT. Considering the nanotube, the consequence of the negative charge transfer that must be taken into account is the decrease of the m-SWNT work function, as calculated by Zhao et al. [156]. The Fermi level at

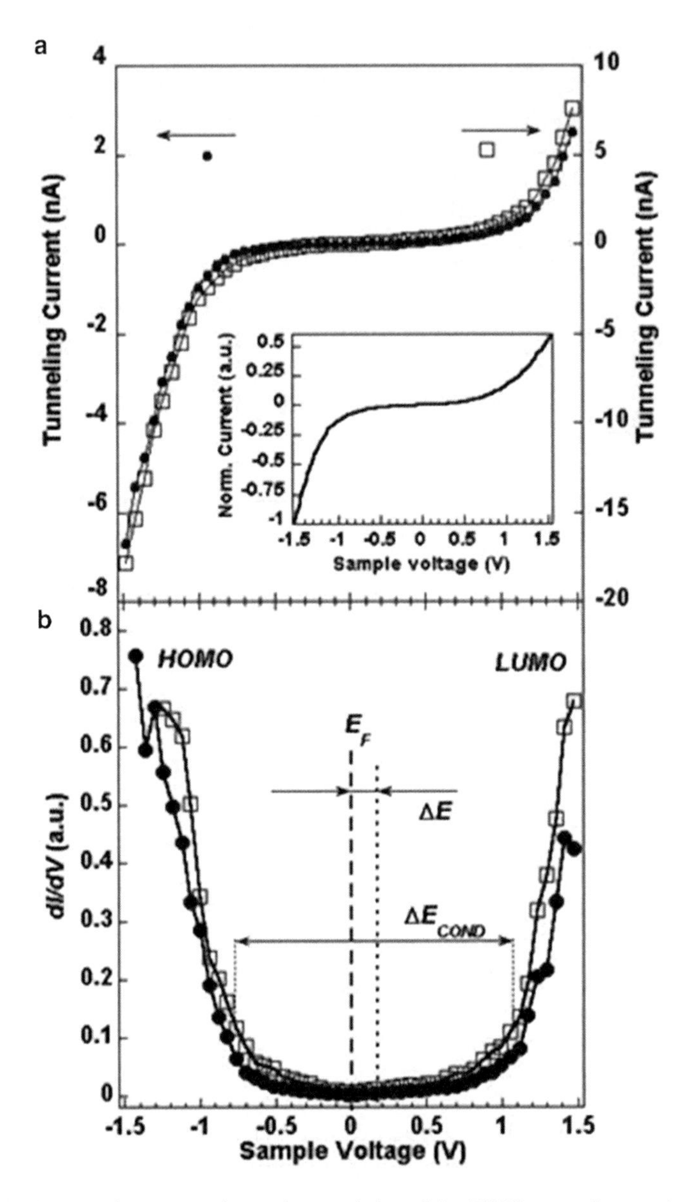

Fig. 34 (**a**) Averaged current–voltage characteristics of the P3HT covered nanotube section of Fig. 32 measured at set point current 0.350 nA and at sample voltages of −800 mV (*dark circles*) and −248 mV(*open squares*). Normalised averaged *IV* curve measured for P3HT deposited on HOPG ($I = 0.500$ nA, $V_{SAMPLE} = -400$ mV) (*inset*). (**b**) Normalised differential conductance curves calculated from *IV* characteristics of (**a**). The Fermi energy is shown as E_F, the differential conductance band gap as ΔE_{COND} and the shift of the Fermi energy with respect to the midgap energy is denoted ΔE

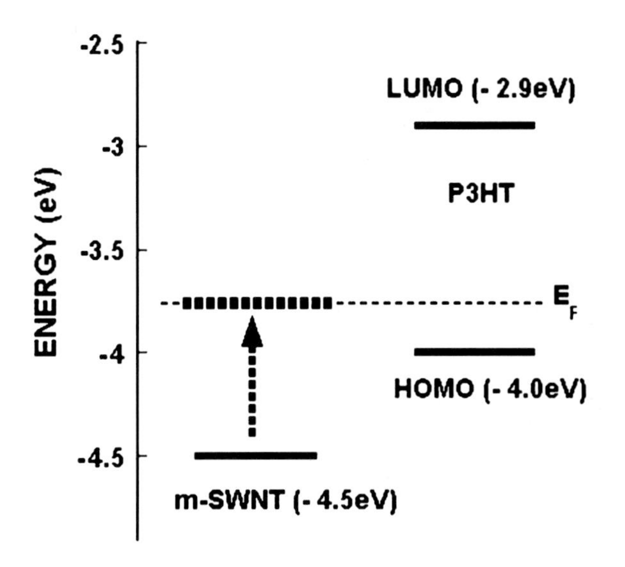

Fig. 35 Energy-level diagram with respect to the vacuum level for isolated metallic single-wall nanotube (m-SWNT) and for P3HT in the DFT/Khon–Sham approach (from [95] and [97], respectively). After the adhesion, the work function of the m-SWNT is reduced (*dashed line*) and aligned with the Fermi energy indicated as E_F

equilibrium results shifted toward the vacuum level and a Schottky barrier is expected to be present at the nanotube surface, as depicted in Fig. 35.

The observed asymmetric behaviour of the *IV* curves consequently arises from the formation of the Schottky barrier, as the electron transfer occurs from the nanotube to the polymer. Although a detailed STS theoretical model is not available, in this case the above considerations appear a reasonable explanation of the STS data.

Figure 36a illustrates another UHV-STM image of a P3HT-covered nanotube. The structure extends for more than 35 nm and presents a high degree of order as the periodic wrapping can be observed over the whole structure.

The period of the polymer structure is evaluated by Fourier transform of the profile along the line shown in Fig. 36a. The power spectrum of the profile line shown in Fig. 36b displays a well-defined peak at the spatial frequency 0.332 nm^{-1}, leading to a period of 3.01 nm. The shape of the polymer wrapping is less evident than in the previous images of Fig. 26 and a more detailed analysis is needed. Figure 37 reports a 3D higher-magnification image on a scale bar of 3 nm and two 2D images acquired along the structure. The images have been filtered with a Gaussian smoothing along the *Y*-axis, and the *Z*-axis colour scale has been adjusted to show structural details of the polymer wrapping.

Figure 37a shows the 3D image of a section of the wrapped nanotube, from which it is possible to have a clear understanding of the polymer coverage. The P3HT self-organisation appears as a helix again but, remarkably, in this example, the helical wrapping shows a variable coiling angle.

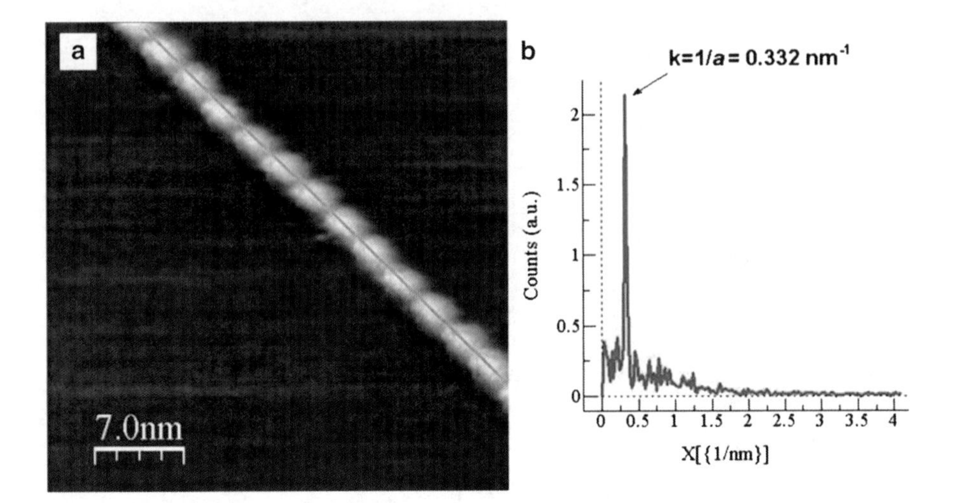

Fig. 36 (**a**) Polymer-covered nanotube imaged by UHV-STM. Scan obtained at $V_{SAMPLE} = -500$ mV and $I = 0.300$ nA. The line of the structure is used for profile analysis. (**b**) Power spectrum of the line profile highlighting a period of 3.01 nm (From Ref. [108])

As can be seen in the inset of Fig. 37a where the structure has been mathematically reconstructed, the periodic variation of the coiling angle eliminates the rotational symmetry of the covered structure. Nevertheless, the inter-coil distance does not vary as can be observed in the examples of Fig. 37.

Rotational symmetry would be present for a constant coiling angle as seen for the case of Fig. 26. Exploring the nanotube along its main axis, it has been possible to image the different patterns of the polymer coverage, since the P3HT structure rotated around along the nanotube. These cases are shown in Fig. 37b, c and demonstrate that the polymer self-organisation appears to have different shapes according to the observation point. The nonconstant coiling angle could be a result of the underlying carbon nanotube's chirality, in this case originating a mismatch with the P3HT backbone structure. At this stage, no further evidence has been collected to support this interpretation.

6.4 High-Resolution Transmission Electron Microscopy (HR-TEM) Analysis

HR-TEM images provide further evidence of the adhesion of polymer on isolated nanotubes and highlight the full inclusion of CNTs in the P3HT matrix. Figure 38a shows a large-diameter five-walled nanotube completely embedded into the polymer matrix. The external diameter is 7.36 nm, whereas the internal is 4.84 nm which leads to an interlayer distance of 0.32 nm. In Fig. 38b–d are displayed isolated nanotubes fully or partially covered by the polymer. It is interesting to

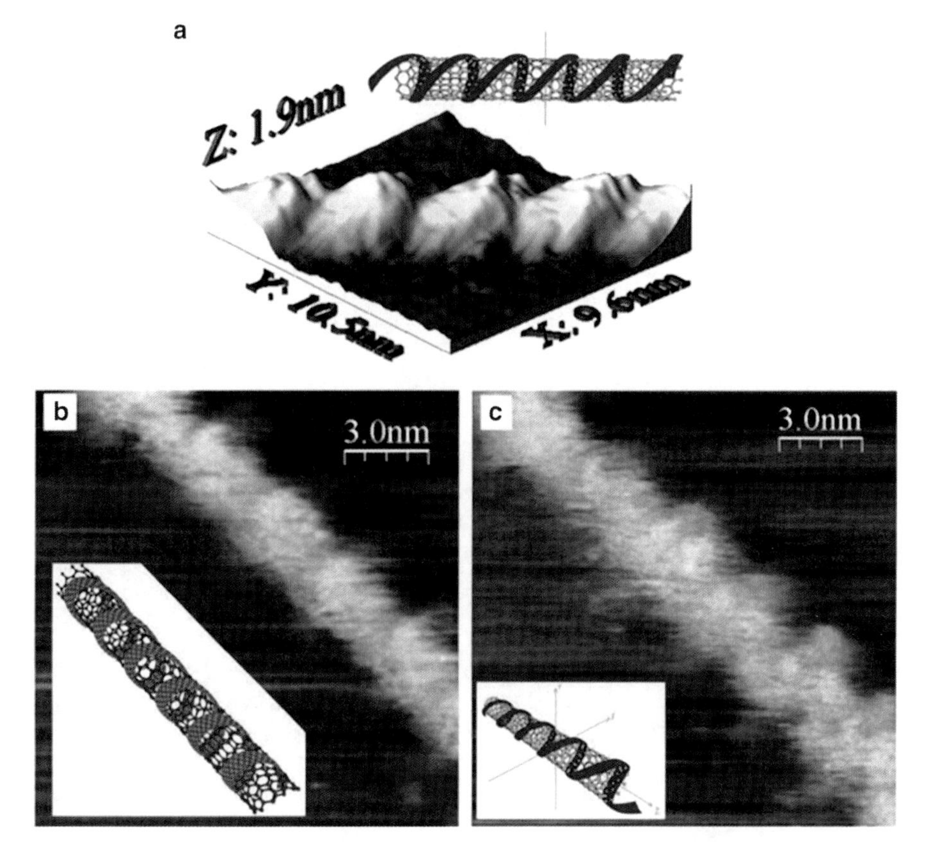

Fig. 37 (**a**) Three-dimensional (3D) image of a P3HT covered nanotube imaged by UHV-STM. (**b**) Corresponding 2D image. Scan obtained at $V_{SAMPLE} = -500$ mV and $I = 0.300$ nA. The image has been filtered with a Gaussian smooth on the *Y*-axis and the *Z* colour scale has been changed to highlight the details of the polymer wrapping. Polymer wrapping generated mathematically. (**c**) The same wrapped nanotube pictured in a different point of the structure (*inset*) (From Ref. [108])

observe that the P3HT layer with a thickness limited to a few nanometres can exhibit a uniform adhesion on the nanotube surface (b, c).

Alternatively, as seen for STM analysis and in Fig. 38d, the nanotube surface can display polymer-free regions. The nanotube in Fig. 38d displays several defects, with internal shells terminated abruptly in correspondence to a kink, marked by an arrow. Although the asymmetric sidewall thickness is clearly visible, the nanotube structure seems to be unaffected by the polymer wrapping. High-resolution transmission electron microscopy can contribute to show the compression of multi-wall nanotubes due to polymer coiling. In the previous paragraphs it has been reported how poly(3-hexylthiophene) interacts with the outer wall structure of carbon nanotubes. The strong adhesion is the result of the electron-rich surface of CNTs and of the polymer π-conjugated structure.

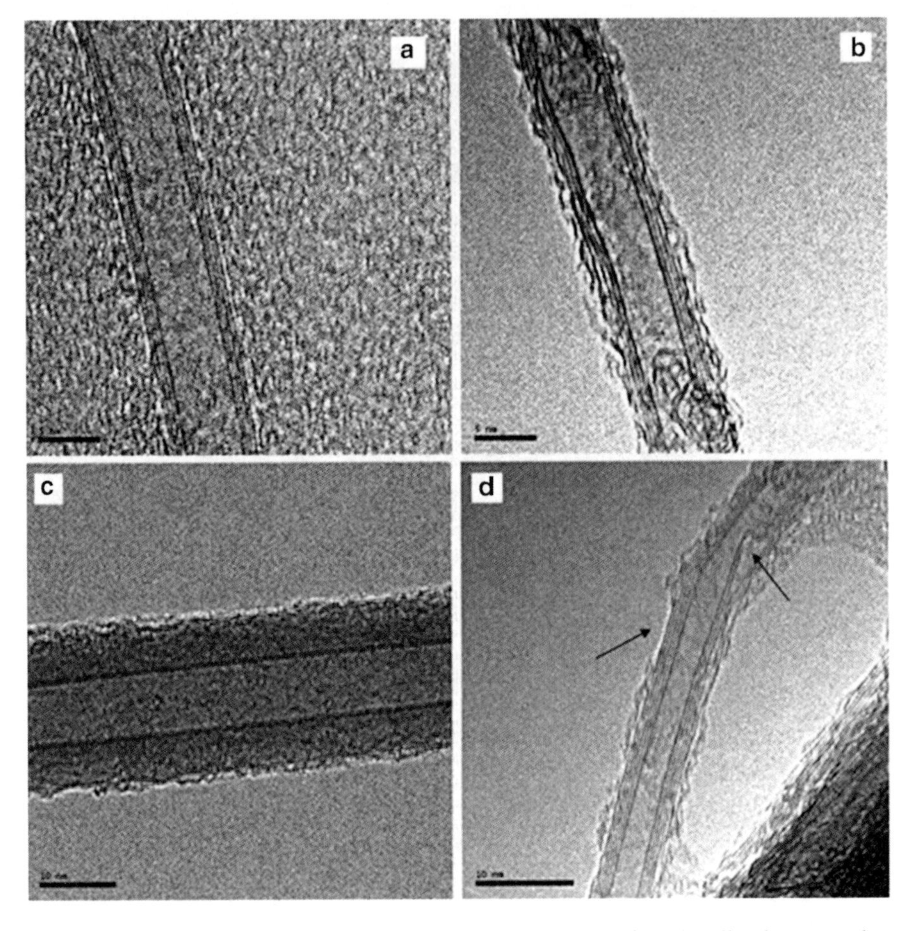

Fig. 38 High-resolution transmission electron microscopy images of multi-wall carbon nanotubes embedded in P3HT matrix (**a**) and fully or partially covered by the polymer (**b–d**). Scale bars are 5 nm (**a, b**) and 10 nm (**c, d**)

A collection of HR-TEM images is reported in Fig. 39 showing the P3HT adhesion onto the multi-wall and single-wall carbon nanotube surface. Figure 39a, b illustrates HR-TEM images of P3HT strands coiled around narrow multi-wall nanotubes of four walls and five walls having diameters of 4.4 nm and 5.4 nm, respectively. The polymer self-assembled structure, which appears like a bulky ribbon coiling around the nanotube, is almost transparent, being visible only through the darker edges of the polymer strands. This effect, which is very similar to the way nanotubes are imaged under TEM [57], is due to the different electron transmission coefficient when crossing vertical or horizontal portion of the polymer ribbon surface.

In both images of Fig. 39a, b, the polymer coiling is irregular with a variable coiling angle with respect to the nanotube axis, ranging from 60° to nearly 90°.

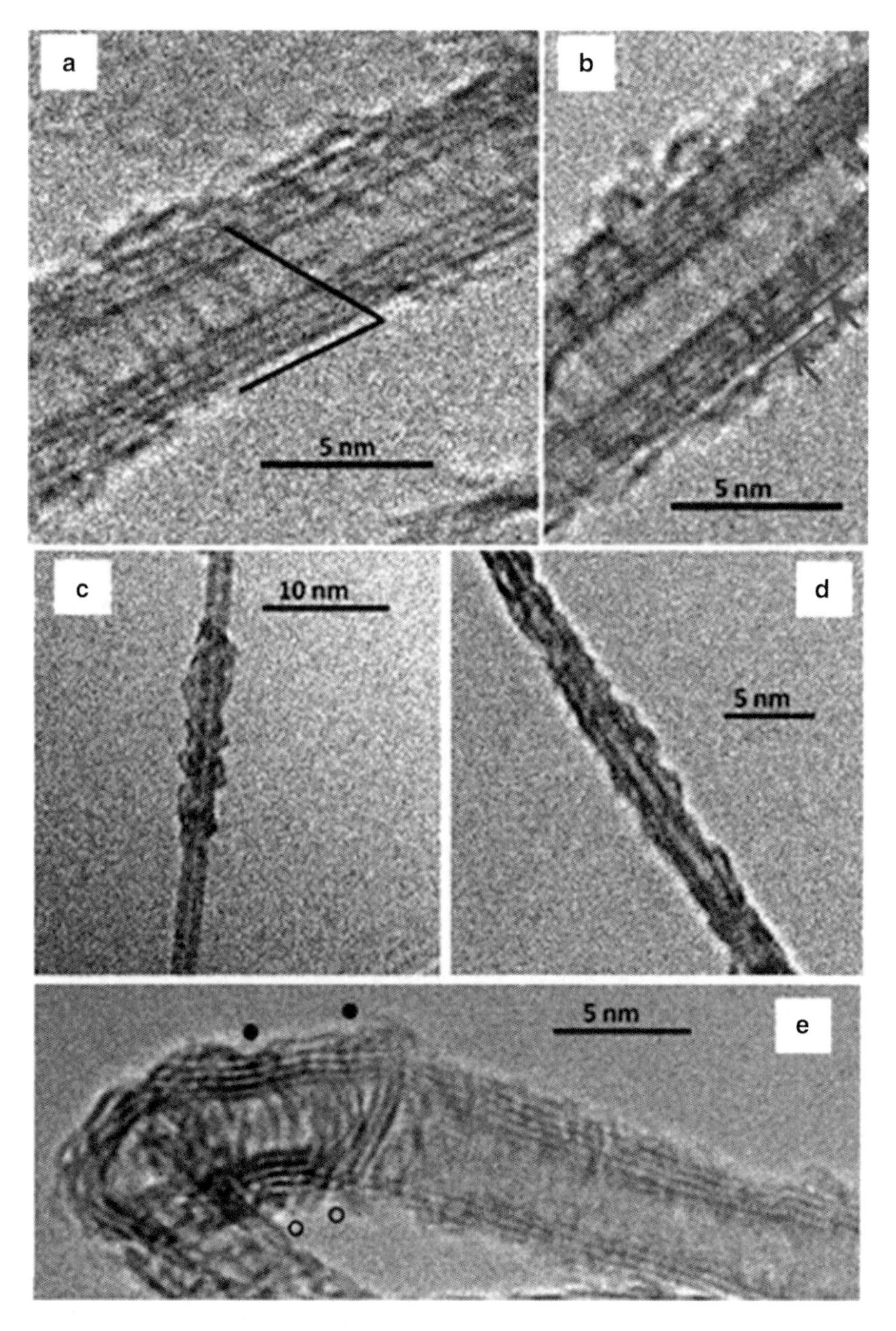

Fig. 39 (**a, b**) High-resolution transmission electron microscopy image of P3HT coiled around multi-wall carbon nanotubes and (**c, d**) around single-wall carbon nanotubes. (**e**) Slightly inclined MWNT showing uniform P3HT coiling fading out of the focus plane (From Ref. [157])

The polymer coil-to-coil distance varies from a few angstroms to more than 3 nm. In Fig. 39b, blue arrows highlight the MWNT's shell interlayer distance of 0.34 nm which is in agreement with values reported by other authors ranging from 0.34 to 0.39 nm [158]. The measured polymer stacking distance is 0.36 nm, matching the stacking distance reported in STM measurements [142].

Figure 39c, d show single-wall carbon nanotubes (SWCNTs) wrapped with P3HT ribbons. As can be observed, the polymer coiling is still present on SWCNTs with small diameter, in this case 2.0 nm and 1.1 nm, respectively. This highlights the flexibility of poly(3-hexylthiophene) structure and demonstrates the bending of the physically adsorbed P3HT around carbon nanotubes with small curvature radii.

Therefore, it can be concluded that P3HT coil wrapping of carbon nanotubes is a general phenomenon, although it may exhibit different arrangements from tube to tube. The wrapping occurs during the interaction in solution phase and persists after the solvent evaporation following the solution cast of the composite material.

Figure 39e illustrates another example of P3HT coiling around a four-wall carbon nanotube slightly bent. Even if the polymer coiling is present on the whole nanotube structure, it is difficult to see (almost invisible) on the right part of the picture, whilst it is easily visible and resolved on the left part. This difference is due to the fact that the right part of the image is out of focus, as the whole structure passes through the focal plane of the TEM. The left part of the structure evidences two clear deformation points, indicated with dark dots which are visible on the upper part, in correspondence to polymer coils. Two more deformation points can be individuated on the lower part (empty circles) but, in this case, it is not possible to exclude that the deformation is due to the buckling of the nanotube. Nevertheless, a squeezing of the carbon nanotube is clearly visible.

In order to further investigate the effects of the polymer coiling on the underlying nanotube structure, MWNTs possessing a relatively large number of inner shells (12–16) wrapped by P3HT have been imaged. In Fig. 40 compression regions and sizable deformations of the outer nanotube radius are clearly visible, as indicated by the short blue arrows. The deformations occur in straight nanotube segments, far away from any bends or kinks, ruling out the possibility that the structural defects imaged originate from long-range nanotube deformation. Radial deformations [159] up to a complete nanotube collapse [160] have been observed in the absence of any covering material and have been attributed to pure van der Waals interaction among nanotube shells. The same interaction, exerted by the polymer wrapping, is likely to be the reason for the sidewall deformation in the observed MWNT.

By observing Fig. 40a, it can be noticed a certain regularity in the deformation of the MWNT, displaying peaks and valleys. Remarkably, on each compression point (valley) it is possible to observe the polymer coil cross section, highlighted by a short blue arrow, whereas the peaks (longer green arrows) appear to coincide with polymer-free regions of the nanotube's external surface. Larger interlayer distances between the MWNT's inner shells are also observed underneath these regions (long black arrows) as a release of the polymer-induced surrounding radial pressure, which appear to coincide with the sidewalls peaks. This result was quite unexpected

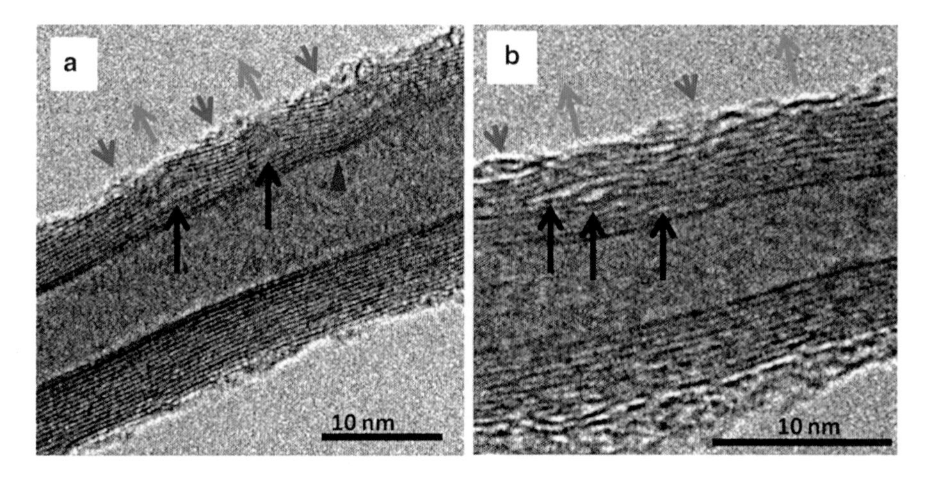

Fig. 40 High HR-TEM images of polymer compressed multi-wall carbon nanotubes. *Short* (*blue*) *arrows* indicate pressure points whereas *long* (*green*) *arrows* highlight the sidewall profile peaks. *Black arrow-tips* indicate the wider interlayer distance consequence of the surrounding pressure points (From Ref. [157])

since it is often assumed that the radial stress is transferred through the nanotube layers, resulting in the MWNT interlayer distance and/or internal radius reduction, a phenomenon observed here only in one point highlighted by the red arrowhead. It should be noted that Fig. 40a displays an asymmetric multi-wall nanotube, where the sidewalls exhibit a different number of inner graphene shells and therefore different thicknesses. The upper and the lower sidewalls of the tube in Fig. 40a are formed by 12 and 16 shells, respectively, with a constant interlayer thickness of 0.38 nm. The measured difference of diameter caused by the deformation is 1.46 nm, whereas the nanotube average diameter is 17.54 nm, resulting in a deformation of 4.2%. The distance between the P3HT coils can be estimated by considering the "polymer compressed zones" of the MWNT structure and results in a polymer repeat distance of 8–9 nm. The multi-wall carbon nanotube reported in Fig. 40b exhibits instead a 12-shell structure, with an interlayer thickness 0.33–0.35 nm. The average diameter is 13.01 nm, whereas the diameter variation due to the compression is 0.98 nm, yielding to deformation of 3.8%. In this case, the increased distance between the inner nanotube walls is more evident. In addition, although this nanotube's cross section is symmetric, the polymer compression effect is visible only on one side of the nanostructure as it was for the previous asymmetric cross-sectioned nanotube. Remarkably, in both cases, the buckling side of the nanotube occurs where the polymer coils, while the nanotube flat side shows a more uniform P3HT coverage.

It has been observed that under induced stress, the radial deformation of nanotubes increases with the radius, but it is inversely proportional to the number of inner shells [161, 162]. Therefore, it can be expected that SWCNTs or DWCNTs with small radii possess very rigid structures and show no visible deformation, while deformations are more likely to be observed for thin-wall MWNTs exhibiting large radii.

To demonstrate this point, it must be assumed in the first instance the pure mechanical capability of a P3HT strand to compress the nanotube, excluding any other mechanism (for example, charge transfer) that could possibly aid the nanotube deformation. First, the value of MWNT elastic radial modulus measured experimentally with atomic force microscope nanoindentation must be assessed. The radial compressive elastic modulus, under asymmetric normal force, can be defined as

$$E_{\mathrm{RAD}} = (F/A)/(\Delta D/D), \tag{19}$$

where F is the applied force, A is the contact area and $\Delta D/D$ is the strain measured on the diameter. Due to the nonlinearity of the phenomena involved, E_{RAD} is better estimated by differentiating (19) or by reducing the extent of the compression. A recent study by Palaci et al. [163] provides an accurate estimation of the MWNT radial elastic modulus. The study was conducted on nanotubes with radii very similar to the case reported here,[1] reducing the extent of the induced deformation and, as for this case, utilising MWNTs synthesised by chemical vapour deposition. For multi-wall nanotubes with external radius larger than 4 nm, the value of E_{RAD} saturates to 30 $\pm$ 10 GPa, close to 36 GPa reported for graphite [164]. Assuming 30 $\pm$ 10 GPa as a valid value for MWNTs, according to (19), for a compression of 4% the estimation of the applied pressure is 1.2 $\pm$ 0.4 GPa. Regarding the polymer, the experimental value for the P3HT's Young modulus has been recently measured by using the buckling-based method [165] and the value reported[2] is 1.33 $\pm$ 0.01 GPa. Therefore, it can be affirmed that the calculated pressure exerted on the MWNT by P3HT coiling can be structurally sustained, and hence applied, by the polymer. It is significant that the observed deformation is close to the maximum that the polymer can withstand.

From the HR-TEM analysis, it is hence possible to assess that one effect of the polymer coiling is to induce structural defects on the nanotube sidewalls such as diameter reduction and interlayer displacements. As shown in several studies, radial deformation is able to modify the nanotube electronic [166, 167] and transport properties [168], with expected significant repercussion on device performance. It must also be noted that the electron transfer from the polymer to the nanotube is suspected to play a fundamental role in the nanotube deformation as observed for potassium-doped double-wall carbon nanotubes [169]. Electron transfer from P3HT to carbon nanotube has been proposed [95], modelled [97] and recently

[1] The multi-wall nanotubes reported in Palaci's study had external radii ranging from 0.2 to 12 nm and constant $R_{\mathrm{ext}}/R_{\mathrm{int}}$ ratio of 2.2 $\pm$ 0.2 nm; in the reported case, the external radii are 6.5 and 8.75 nm and the $R_{\mathrm{ext}}/R_{\mathrm{int}}$ ratio is 2.23 and 2.46.

[2] The value provided for the Young's modulus of P3HT has been measured for polymer films. The authors believe that the tensile strength of the single polymer backbone can be higher, confirming the possibility of carbon nanotube deformation by P3HT strands.

experimentally observed by Motta's group with scanning tunnelling spectroscopy [152]. The local charge transfer on the nanotube surface is expected to weaken the C–C bond on the nanotube surface and to soften the vibrational modes [117].

6.5 Raman Spectra of P3HT/MWNTs

In order to support the previous results and provide further evidence of polymer adhesion-induced nanotube compression, MWNT/P3HT samples were investigated with vibrational spectroscopy. Raman spectroscopy studies have demonstrated that frequency displacement of CNT first-order vibrational modes often occurs when the carbon nanotubes are incorporated in composites. Specifically, G-band peak frequency upshifts and downshifts occur and have been attributed to stress-induced compression [114, 139], molecular adsorption on the nanotube surface [118] and charge transfer [117]. Studies on single-wall nanotubes [170–172] proved that the G-band peak position in Raman spectra, related to transversal modes, increases almost linearly with the external pressure applied, before attaining a nonlinear zone with a reduced increase rate that can be related to reversible nanotube collapse. For MWNTs, the same trend was observed [173] with a reported increase rate of 4.3 cm^{-1} GPa^{-1}.

A series of first-order Raman spectra starting from pristine MWNT samples and moving to increasing P3HT relative weight content up to 1:1 ratio were collected for this study. Figure 41a shows the first-order Raman spectra of pristine multi-wall nanotubes and 1:1 w/w MWNT/P3HT samples. The pristine MWNT spectrum between 1,000 and 1,800 cm^{-1} features the defect-induced D-band [174] centred at 1,337 cm^{-1} and the characteristic tangential mode-related G-band [65] centred at 1,568 cm^{-1}. Conversely, the 1:1 w/w MWNT/P3HT sample's spectrum exhibits vibrational bands originating from the polymer's presence. The highest-intensity P3HT-related peak occurs at 1,446 cm^{-1} which is attributed to $C_\alpha{=}C_\beta$ stretching modes, whereas weak vibrational modes at 1,377 cm^{-1}, related to C_β–$C_{\beta'}$ stretching and $C_\beta{}^+$–H bending, are barely visible because they overlapped the MWNT D-band. Other vibrational modes for the polymer occurring at 1,090 and 1,210 cm^{-1} are discussed elsewhere [115]. The G-band of the 1:1 w/w MWNT/P3HT composite displays an upshift of 8 cm^{-1}, with the experimental maximum centred at 1,576 cm^{-1}. The peak is also broadened, showing an FWHM of 55 cm^{-1}, 6 cm^{-1} larger than the peak of the pristine MWNT sample.

Figure 41b shows the G-band Raman upshift as the polymer content in the compound is raised. At low polymer content, the contribution of compressed nanotubes is negligible with little or no upshift observed in the spectrum. As the P3HT quantity is raised, the upshift increases up to a maximum of 11 cm^{-1} for the 0.5 w/w sample. A relative downshift of 3 cm^{-1} is observed for 1:1 w/w sample reported in Fig. 41b.

The origin of the relative downshift at high polymer loads can be attributed to charge migration from the P3HT. It has been reported that negative charge-doped

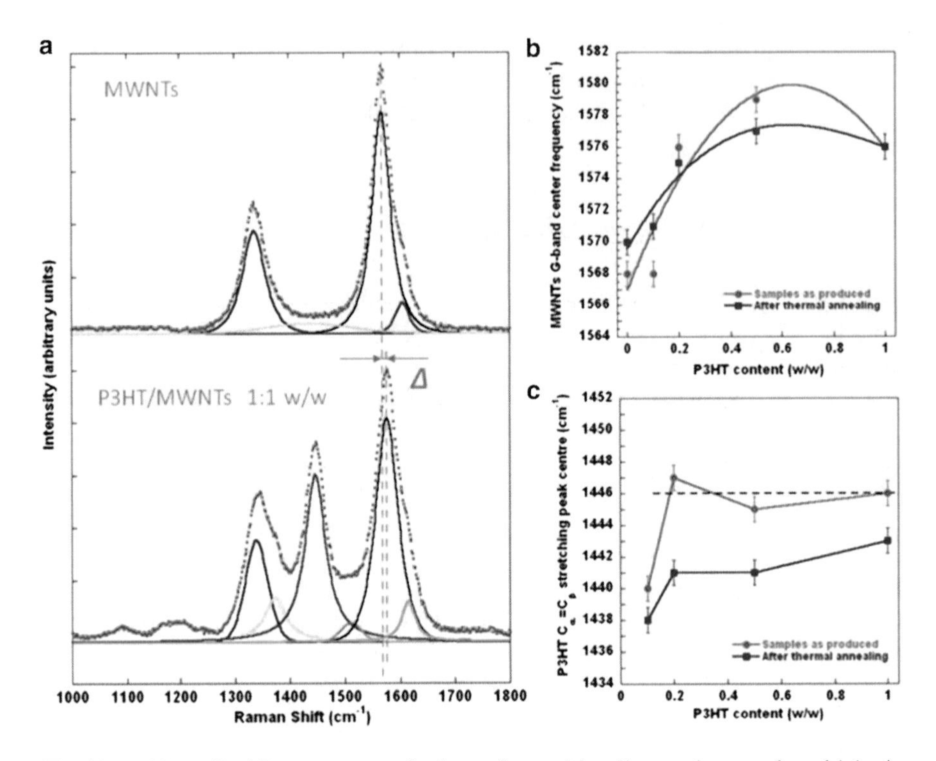

Fig. 41 (**a**) Normalised Raman spectra of polymer-free multi-wall nanotubes sample and 1:1 w/w MWNT/P3HT sample evidencing the compression-induced G-band peak upshift, indicated with Δ. The G-band peak for pristine MWNT samples is centred at 1,568 cm^{-1}. (**b**) Wave-number shift of multi-wall carbon nanotubes G-band versus P3HT relative content. (**c**) P3HT highest intensity peak shift as the compound relative content of the polymer is raised (From Ref. [157])

nanotubes exhibit a G-band displacement to lower wave numbers [117]. It can be proposed that the compression effect is predominant until the nanotubes are completely wrapped or the amount of charge transferred (proportional to the polymer content) equilibrates, eventually reversing the peak shift. Considering that both compression and doping are related to P3HT adsorption on the nanotube surface, it can be concluded that the resultant G-band shift is the result of these two phenomena operating in competition. Therefore, the electron migration in the reported measurements placed a limit on the compression-related G-band peak upshift that could have otherwise resulted in a value higher than the value measured.[3] Although it can be assessed that a completely covered MWNTs is characterised by an upshifted G-band, it must be observed that a partially coiled nanotube, like the ones reported

[3] It must be noticed that a charge migration from the nanotube to the polymer could have supported upshifts higher than expected, but would have failed to explain the relative downshift at higher polymer contents.

previously in the text, is likely to show a Raman spectrum which could be more affected by the polymer doping rather than the compression. Further studies on limited number of nanotube spectra are being pursued at the moment to elucidate this point.

Raman spectra of the same samples were next collected after a low-temperature annealing (120°C). The annealing step was expected to lower the compressive forces on the carbon nanotubes due to polymer's reorganisation and thermal relaxation. As can be seen in Fig. 41b, after the thermal treatment, the maximum upshift observed is reduced to 9 cm^{-1}. Focussing on the P3HT $C_{\alpha}{=}C_{\beta}$ stretching vibration mode in the Raman spectra, Fig. 41c shows the measured central frequency of this feature for the different composites. As can be observed, the vibration mode frequency measured at the expected value of 1,446 cm^{-1} is downshifted for the lowest P3HT content sample (1:0.1 w/w). It can be noted that the 6 cm^{-1} observed downshift occurs for the composite with the maximum carbon nanotube surface available for polymer adhesion. Heller et al. [175] reported P3HT C=C stretching modes downshift that coincided with increased values of the polymer's effective conjugation length. It can be therefore considered that the electron-rich MWNT surface plays a significant role in the P3HT self-organisation, leading to polymer segments with order-enhanced structures. In order to confirm this observation, the same analysis was performed on the samples after the thermal annealing. Even at the highest content ratio, the P3HT main peak is downshifted by a few wave numbers that can be interpreted as evidence for the contribution of P3HT segments possessing extended conjugation lengths interacting with nanotubes.[4]

Further support of the effect of thermal annealing on P3HT self-assembly onto MWNTs is provided by continuously acquired HR-TEM images of a polymer-covered nanotube under high-energy electron beam.[5] Figure 42 shows four images collected at different times clearly showing the polymer reorganisation onto the nanotube surface.

If Fig. 42a is taken as reference, in Fig. 42a b new polymer-induced defect (indicated with α) is observed on the MWNT structure after more than 11 min electron beam exposure. The following image (c) shows two other defective features, indicated with β and γ. In the same figure, the compressed zone α is more evident and generates a sizable, localised inner diameter reduction. After almost 24 min of electron beam exposure, the polymer is still rearranging its structure on the nanotube surface while the defect β is barely visible and the induced elastic compression in α is no longer

[4] Although thermal annealing can enhance the polymer crystallisation order, Raman spectra collected on pristine P3HT samples before and after the thermal treatment did not show any downshift. This confirms that the nanotube sidewall structure can act as template to promote a more ordered polymer phase.

[5] With high-energy electron beam is intended a radiation able to raise the temperature of the sample under analysis. Although there is no control of the temperature of the sample in the time frame under analysis, damages to the nanotube can be excluded. The damage of the sample occurred after a long-time exposure (not shown).

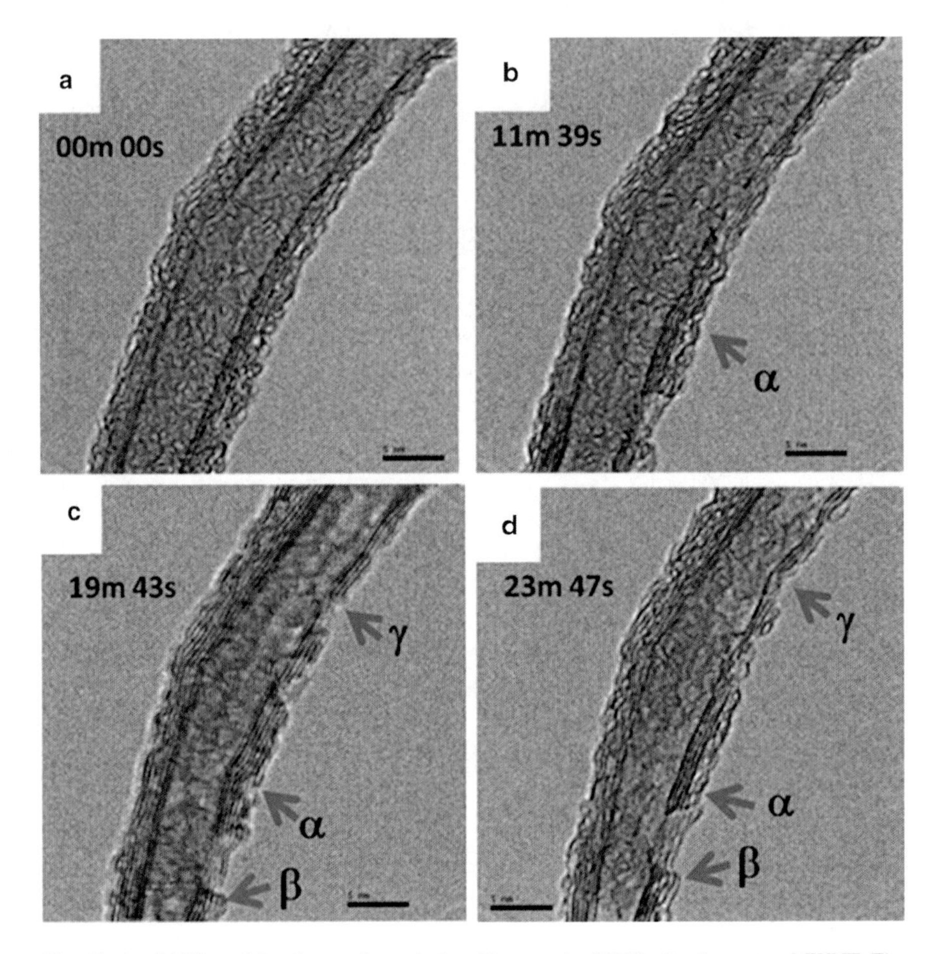

Fig. 42 (**a–d**) Effect of the electron beam induced heat on the P3HT adhesion onto a MWNT. The sizable effect of the polymer-induced compression is shown with consecutive images collected at different times on the same structure. Scale bar is 5 nm for all images (From Ref. [157])

perceptible. Conversely, compression point γ becomes more obvious and the nanotube diameter reduction appears at another location of the structure. The elastic restoration of structural defect α proves that up to 24 min of analysis after the reference time, no irreversible nanotube deformation or structural damage has been induced by the electron beam.

7 Molecular Dynamics Simulation

Classical molecular dynamics (MD) simulations can be successfully used to give a clear understanding of stable polymer conformation and to provide useful information on the nanoscale self-assembly processes. Two recent studies have been proposed on the P3HT/SWNT self-assembly.

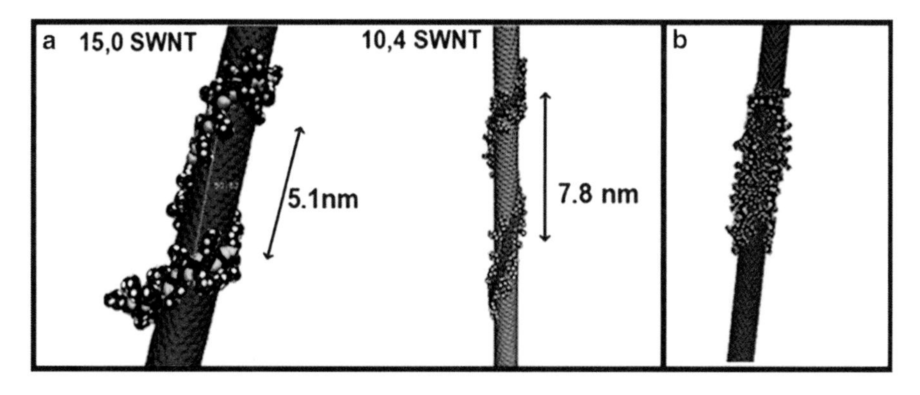

Fig. 43 (**a**) P3HT equilibrium self-assembly around (15,0) and (10,4) SWNTs. The distance between the polymer coils is in good agreement with experimental data observed in the previous section. (**b**) A more disordered stable conformation of a (15,0) SWNT wrapped by P3HT. The image is very close to the observed P3HT covered SWNT reported in Fig. 39c. From [24]

Bernardi et al. [24] showed that P3HT stable helical wrapping of single-wall carbon nanotubes can occur as depicted in Fig. 43a. As it can be observed, during the interaction, helical structures of P3HT can self-assemble on the nanotube surface. The result obtained has been originated starting from a P3HT 20-mer chain which was initialised at a distance of about 1 nm from the SWNT surface and then relaxed before the simulation run. It has been observed that after the equilibration, it is the competition between the π–π stacking interaction and the torsional rigidity of the polymer that drives the chain dynamics. In Fig. 43a, it is evident how the tendency of P3HT to coil the SWNT is not affected by the chirality of the nanotube which can be either zigzag (15,0) or armchair (10,4). It can be observed that the distance between the coils is similar to the one experimentally measured via scanning tunnelling microscopy as reported in Fig. 28. It is also interesting to consider that bundled conformations of the polymer can originate in particular conditions. Figure 43b shows a 50-mer P3HT chain kinetically trapped in a disordered self-assembled structure. This conformation is incredibly similar to Fig. 39c in which a P3HT chain is shown wrapping a SWNT in a disordered manner.

The investigation of the polymer chain structure adsorbed on the nanotube surface can provide interesting information on the resultant conjugation length. In particular, as anticipated by experimental results, MD simulations can demonstrate if the ordered, one-dimensional (1D) structure of the carbon nanotube can limit the torsional disorder of the P3HT upon adsorption. To explore this effect, a large number of simulation runs were performed considering P3HT chains in the presence of carbon nanotubes. Both concentrations and the length of the SWNTs involved in the interaction were varied. After the P3HT adsorption process was concluded, the results for each run were averaged. The analysis for all the cases explored is reported in Fig. 44 where the fraction of conjugation-breaking torsional angles (>42.5 [176, 177]) indicated with f is calculated for different SWNT weight contents and lengths. It must be noted that $1/f$ is the resulting average conjugation

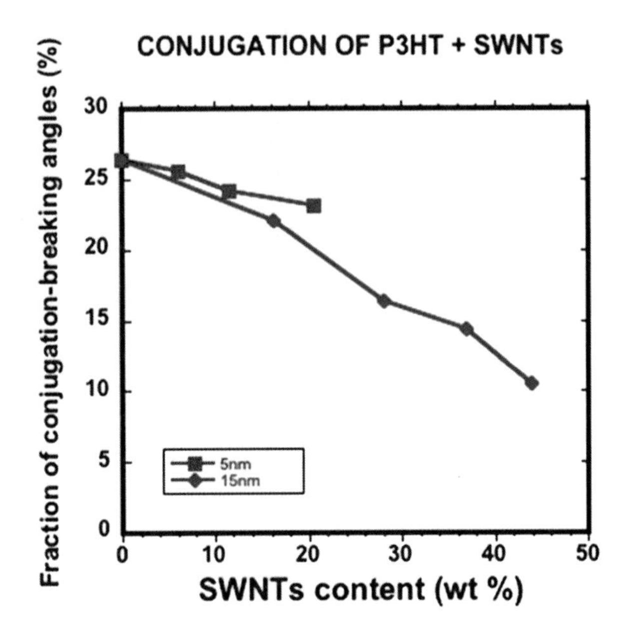

Fig. 44 Fraction of conjugation-breaking angles plotted against the SWNT concentration. The value decreases monotonically as the SWNT concentration is increased. When longer SWNTs are considered, the decrease rate is higher due to the larger SWNT surface available. From [24]

length of the polymer expressed in terms of thiophene rings. At the beginning, it is considered the case of 0% SWNT load (pure P3HT in the form of 20-mer chains at 1/40 the condensed phase density). In these conditions, it is found that the f is 26.5%, which means that the conjugation length is approximately four thiophene rings. Since for a single chain the value found is the same, it can be observed that the P3HT–P3HT interactions do not affect the intrinsic degree of freedom of the polymer chains. Alternatively, when the weight load of SWNTs is increased, a monotonic decrease of f is found as depicted in Fig. 44 for 5- and 15-nm-long nanotubes. It is evident that the introduction of extended length SWNTs can improve more effectively the degree of conjugation of the polymer. This can be interpreted as a direct consequence of the increased templating SWNT surface available for the polymer. In fact, the reduced torsional disorder of the polymer chains is due to the SWNT external surface on which the P3HT chains stack by π–π coplanar interaction. It is interesting to point out that if 15-nm-long SWNTs are considered, the average conjugation length of the polymer chains increases from approximately four thiophene rings for 0% SWNT content to ten rings for a concentration of 40%. This result confirms the observation made previously in UV–Vis and Raman spectra.

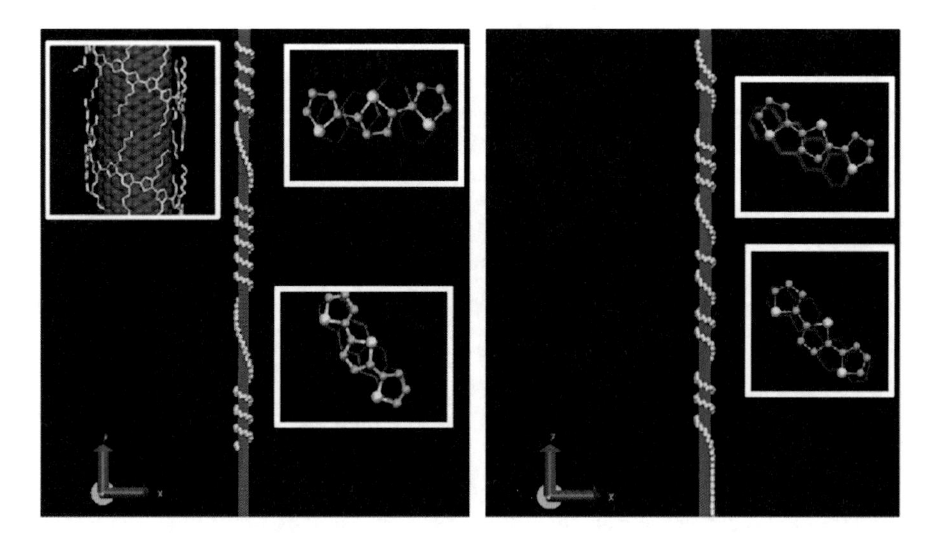

Fig. 45 P3HT helical structure on (15,0) zig-zag (*left*) and (9,9) armchair (*right*) nanotubes after 0.5 ns at 300 K. Only the sulphur atoms are reported for clarity as *yellow balls*. The *insets* show the details of the interdigitated structure of the hexyl chains and the tiophene/nanotube lattice matching. From [25]

Another interesting result has been obtained by Caddeo et al. [25] which explored the various P3HT conformations observed on zigzag and armchair nanotubes. In Fig. 45 there are two out-of-equilibrium P3HT structures coiling a (15,0) and a (9,9) SWNTs, exhibiting different polymer structure morphologies. Molecular dynamics simulations show that during a 2-ns annealing run at 300 K, both structures unwrap, following dissimilar evolution paths, differing in terms of coiling angles and their relative counts. In the zigzag case, the polymer structure is constituted by approximately 13 coils at $\sim 60^\circ$ and 2 coils at $\sim 30^\circ$. In the armchair case, approximately four coils show an angle of $\sim 45^\circ$, whereas approximately eight coils exhibit angles of $\sim 30^\circ$. The first important observation is that the SWNT chirality affects the polymer self-assembly morphology. The figure insets report the relative positions of the thiophene rings and nanotube lattice for the two cases. It can be noticed that the angles observed in the MD simulations are in good agreement to the measured angles.

A final observation can be made on the P3HT self-assembled structure on the (15,0) nanotube surface. It can be noticed that the coiled polymer forms periodic clusters of two or more neighbouring coils along the nanotube structure. This behaviour is due to the polymer–polymer interchain interaction, which originates the interdigitated structure reported in the inset of Fig. 45.

8 Summary

In this chapter, we presented a review of the structural and electronic properties of P3HT and CNTs composites. Observations made by using bulk properties techniques (AFM, TEM, UV–Vis, Raman) have been extended by atomic surface analysis (STM, STS) and by molecular dynamic simulations:

- Low-resolution AFM and TEM images show that P3HT/CNT form homogeneous composites in which nanotubes are fully embedded in the polymer structure. These structural findings are confirmed by the changes observed in the CNT Raman vibrational modes of the mixed samples.
- Optical and morphological analyses show that the polymer order is enhanced after the introduction of the nanotubes in the final composite. The interaction with nanotubes increases the polymer ordering as evidenced by the characteristic red shift of the UV–Vis composites spectra and by the P3HT $C_{\alpha}=C_{\beta}$ stretching vibration modes in the Raman spectra.
- Atomic scale microscopy investigations (STM and HR-TEM) unveiled the details of the P3HT self-assembly around the nanotube, showing the polymer backbone tracing helical structures along the nanotube main axis. In the most ordered structures, it has been possible to relate the coiling frequency to the nanotube diameter and the coiling angle. These images show the polymer structure stacking on the nanotube at an average distance of 0.36 nm, which is compatible with previous measurements on graphite and with the expected nanotube–polymer π–π interaction. This is the bond believed to be responsible for the chirality-driven polymer adhesion, evidenced by the coincidence of the coiling angle with the main axes of the nanotube surface.
- Large-diameter carbon nanotubes are compressed by the polymer wrapping; this deformation is especially visible in the outer shells by HR-TEM. The ability of the polymer to perturb the nanotube structure has been demonstrated by using literature data regarding the nanotube compressibility and the polymer Young's module. The polymer compression leads to a considerable upshift of the G-band peak position in the Raman spectra. The expected shift of 7 cm^{-1} has been observed experimentally, although mixed with other shifts related to electron transfer from polymer to nanotube in the mixture, confirming the effect of the polymer adhesion on nanotubes.
- High-resolution scanning tunnelling spectroscopy acquired on bare sections of carbon nanotubes show a density of states in good agreement with independent theoretical calculation for isolated tubes. Remarkably, the DOS is not affected by the polymer presence surrounding the point under observation, while the Fermi level results shifted on polymer-covered sections. This leads to the conclusion that poly(3-hexylthiophene) transfers electrons to metallic nanotubes, shifting the Fermi level of the tubes toward the vacuum level and lowering the likelihood of charge separation in photovoltaic devices.
- All the microscopic observations have been confirmed by two theoretical works, the first showing the behaviour of an ensemble of nanotubes and P3HT chains in

solution by Monte Carlo simulation, and the other evaluating the minimum energy configurations of a P3HT strand on nanotubes with various chirality.

Acknowledgements The authors are thankful to Prof. M. De Crescenzi and Prof. J.M. Bell for helpful discussions and direction, Dr. E.R. Waclawik, Dr M. Scarselli, Dr. P. Castrucci, Dr. M. Diociaiuti, Dr. S. Casciardi for helping in the measurements and for their contribution to the discussions, Prof. J.C. Grossman, M. Bernardi for providing their expertise in the molecular dynamics calculations. The authors also acknowledge the financial support of the Queensland Government through the NIRAP project "Solar Powered Nanosensors."

References

1. Kymakis, E., Alexandou, I., Amaratunga, G.A.J.: Single-walled carbon nanotube-polymer composites: electrical, optical and structural investigation. Synth. Met. **127**(1–3), 59–62 (2002)
2. Kymakis, E., Alexandrou, I., Amaratunga, G.A.J.: High open-circuit voltage photovoltaic devices from carbon-nanotube-polymer composites. J. Appl. Phys. **93**(3), 1764–1768 (2003)
3. Kymakis, E., Amaratunga, G.A.J.: Single-wall carbon nanotube/conjugated polymer photovoltaic devices. Appl. Phys. Lett. **80**(1), 112–114 (2002)
4. Pyo, M., Bae, E.G., Cho, Y., Jung, Y.S., Zong, K.: Composites of low bandgap conducting polymer-wrapped MWNT and poly(methyl methacrylate) for low percolation and high transparency. Synth. Met. **160**(19–20), 2224–2227 (2010). doi:10.1016/j.synthmet.2010.07.032
5. Coleman, J.N., Cadek, M., Blake, R., Nicolosi, V., Ryan, K.P., Belton, C., Fonseca, A., Nagy, J.B., Gun'ko, Y.K., Blau, W.J.: High-performance nanotube-reinforced plastics: understanding the mechanism of strength increase. Adv. Funct. Mater. **14**(8), 791–798 (2004)
6. Wall, A., Coleman, J.N., Ferreira, M.S.: Physical mechanism for the mechanical reinforcement in nanotube-polymer composite materials. Phys. Rev. B **71**(12), 125421 (2005)
7. O'Connell, M.J., Boul, P., Ericson, L.M., Huffman, C., Wang, Y.H., Haroz, E., Kuper, C., Tour, J., Ausman, K.D., Smalley, R.E.: Reversible water-solubilization of single-walled carbon nanotubes by polymer wrapping. Chem. Phys. Lett. **342**(3–4), 265–271 (2001)
8. Star, A., Steuerman, D.W., Heath, J.R., Stoddart, J.F.: Starched carbon nanotubes. Angew. Chem. **114**(14), 2618–2622 (2002). doi:10.1002/1521-3757(20020715)114:14<2618::aid-ange2618>3.0.co;2-4
9. Kim, O.-K., Je, J., Baldwin, J.W., Kooi, S., Pehrsson, P.E., Buckley, L.J.: Solubilization of single-wall carbon nanotubes by supramolecular encapsulation of helical amylose. J. Am. Chem. Soc. **125**(15), 4426–4427 (2003). doi:10.1021/ja029233b
10. Chen, J., Dyer, M.J., Yu, M.-F.: Cyclodextrin-mediated soft cutting of single-walled carbon nanotubes. J. Am. Chem. Soc. **123**(25), 6201–6202 (2001). doi:10.1021/ja015766t
11. Chambers, G., Carroll, C., Farrell, G.F., Dalton, A.B., McNamara, M., In het Panhuis, M., Byrne, H.J.: Characterization of the interaction of gamma cyclodextrin with single-walled carbon nanotubes. Nano Lett. **3**(6), 843–846 (2003). doi:10.1021/nl034181p
12. Dodziuk, H., Ejchart, A., Anczewski, W., Ueda, H., Krinichnaya, E., Dolgonos, G., Kutner, W.: Water solubilization, determination of the number of different types of single-wall carbon nanotubes and their partial separation with respect to diameters by complexation with [small eta]-cyclodextrin. Chem. Commun. **8**, 986–987 (2003)
13. Dieckmann, G.R., Dalton, A.B., Johnson, P.A., Razal, J., Chen, J., Giordano, G.M., Muñoz, E., Musselman, I.H., Baughman, R.H., Draper, R.K.: Controlled assembly of carbon nanotubes by designed amphiphilic peptide helices. J. Am. Chem. Soc. **125**(7), 1770–1777 (2003). doi:10.1021/ja029084x

14. Chen, R.J., Zhang, Y., Wang, D., Dai, H.: Noncovalent sidewall functionalization of single-walled carbon nanotubes for protein immobilization. J. Am. Chem. Soc. **123**(16), 3838–3839 (2001). doi:10.1021/ja010172b
15. Zheng, M., Jagota, A., Semke, E.D., Diner, B.A., McLean, R.S., Lustig, S.R., Richardson, R.E., Tassi, N.G.: DNA-assisted dispersion and separation of carbon nanotubes. Nat. Mater. **2**(5), 338–342 (2003)
16. Zheng, M., Jagota, A., Strano, M.S., Santos, A.P., Barone, P., Chou, S.G., Diner, B.A., Dresselhaus, M.S., McLean, R.S., Onoa, G.B., Samsonidze, G.G., Semke, E.D., Usrey, M., Walls, D.J.: Structure-based carbon nanotube sorting by sequence-dependent DNA assembly. Science **302**(5650), 1545–1548 (2003)
17. Kang, Y.K., Lee, O.-S., Deria, P., Kim, S.H., Park, T.-H., Bonnell, D.A., Saven, J.G., Therien, M.J.: Helical wrapping of single-walled carbon nanotubes by water soluble poly (p-phenyleneethynylene). Nano Lett. **9**(4), 1414–1418 (2009). doi:10.1021/nl8032334
18. Naito, M., Nobusawa, K., Onouchi, H., Nakamura, M., Yasui, K.-I., Ikeda, A., Fujiki, M.: Stiffness- and conformation-dependent polymer wrapping onto single-walled carbon nanotubes. J. Am. Chem. Soc. **130**(49), 16697–16703 (2008). doi:10.1021/ja806109z
19. Yi, W., Malkovskiy, A., Chu, Q., Sokolov, A.P., Colon, M.L., Meador, M., Pang, Y.: Wrapping of single-walled carbon nanotubes by a π-conjugated polymer: the role of polymer conformation-controlled size selectivity. J. Phys. Chem. B **112**(39), 12263–12269 (2008). doi:10.1021/jp804083n
20. Curran, S.A., Ajayan, P.M., Blau, W.J., Carroll, D.L., Coleman, J.N., Dalton, A.B., Davey, A.P., Drury, A., McCarthy, B., Maier, S., Strevens, A.: A composite from poly(m-phenylene-vinylene-co-2,5-dioctoxy-p-phenylenevinylene) and carbon nanotubes: a novel material for molecular optoelectronics. Adv. Mater. **10**(14), 1091 (1998)
21. Star, A., Stoddart, J.F., Steuerman, D., Diehl, M., Boukai, A., Wong, E.W., Yang, X., Chung, S.-W., Choi, H., Heath, J.R.: Preparation and properties of polymer-wrapped single-walled carbon nanotubes. Angew. Chem. **113**(9), 1771–1775 (2001)
22. Star, A., Stoddart, J.F.: Dispersion and solubilization of single-walled carbon nanotubes with a hyperbranched polymer. Macromolecules **35**(19), 7516–7520 (2002). doi:10.1021/ma0204150
23. Keogh, S.M., Hedderman, T.G., Lynch, P., Farrell, G.F., Byrne, H.J.: Bundling and diameter selectivity in HiPco SWNTs poly(p-phenylene vinylene-co-2,5-dioctyloxy-m-phenylene vinylene) composites. J. Phys. Chem. B **110**(39), 19369–19374 (2006). doi:10.1021/jp056321k
24. Bernardi, M., Giulianini, M., Grossman, J.C.: Self-assembly and its impact on interfacial charge transfer in carbon nanotube/P3HT solar cells. ACS Nano **4**(11), 6599–6606 (2010). doi:10.1021/nn1018297
25. Caddeo, C., Melis, C., Colombo, L., Mattoni, A.: Understanding the helical wrapping of poly (3-hexylthiophene) on carbon nanotubes. J. Phys. Chem. C **114**(49), 21109–21113 (2010). doi:10.1021/jp107370v
26. Tallury, S.S., Pasquinelli, M.A.: Molecular dynamics simulations of flexible polymer chains wrapping single-walled carbon nanotubes. J. Phys. Chem. B **114**(12), 4122–4129 (2010). doi:10.1021/jp908001d
27. Liu, Y., Chipot, C., Shao, X., Cai, W.: Solubilizing carbon nanotubes through noncovalent functionalization. Insight from the reversible wrapping of alginic acid around a single-walled carbon nanotube. J. Phys. Chem. B **114**(17), 5783–5789 (2010). doi:10.1021/jp9110772
28. Gurevitch, I., Srebnik, S.: Monte Carlo simulation of polymer wrapping of nanotubes. Chem. Phys. Lett. **444**(1–3), 96–100 (2007). doi:10.1016/j.cplett.2007.06.112
29. Gurevitch, I., Srebnik, S.: Conformational behavior of polymers adsorbed on nanotubes. J. Chem. Phys. **128**(14), 144901–144908 (2008). doi:10.1063/1.2894842
30. Srebnik, S.: Physical association of polymers with nanotubes. J. Polym. Sci. B Polym. Phys. **46**(24), 2711–2718 (2008). doi:10.1002/polb.21605

31. Bolto, B.A., McNeill, R., Weiss, D.E.: Electronic conduction in polymers. III. Electronic properties of polypyrrole. Aust. J. Chem. **16**(6), 1090–1103 (1963)
32. Shirakawa, H., Louis, E.J., MacDiarmid, A.G., Chiang, C.K., Heeger, A.J.: Synthesis of electrically conducting organic polymers: halogen derivatives of polyacetylene, (CH)x. J. Chem. Soc. Chem. Commun. **16**, 578–580 (1977)
33. Malliaras, G., Friend, R.: An organic electronics primer. Phys. Today **58**(5), 53–58 (2005)
34. Roncali, J.: Conjugated poly(thiophenes): synthesis, functionalization, and applications. Chem. Rev. **92**(4), 711–738 (2002)
35. Naarmann, H.: Polymers, electrically conducting. In: Ullmann's Encyclopedia of Industrial Chemistry. Wiley-VCH, Weinheim (2002). doi:10.1002/14356007.a21_429
36. Pope, M., Swenberg, C.E.: Electronic Processes in Organic Crystals and Polymers, 2nd edn. Oxford University Press, New York (1999)
37. Moliton, A.: Optoelectronics of Molecules and Polymers. Springer, New York (2006)
38. Somoza, M.M.: Depiction of Franck–Condon principle in absorption and fluorescence. http://en.wikipedia.org/wiki/File:Franck-Condon-diagram.png (2006)
39. Somoza, M.M.: Depiction of absorption and fluorescence progression due to changes in vibrational levels during electronic transition. http://en.wikipedia.org/wiki/File:Vibration-fluor-abs.png (2006)
40. Peumans, P., Bulovic, V., Forrest, S.R.: Efficient, high-bandwidth organic multilayer photodetectors. Appl. Phys. Lett. **76**(26), 3855–3857 (2000)
41. Jerome, D., Bechgaard, K.: Condensed-matter physics: superconducting plastic. Nature **410** (6825), 162–163 (2001)
42. Hugger, S., Thomabb, R., Heinzel, T., Thurn-Albrecht, T.: Semicrystalline morphology in thin films of poly(3-hexylthiophene). Colloid Polym. Sci. **282**, 932–938 (2004)
43. Prosa, T.J., Winokur, M.J., Moulton, J., Smith, P., Heeger, A.J.: X-ray structural studies of poly(3-alkylthiophenes): an example of an inverse comb. Macromolecules **25**(17), 4364–4372 (1992)
44. Chung, T.C., Kaufman, J.H., Heeger, A.J., Wudl, F.: Charge storage in doped poly(thiophene): optical and electrochemical studies. Phys. Rev. B **30**(2), 702 (1984)
45. Mintmire, J.W., White, C.T., Elert, M.L.: Heteroatom effects in heterocyclic ring chain polymers. Synth. Met. **16**(2), 235–243 (1986)
46. Kaneto, K., Kohno, Y., Yoshino, K.: Absorption spectra induced by photoexcitation and electrochemical doping in polythiophene. Solid State Commun. **51**(5), 267–269 (1984)
47. Harbeke, G., Meier, E., Kobel, W., Egli, M., Kiess, H., Tosatti, E.: Spectroscopic evidence for polarons in poly(3-methylthiophene). Solid State Commun. **55**(5), 419–422 (1985)
48. Sirringhaus, H., Wilson, R.J., Friend, R.H., Inbasekaran, M., Wu, W., Woo, E.P., Grell, M., Bradley, D.D.C.: Mobility enhancement in conjugated polymer field-effect transistors through chain alignment in a liquid-crystalline phase. Appl. Phys. Lett. **77**(3), 406–408 (2000)
49. Nalwa, H.S.: Handbook of Organic Conductive Molecules and Polymers, vol. 3. Wiley, New York (1997)
50. Kline, R.J., McGehee, M.D., Kadnikova, E.N., Liu, J., Frechet, J.M.J.: Controlling the field-effect mobility of regioregular polythiophene by changing the molecular weight. Adv. Mater. **15**(18), 1519–1522 (2003)
51. Joung, M.J., Kim, C.A., Kang, S.Y., Baek, K.-H., Kim, G.H., Ahn, S.D., You, I.K., Ahn, J.H., Suh, K.S.: The application of soluble and regioregular poly(3-hexylthiophene) for organic thin-film transistors. Synth. Met. **149**(1), 73–77 (2005)
52. Goh, C., Kline, R.J., McGehee, M.D., Kadnikova, E.N., Frechet, J.M.J.: Molecular-weight-dependent mobilities in regioregular poly(3-hexyl-thiophene) diodes. Appl. Phys. Lett. **86** (12), 122110–122113 (2005)
53. Hotta, S., Soga, M., Sonoda, N.: Novel organosynthetic routes to polythiophene and its derivatives. Synth. Met. **26**(3), 267–279 (1988)

54. Kaminorz, Y., Smela, E., Inganäs, O., Brehmer, L.: Sensitivity of polythiophene planar light-emitting diodes to oxygen. Adv. Mater. **10**(10), 765–769 (1998)
55. Abdou, M.S.A., Orfino, F.P., Xie, Z.W., Deen, M.J., Holdcroft, S.: Reversible charge-transfer complexes between molecular-oxygen and poly(3-alkylthiophene)S. Adv. Mater. **6**(11), 838–841 (1994)
56. Dennler, G., Lungenschmied, C., Neugebauer, H., Sariciftci, N.S., Latrèche, M., Czeremuszkin, G., Wertheimer, M.R.: A new encapsulation solution for flexible organic solar cells. Thin Solid Films **511–512**, 349–353 (2006)
57. Iijima, S.: Helical microtubules of graphitic carbon. Nature **354**(6348), 56–58 (1991)
58. Yu, M.-F., Lourie, O., Dyer, M.J., Moloni, K., Kelly, T.F., Ruoff, R.S.: Strength and breaking mechanism of multiwalled carbon nanotubes under tensile load. Science **287**(5453), 637–640 (2000). doi:10.1126/science.287.5453.637
59. Dalton, A.B., Collins, S., Munoz, E., Razal, J.M., Ebron, V.H., Ferraris, J.P., Coleman, J.N., Kim, B.G., Baughman, R.H.: Super-tough carbon-nanotube fibres. Nature **423**(6941), 703 (2003)
60. Iijima, S., Brabec, C., Maiti, A., Bernholc, J.: Structural flexibility of carbon nanotubes. J. Chem. Phys. **104**(5), 2089–2092 (1996)
61. Kim, P., Shi, L., Majumdar, A., McEuen, P.L.: Thermal transport measurements of individual multiwalled nanotubes. Phys. Rev. Lett. **87**(21), 215502 (2001)
62. Baughman, R.H., Cui, C., Zakhidov, A.A., Iqbal, Z., Barisci, J.N., Spinks, G.M., Wallace, G.G., Mazzoldi, A., De Rossi, D., Rinzler, A.G., Jaschinski, O., Roth, S., Kertesz, M.: Carbon nanotube actuators. Science **284**(5418), 1340–1344 (1999). doi:10.1126/science.284.5418.1340
63. Ströck, M.: Allotropes of carbon. http://en.wikipedia.org/wiki/File:Eight_Allotropes_of_Carbon.png (2006)
64. Ebbesen, T.W., Lezec, H.J., Hiura, H., Bennett, J.W., Ghaemi, H.F., Thio, T.: Electrical conductivity of individual carbon nanotubes. Nature **382**(6586), 54–56 (1996)
65. Saito, R., Dresselhaus, M.S., Dresselhaus, G.: Physical Properties of Carbon Nanotubes. Imperial College Press, London (1998)
66. Collins, P.G., Bradley, K., Ishigami, M., Zettl, A.: Extreme oxygen sensitivity of electronic properties of carbon nanotubes. Science **287**(5459), 1801–1804 (2000). doi:10.1126/science.287.5459.1801
67. Cantalini, C., Valentini, L., Armentano, I., Kenny, J.M., Lozzi, L., Santucci, S.: Carbon nanotubes as new materials for gas sensing applications. J. Eur. Ceram. Soc. **24**(6), 1405–1408 (2004)
68. Kong, J., Franklin, N.R., Zhou, C., Chapline, M.G., Peng, S., Cho, K., Dai, H.: Nanotube molecular wires as chemical sensors. Science **287**(5453), 622–625 (2000). doi:10.1126/science.287.5453.622
69. Zhu, W., Bower, C., Zhou, O., Kochanski, G., Jin, S.: Large current density from carbon nanotube field emitters. Appl. Phys. Lett. **75**(6), 873–875 (1999)
70. Murakami, H., Hirakawa, M., Tanaka, C., Yamakawa, H.: Field emission from well-aligned, patterned, carbon nanotube emitters. Appl. Phys. Lett. **76**(13), 1776–1778 (2000)
71. Kim, P., Lieber, C.M.: Nanotube nanotweezers. Science **286**(5447), 2148–2150 (1999). doi:10.1126/science.286.5447.2148
72. Hafner, J.H., Cheung, C.L., Lieber, C.M.: Growth of nanotubes for probe microscopy tips. Nature **398**(6730), 761–762 (1999)
73. Hinds, B.J., Chopra, N., Rantell, T., Andrews, R., Gavalas, V., Bachas, L.G.: Aligned multiwalled carbon nanotube membranes. Science **303**(5654), 62–65 (2004). doi:10.1126/science.1092048
74. O'Connell, M.: Carbon Nanotubes Properties and Applications. Taylor & Francis, Boca Raton (2006)
75. Iijima, S., Ichihashi, T.: Single-shell carbon nanotubes of 1-nm diameter. Nature **363**(6430), 603–605 (1993)

76. Bethune, D.S., Klang, C.H., de Vries, M.S., Gorman, G., Savoy, R., Vazquez, J., Beyers, R.: Cobalt-catalysed growth of carbon nanotubes with single-atomic-layer walls. Nature **363** (6430), 605–607 (1993)
77. Journet, C., Maser, W.K., Bernier, P., Loiseau, A., de la Chapelle, M.L., Lefrant, S., Deniard, P., Lee, R., Fischer, J.E.: Large-scale production of single-walled carbon nanotubes by the electric-arc technique. Nature **388**(6644), 756–758 (1997)
78. Waldorff, E.I., Waas, A.M., Friedmann, P.P., Keidar, M.: Characterization of carbon nanotubes produced by arc discharge: effect of the background pressure. J. Appl. Phys. **95** (5), 2749–2754 (2004)
79. Thess, A., Lee, R., Nikolaev, P., Dai, H., Petit, P., Robert, J., Xu, C., Lee, Y.H., Kim, S.G., Rinzler, A.G., Colbert, D.T., Scuseria, G.E., Tomanek, D., Fischer, J.E., Smalley, R.E.: Crystalline ropes of metallic carbon nanotubes. Science **273**(5274), 483–487 (1996). doi:10.1126/science.273.5274.483
80. Liu, J., Rinzler, A.G., Dai, H., Hafner, J.H., Bradley, R.K., Boul, P.J., Lu, A., Iverson, T., Shelimov, K., Huffman, C.B., Rodriguez-Macias, F., Shon, Y.-S., Lee, T.R., Colbert, D.T., Smalley, R.E.: Fullerene pipes. Science **280**(5367), 1253–1256 (1998). doi:10.1126/science.280.5367.1253
81. Ren, Z.F., Huang, Z.P., Xu, J.W., Wang, J.H., Bush, P., Siegal, M.P., Provencio, P.N.: Synthesis of large arrays of well-aligned carbon nanotubes on glass. Science **282**(5391), 1105–1107 (1998). doi:10.1126/science.282.5391.1105
82. Ren, Z.F., Huang, Z.P., Wang, D.Z., Wen, J.G., Xu, J.W., Wang, J.H., Calvet, L.E., Chen, J., Klemic, J.F., Reed, M.A.: Growth of a single freestanding multiwall carbon nanotube on each nanonickel dot. Appl. Phys. Lett. **75**(8), 1086–1088 (1999)
83. Dai, H.: Carbon nanotubes: opportunities and challenges. Surf. Sci. **500**(1–3), 218–241 (2002)
84. Venema, L.C.: Electronic Structure of Carbon Nanotubes. IOS Press, Amsterdam (2000)
85. Avouris, P.: Carbon nanotube electronics. Chem. Phys. **281**(2–3), 429–445 (2002)
86. Venema, L.C., Meunier, V., Lambin, P., Dekker, C.: Atomic structure of carbon nanotubes from scanning tunneling microscopy. Phys. Rev. B **61**(4), 2991 (2000)
87. Mintmire, J.W., White, C.T.: Universal density of states for carbon nanotubes. Phys. Rev. Lett. **81**(12), 2506 (1998)
88. Mintmire, J.W., White, C.T.: Electronic and structural properties of carbon nanotubes. Carbon **33**(7), 893–902 (1995)
89. Akai, Y., Saito, S.: Electronic structure, energetics and geometric structure of carbon nanotubes: a density-functional study. Physica E Low Dimens. Syst. Nanostruct. **29**(3–4), 555–559 (2005)
90. Brown, E., Hao, L., Gallop, J.C., Macfarlane, J.C.: Ballistic thermal and electrical conductance measurements on individual multiwall carbon nanotubes. Appl. Phys. Lett. **87**(2), 023107 (2005)
91. Kymakis, E., Koudoumas, E., Franghiadakis, I., Amaratunga, G.A.J.: Post-fabrication annealing effects in polymer-nanotube photovoltaic cells. J. Phys. D Appl. Phys. **39**(6), 1058–1062 (2006)
92. Miller, A.J., Hatton, R.A., Silva, S.R.P.: Water-soluble multiwall-carbon-nanotube-polythiophene composite for bilayer photovoltaics. Appl. Phys. Lett. **89**(12), 123115 (2006)
93. Miller, A.J., Hatton, R.A., Silva, S.R.P.: Interpenetrating multiwall carbon nanotube electrodes for organic solar cells. Appl. Phys. Lett. **89**(13), 133117 (2006)
94. Patyk, R.L., Lomba, B.S., Nogueira, A.F., Furtado, C., Santos, A.P., Mello, R.M.Q., Micaroni, L., Hümmelgen, I.A.: Carbon nanotube-polybithiophene photovoltaic devices with high open-circuit voltage. Phys. Stat. Sol. Rapid Res. Lett. **1**(1), R43–R45 (2007)
95. Geng, J., Zeng, T.: Influence of single-walled carbon nanotubes induced crystallinity enhancement and morphology change on polymer photovoltaic devices. J. Am. Chem. Soc. **128**(51), 16827–16833 (2006). doi:10.1021/ja065035z

96. Kanai, Y., Grossman, J.C.: Insights on interfacial charge transfer across P3HT/fullerene photovoltaic heterojunction from ab initio calculations. Nano Lett. **7**(7), 1967–1972 (2007). doi:10.1021/nl0707095
97. Kanai, Y., Grossman, J.C.: Role of semiconducting and metallic tubes in P3HT/carbon-nanotube photovoltaic heterojunctions: density functional theory calculations. Nano Lett. **8** (3), 908–912 (2008). doi:10.1021/nl0732777
98. Du Pasquier, A., Unalan, H.E., Kanwal, A., Miller, S., Chhowalla, M.: Conducting and transparent single-wall carbon nanotube electrodes for polymer-fullerene solar cells. Appl. Phys. Lett. **87**(20), 203511 (2005)
99. Rowell, M.W., Topinka, M.A., McGehee, M.D., Prall, H.-J., Dennler, G., Sariciftci, N.S., Hu, L., Gruner, G.: Organic solar cells with carbon nanotube network electrodes. Appl. Phys. Lett. **88**(23), 233506 (2006)
100. van de Lagemaat, J., Barnes, T.M., Rumbles, G., Shaheen, S.E., Coutts, T.J., Weeks, C., Levitsky, I., Peltola, J., Glatkowski, P.: Organic solar cells with carbon nanotubes replacing In[sub 2]O[sub 3]:Sn as the transparent electrode. Appl. Phys. Lett. **88**(23), 233503 (2006)
101. Berson, S., de Bettignies, R., Bailly, S., Guillerez, S., Jousselme, B.: Elaboration of P3HT/CNT/PCBM composites for organic photovoltaic cells. Adv. Funct. Mater. **17**(16), 3363–3370 (2007)
102. Pradhan, B., Batabyal, S.K., Pal, A.J.: Functionalized carbon nanotubes in donor/acceptor-type photovoltaic devices. Appl. Phys. Lett. **88**(9), 093106 (2006)
103. Wu, M.-C., Lin, Y.-Y., Chen, S., Liao, H.-C., Wu, Y.-J., Chen, C.-W., Chen, Y.-F., Su, W.-F.: Enhancing light absorption and carrier transport of P3HT by doping multi-wall carbon nanotubes. Chem. Phys. Lett. **468**(1–3), 64–68 (2009)
104. Kymakis, E., Kornilios, N., Koudoumas, E.: Carbon nanotube doping of P3HT: PCBM photovoltaic devices. J. Phys. D Appl. Phys. **41**(16), 165110 (2008)
105. Viswanathan, G., Chakrapani, N., Yang, H., Wei, B., Chung, H., Cho, K., Ryu, C.Y., Ajayan, P.M.: Single-step in situ synthesis of polymer-grafted single-wall nanotube composites. J. Am. Chem. Soc. **125**(31), 9258–9259 (2003). doi:10.1021/ja0354418
106. Shim, M., Shi Kam, N.W., Chen, R.J., Li, Y., Dai, H.: Functionalization of carbon nanotubes for biocompatibility and biomolecular recognition. Nano Lett. **2**(4), 285–288 (2002). doi:10.1021/nl015692j
107. Barber, A.H., Cohen, S.R., Wagner, H.D.: Measurement of carbon nanotube–polymer interfacial strength. Appl. Phys. Lett. **82**(23), 4140–4142 (2003)
108. Giulianini, M., Waclawik, E.R., Bell, J.M., Scarselli, M., Castrucci, P., De Crescenzi, M. et al.: Microscopic and spectroscopic investigation of Poly(3-hexylthiophene) interaction with carbon nanotubes. Polymers **3**(3), 1433–1446 (2011)
109. Trznadel, M., Pron, A., Zagorska, M., Chrzaszcz, R., Pielichowski, J.: Effect of molecular weight on spectroscopic and spectroelectrochemical properties of regioregular poly(3-hexylthiophene). Macromolecules **31**(15), 5051–5058 (1998). doi:10.1021/ma970627a
110. Brown, P.J., Thomas, D.S., Kohler, A., Wilson, J.S., Kim, J.S., Ramsdale, C.M., Sirringhaus, H., Friend, R.H.: Effect of interchain interactions on the absorption and emission of poly(3-hexylthiophene). Phys. Rev. B **67**(6), 064203 (2003)
111. Hotta, S., Rughooputh, S.D.D.V., Heeger, A.J., Wudl, F.: Spectroscopic studies of soluble poly(3-alkylthienylenes). Macromolecules **20**(1), 212–215 (2002). doi:10.1021/ma00167a038
112. Wu, H.-X., Qiu, X.-Q., Cao, W.-M., Lin, Y.-H., Cai, R.-F., Qian, S.-X.: Polymer-wrapped multiwalled carbon nanotubes synthesized via microwave-assisted in situ emulsion polymerization and their optical limiting properties. Carbon **45**(15), 2866–2872 (2007)
113. Dresselhaus, M.S., Dresselhaus, G., Hofmann, M.: The big picture of Raman scattering in carbon nanotubes. Vib. Spectrosc. **45**(2), 71–81 (2007)
114. McNally, T., Potschke, P., Halley, P., Murphy, M., Martin, D., Bell, S.E.J., Brennan, G.P., Bein, D., Lemoine, P., Quinn, J.P.: Polyethylene multiwalled carbon nanotube composites. Polymer **46**, 8222–8232 (2005)

115. Baibarac, M., Lapkowski, M., Pron, A., Lefrant, S., Baltog, I.: SERS spectra of poly(3-hexylthiophene) in oxidized and unoxidized states. J. Raman Spectrosc. **29**(9), 825–832 (1998)
116. Claye, A., Rahman, S., Fischer, J.E., Sirenko, A., Sumanasekera, G.U., Eklund, P.C.: In situ Raman scattering studies of alkali-doped single wall carbon nanotubes. Chem. Phys. Lett. **333**(1–2), 16–22 (2001)
117. Rao, A.M., Eklund, P.C., Bandow, S., Thess, A., Smalley, R.E.: Evidence for charge transfer in doped carbon nanotube bundles from Raman scattering. Nature **388**(6639), 257–259 (1997)
118. Baskaran, D., Mays, J.W., Bratcher, M.S.: Noncovalent and nonspecific molecular interactions of polymers with multiwalled carbon nanotubes. Chem. Mater. **17**(13), 3389–3397 (2005)
119. Czerw, R., Guo, Z., Ajayan, P.M., Sun, Y.-P., Carroll, D.L.: Organization of polymers onto carbon nanotubes: a route to nanoscale assembly. Nano Lett. **1**(8), 423–427 (2001). doi:10.1021/nl015548y
120. Shan, B., Cho, K.: First-principles study of work functions of double-wall carbon nanotubes. Phys. Rev. B **73**(8), 081401–081404 (2006)
121. Chirvase, D., Chiguvare, Z., Knipper, M., Parisi, J., Dyakonov, V., Hummelen, J.C.: Electrical and optical design and characterisation of regioregular poly(3-hexylthiophene-2,5diyl)/fullerene-based heterojunction polymer solar cells. Synth. Met. **138**(1–2), 299–304 (2003)
122. Gotovac, S., Honda, H., Hattori, Y., Takahashi, K., Kanoh, H., Kaneko, K.: Effect of nanoscale curvature of single-walled carbon nanotubes on adsorption of polycyclic aromatic hydrocarbons. Nano Lett. **7**(3), 583–587 (2007)
123. Lu, J., Nagase, S., Zhang, X., Wang, D., Ni, M., Maeda, Y., Wakahara, T., Nakahodo, T., Tsuchiya, T., Akasaka, T., Gao, Z., Yu, D., Ye, H., Mei, W.N., Zhou, Y.: Selective interaction of large or charge-transfer aromatic molecules with metallic single-wall carbon nanotubes: critical role of the molecular size and orientation. J. Am. Chem. Soc. **128**(15), 5114–5118 (2006). doi:10.1021/ja058214+
124. Chen, F., Wang, B., Chen, Y., Li, L.-J.: Toward the extraction of single species of single-walled carbon nanotubes using fluorene-based polymers. Nano Lett. **7**(10), 3013–3017 (2007)
125. Coleman, J.N., Ferreira, M.S.: Geometric constraints in the growth of nanotube-templated polymer monolayers. Appl. Phys. Lett. **84**(5), 798–800 (2004)
126. Hugelmann, M., Schindler, W.: Schottky diode characteristics of electrodeposited Au/n-Si (111) nanocontacts. Appl. Phys. Lett. **85**(16), 3608–3610 (2004)
127. Gheber, L.A., Hershfinkel, M., Gorodetsky, G., Volterra, V.: Scanning tunneling spectroscopy studies of the Au-H-terminated Si interface. Appl. Phys. Lett. **69**(3), 400 (1996)
128. Wildöer, J.W.G., Venema, L.C., Rinzler, A.G., Smalley, R.E., Dekker, C.: Electronic structure of atomically resolved carbon nanotubes. Nature **391**(6662), 59–62 (1998)
129. Reed, M.A., Zhou, C., Muller, C.J., Burgin, T.P., Tour, J.M.: Conductance of a molecular junction. Science **278**(5336), 252–254 (1997). doi:10.1126/science.278.5336.252
130. Akai-Kasaya, M., Shimizu, K., Watanabe, Y., Saito, A., Aono, M., Kuwahara, Y.: Electronic structure of a polydiacetylene nanowire fabricated on highly ordered pyrolytic graphite. Phys. Rev. Lett. **91**(25), 255501 (2003)
131. Terada, Y., Choi, B.-K., Heike, S., Fujimori, M., Hashizume, T.: Injection of molecules onto hydrogen-terminated Si(100) surfaces via a pulse valve. J. Appl. Phys. **93**(12), 10014–10017 (2003)
132. Cadek, M., Coleman, J.N., Barron, V., Hedicke, K., Blau, W.J.: Morphological and mechanical properties of carbon-nanotube-reinforced semicrystalline and amorphous polymer composites. Appl. Phys. Lett. **81**(27), 5123–5125 (2002)
133. Schadler, L.S., Giannaris, S.C., Ajayan, P.M.: Load transfer in carbon nanotube epoxy composites. Appl. Phys. Lett. **73**(26), 3842–3844 (1998)

134. Zhan, G.-D., Kuntz, J.D., Wan, J., Mukherjee, A.K.: Single-wall carbon nanotubes as attractive toughening agents in alumina-based nanocomposites. Nat. Mater. **2**(1), 38–42 (2003)
135. Bower, C., Rosen, R., Jin, L., Han, J., Zhou, O.: Deformation of carbon nanotubes in nanotube–polymer composites. Appl. Phys. Lett. **74**(22), 3317–3319 (1999)
136. Lourie, O., Cox, D.M., Wagner, H.D.: Buckling and collapse of embedded carbon nanotubes. Phys. Rev. Lett. **81**(8), 1638 (1998)
137. Wagner, H.D., Lourie, O., Feldman, Y., Tenne, R.: Stress-induced fragmentation of multiwall carbon nanotubes in a polymer matrix. Appl. Phys. Lett. **72**(2), 188–190 (1998)
138. Hadjiev, V.G., Iliev, M.N., Arepalli, S., Nikolaev, P., Files, B.S.: Raman scattering test of single-wall carbon nanotube composites. Appl. Phys. Lett. **78**(21), 3193 (2001)
139. Musumeci, A.W., Silva, G.G., Liu, J.-W., Martens, W.N., Waclawik, E.R.: Structure and conductivity of multi-walled carbon nanotube/poly(3-hexylthiophene) composite films. Polymer **48**(6), 1667–1678 (2007)
140. Horcas, I., Fernandez, R., Gomez-Rodriguez, J.M., Colchero, J., Gomez-Herrero, J., Baro, A.M.: WSXM: a software for scanning probe microscopy and a tool for nanotechnology. Rev. Sci. Instrum. **78**(1), 013705–013708 (2007)
141. Note: Polymer 3D structure generated with Archim v2.1. http://www.archimy.com
142. Giulianini, M., Waclawik, E.R., Bell, J.M., De Crescenzi, M., Castrucci, P., Scarselli, M., Motta, N.: Regioregular poly(3-hexyl-thiophene) helical self-organization on carbon nanotubes. Appl. Phys. Lett. 95(1), 013304 (2009)
143. Mena-Osteritz, E.: Superstructures of self-organizing thiophenes. Adv. Mater. **14**(8), 609–616 (2002)
144. Goh, R.G.S., Motta, N., Bell, J.M., Waclawik, E.R.: Effects of substrate curvature on the adsorption of poly(3-hexylthiophene) on single-walled carbon nanotubes. Appl. Phys. Lett. **88**(5), 053101–053103 (2006)
145. Grevin, B., Rannou, P., Payerne, R., Pron, A., Travers, J.P.: Scanning tunneling microscopy investigations of self-organized poly(3-hexylthiophene) two-dimensional polycrystals. Adv. Mater. **15**(11), 881–884 (2003)
146. Mena-Osteritz, E., Meyer, A., Langeveld-Voss, B.M.W., Janssen, R.A.J., Meijer, E.W., Bauerle, P.: Two-dimensional crystals of poly(3-alkylthiophene)s: direct visualization of polymer folds in submolecular resolution. Angew. Chem. **39**(15), 2680–2684 (2000)
147. Brinkmann, M., Wittmann, J.C.: Orientation of regioregular poly(3-hexylthiophene) by directional solidification: a simple method to reveal the semicrystalline structure of a conjugated polymer. Adv. Mater. **18**(7), 860–863 (2006)
148. Giulianini, M., Capasso, A., Waclawik, E., Bell, J., Scarselli, M., Castrucci, P., De Crescenzi, M., Motta, N.: UHV-STM study of P3HT adhesion on carbon nanotubes for solar cells application. Paper presented at the Proceedings of Nanophotonics Down Under 2009 Devices and Applications
149. McCarthy, B., Coleman, J.N., Czerw, R., Dalton, A.B., Panhuis, M.I.H., Maiti, A., Drury, A., Bernier, P., Nagy, J.B., Lahr, B., Byrne, H.J., Carroll, D.L., Blau, W.J.: A microscopic and spectroscopic study of interactions between carbon nanotubes and a conjugated polymer. J. Phys. Chem. B **106**(9), 2210–2216 (2002)
150. Wei, C.: Radius and chirality dependent conformation of polymer molecule at nanotube interface. Nano Lett. **6**(8), 1627–1631 (2006)
151. Kraabel, B., Moses, D., Heeger, A.J.: Direct observation of the intersystem crossing in poly (3-octylthiophene). J. Chem. Phys. **103**(12), 5102–5108 (1995)
152. Giulianini, M., Waclawik, E.R., Bell, J.M., Scarselli, M., Castrucci, P., De Crescenzi, M., Motta, N.: Poly(3-hexyl-thiophene) coil-wrapped single wall carbon nanotube investigated by scanning tunneling spectroscopy. Appl. Phys. Lett. 95(14), 143116 (2009)
153. Ago, H., Kugler, T., Cacialli, F., Salaneck, W.R., Shaffer, M.S.P., Windle, A.H., Friend, R. H.: Work functions and surface functional groups of multiwall carbon nanotubes. J. Phys. Chem. B **103**(38), 8116–8121 (1999). doi:10.1021/jp991659y

154. Ouyang, M., Huang, J.L., Cheung, C.L., Lieber, C.M.: Atomically resolved single-walled carbon nanotube intramolecular junctions. Science **291**(5501), 97–100 (2001)
155. Scifo, L., Dubois, M., Brun, M., Rannou, P., Latil, S., Rubio, A., Grevin, B.: Probing the electronic properties of self-organized poly(3-dodecylthiophene) monolayers by two-dimensional scanning tunneling spectroscopy imaging at the single chain scale. Nano Lett. **6**(8), 1711–1718 (2006)
156. Zhao, J., Han, J., Lu, J.P.: Work functions of pristine and alkali-metal intercalated carbon nanotubes and bundles. Phys. Rev. B **65**(19), 193401 (2002)
157. Giulianini, M., Waclawik, E.R., Bell, J.M., Crescenzi, M.D., Castrucci, P., Scarselli, M., Diociauti, M., Casciardi, S., Motta, N.: Evidence of Multiwall Carbon Nanotube Deformation Caused by Poly(3-hexylthiophene) Adhesion. The Journal of Physical Chemistry C 115(14), 6324-6330 (2011). doi:10.1021/jp2000267
158. Kiang, C.H., Endo, M., Ajayan, P.M., Dresselhaus, G., Dresselhaus, M.S.: Size effects in carbon nanotubes. Phys. Rev. Lett. **81**(9), 1869 (1998)
159. Ruoff, R.S., Tersoff, J., Lorents, D.C., Subramoney, S., Chan, B.: Radial deformation of carbon nanotubes by van der Waals forces. Nature **364**(6437), 514–516 (1993)
160. Chopra, N.G., Benedict, L.X., Crespi, V.H., Cohen, M.L., Louie, S.G., Zettl, A.: Fully collapsed carbon nanotubes. Nature **377**(6545), 135–138 (1995)
161. Hertel, T., Walkup, R.E., Avouris, P.: Deformation of carbon nanotubes by surface van der Waals forces. Phys. Rev. B **58**(20), 13870 (1998)
162. Park, M.H., Jang, J.W., Lee, C.E., Lee, C.J.: Interwall support in double-walled carbon nanotubes studied by scanning tunneling microscopy. Appl. Phys. Lett. **86**(2), 023110–023113 (2005). doi:10.1063/1.1851615
163. Palaci, I., Fedrigo, S., Brune, H., Klinke, C., Chen, M., Riedo, E.: Radial elasticity of multiwalled carbon nanotubes. Phys. Rev. Lett. **94**(17), 175502 (2005)
164. Kelly, B.T.: The Physics of Graphite. Applied Science Publishers, London (2006)
165. Tahk, D., Lee, H.H., Khang, D.-Y.: Elastic moduli of organic electronic materials by the buckling method. Macromolecules **42**(18), 7079–7083 (2009). doi:10.1021/ma900137k
166. Gülseren, O., Yildirim, T., Ciraci, S., Kılıç, Ç.: Reversible band-gap engineering in carbon nanotubes by radial deformation. Phys. Rev. B **65**(15), 155410 (2002)
167. Heo, J., Bockrath, M.: Local electronic structure of single-walled carbon nanotubes from electrostatic force microscopy. Nano Lett. **5**(5), 853–857 (2005). doi:10.1021/nl0501765
168. Bozovic, D., Bockrath, M., Hafner, J.H., Lieber, C.M., Park, H., Tinkham, M.: Electronic properties of mechanically induced kinks in single-walled carbon nanotubes. Appl. Phys. Lett. **78**(23), 3693–3695 (2001)
169. Chun, K.-Y., Lee, C.J.: Potassium doping in the double-walled carbon nanotubes at room temperature. J. Phys. Chem. C **112**(12), 4492–4497 (2008). doi:10.1021/jp077453b
170. Wood, J.R., Frogley, M.D., Meurs, E.R., Prins, A.D., Peijs, T., Dunstan, D.J., Wagner, H.D.: Mechanical response of carbon nanotubes under molecular and macroscopic pressures. J. Phys. Chem. B **103**(47), 10388–10392 (1999). doi:10.1021/jp992136t
171. Venkateswaran, U.D., Rao, A.M., Richter, E., Menon, M., Rinzler, A., Smalley, R.E., Eklund, P.C.: Probing the single-wall carbon nanotube bundle: Raman scattering under high pressure. Phys. Rev. B **59**(16), 10928 (1999)
172. Peters, M.J., McNeil, L.E., Lu, J.P., Kahn, D.: Structural phase transition in carbon nanotube bundles under pressure. Phys. Rev. B **61**(9), 5939 (2000)
173. Thomsen, C., Reich, S., Jantoljak, H., Loa, I., Syassen, K., Burghard, M., Duesberg, G.S., Roth, S.: Raman spectroscopy on single- and multi-walled nanotubes under high pressure. Appl. Phys. A Mater. Sci. Process. **69**(3), 309–312 (1999)
174. Rao, A.M., Jorio, A., Pimenta, M.A., Dantas, M.S.S., Saito, R., Dresselhaus, G., Dresselhaus, M.S.: Polarized Raman study of aligned multiwalled carbon nanotubes. Phys. Rev. Lett. **84** (8), 1820 (2000)
175. Heller, C., Leising, G., Godon, C., Lefrant, S., Fischer, W., Stelzer, F.: Raman excitation profiles of conjugated segments in solution. Phys. Rev. B **51**(13), 8107 (1995)

176. Bredas, J.L., Street, G.B., Themans, B., Andre, J.M.: Organic polymers based on aromatic rings (polyparaphenylene, polypyrrole, polythiophene): evolution of the electronic properties as a function of the torsion angle between adjacent rings. J. Chem. Phys. **83**(3), 1323–1329 (1985). doi:10.1063/1.449450
177. Vukmirović, N., Wang, L.-W.: Electronic structure of disordered conjugated polymers: polythiophenes. J. Phys. Chem. B **113**(2), 409–415 (2008). doi:10.1021/jp808360y

InAs Epitaxy on GaAs(001): A Model Case of Strain-Driven Self-assembling of Quantum Dots

E. Placidi, F. Arciprete, R. Magri, M. Rosini, A. Vinattieri, L. Cavigli, M. Gurioli, E. Giovine, L. Persichetti, M. Fanfoni, F. Patella, and A. Balzarotti

Abstract We review basic topics of the self-aggregation process of InAs quantum dots on the GaAs(001) surface with reference to our recent experimental and theoretical studies. Atomic-force and scanning-tunnelling microscopy, and reflection high-energy electron diffraction measurements are presented for discussing issues such as formation and composition of the wetting layer, evolution of the 2D to 3D transition, size distribution and equilibrium shape of the islands. Single-dot emission is demonstrated by micro-photoluminescence spectra of samples where quantum dots were confined on selected nanoscale areas of the surface by molecular-beam epitaxial growth on lithographed substrates. Theoretical ab initio studies of the In diffusion on the wetting layer, and simulations with the finite element method of the elastic energy relaxation of the island–substrate system are also discussed.

E. Placidi (✉)
CNR-ISM, via Fosso del Cavaliere 100, 00133 Rome, Italy

Dipartimento di Fisica, Università di Roma Tor Vergata, via della Ricerca Scientifica 1, 00133 Rome, Italy
e-mail: ernesto.placidi@roma2.infn.it

F. Arciprete • L. Persichetti • M. Fanfoni • F. Patella • A. Balzarotti
Dipartimento di Fisica, Università di Roma Tor Vergata, via della Ricerca Scientifica 1, 00133 Rome, Italy

R. Magri • M. Rosini
Dipartimento di Fisica, Università degli Studi di Modena e Reggio Emilia, via Campi 213/A, 41100 Modena, Italy

Centro S3 CNR-Istituto di Nanoscienze, via Campi 213/A, 41125 Modena, Italy

A. Vinattieri • L. Cavigli • M. Gurioli
Dipartimento di Fisica, Università di Firenze, via G. Sansone 1, 59100 Sesto Fiorentino, Italy

E. Giovine
Istituto di Fotonica e Nanotecnologie, Consiglio Nazionale delle Ricerche, via Cineto Romano, Rome, Italy

S. Bellucci (ed.), *Self-Assembly of Nanostructures: The INFN Lectures, Vol. III*, Lecture Notes in Nanoscale Science and Technology 12,
DOI 10.1007/978-1-4614-0742-3_2,

1 Introduction

Due to carrier localisation in all three dimensions, quantum dots (QDs) own discrete energy level structure having the potential for use in a variety of novel electronic devices. QDs of III–V compounds, and in particular, InAs QDs on GaAs (001), are widely employed, at present, in conventional devices such as light emitting diodes, photovoltaic cells and quantum semiconductor lasers. Moreover, they offer appealing perspectives for more sophisticated applications in new generation devices such as single-photon emitters for nano-photonics and quantum computing [1–3]. The latter applications require the close control of both the size and position of the QD on the substrate surface.

The size being of the order of nanometres, QDs are difficult to manufacture using standard lithographic techniques; thus, various alternative methods have actually been devised to produce them. A particularly effective method is to directly grow the dots by molecular beam epitaxy (MBE), depositing under appropriate growth conditions a thin film layer of the material on the substrate of a different material (heteroepitaxy). Due to the difference in the lattice parameters, lattice-mismatch occurs that can lead to the break-up of the film into coherently strained islands. The driving force behind the process is the strain energy stored in the lattice-mismatched epitaxial thin film that makes it unstable. Though the formation of islands increases the surface area, and thus the free energy, the reduction in elastic energy when the film relaxes overwhelms the surface effect, and the total energy of the system is actually lowered.

There are quite a number of issues in the fabrication process of QDs by heteroepitaxy that must be understood in great detail in order to grow QDs arrays of the desired structural and electronic properties suitable for applications.

This review reports mainly on the work we have done in last years on the InAs/GaAs system, which tackle most of the important aspects of the structural and electronic properties of the system. The investigated samples were all grown by MBE using different growth procedures, and analysed by an ample set of experimental and theoretical tools. The structural characterisation of the two-dimensional (2D) and three-dimensional (3D) phase of the system was performed by atomic-force and scanning-tunnelling microscopy (AFM and STM) and by reflection high-energy electron diffraction (RHEED) during growth. Experimental results on composition of the wetting layer (WL), evolution of the 2D–3D transition, size distribution and shape of island are discussed in detail in Sects. 3 and 4. Theoretical ab initio studies of the In diffusion on the differently reconstructed wetting layer are reported in Sect. 3, while simulations of the elastic energy relaxation inside InAs islands and GaAs substrate with the finite element method (FEM) are reported in Sect. 4. Micro-photoluminescence (μPL) spectra are shown in Sect. 5, which evidence single-dot emission of InAs islands spatially confined to the nanoscale by MBE selective-epitaxy on e-beam lithographed SiO_2/GaAs substrates.

The review starts with Sect. 2 that reminds background concepts on heteroepitaxy.

2 Background Concepts on Heteroepitaxy

2.1 Macroscopic Regimes of Heteroepitaxial Growth

Heteroepitaxy refers to the epitaxial deposition of a film of one material on a monocrystalline substrate of another material. In molecular beam epitaxy (MBE), films are grown from gaseous precursors. The substrate acts as a seed crystal providing a template for positioning the first impinging atoms of the film, and each atomic layer has the same function for the next layer. In the heteroepitaxy of a material "A" on a substrate "B," the growth morphology is mainly determined by the surface energies of the overlayer γ_A, of the substrate γ_B and of the interface γ_{AB}. The inequality, $\gamma_B > \gamma_A + \gamma_{AB}$, sets the condition for the epitaxial layer to wet the substrate; in this case, growth is called of the Frank–Van der Merwe type. If the inequality has the opposite sign, one usually obtains Volmer–Weber growth, i.e., direct islanding of "A" on substrate "B." The Stranski–Krastanov growth is an intermediate case that occurs when there is initial wetting of the substrate but, with increasing overlayer thickness, surface energies are changed by surface stress or by interface mixing and/or segregation, so that at a critical thickness of "A" the initial wetting condition does not hold anymore and islands start forming from then on. This latter case is typical when the film lattice parameter differs significantly from that of the substrate.

In terms of lattice parameters, the mismatch of film "A" on substrate "B" is

$$\varepsilon = \frac{b-a}{a}, \tag{1}$$

where b and a denote the substrate and film lattice constants, respectively. The growth is stress free, or incoherent, when the substrate and film keep their different bulk lattice constants. The growth is pseudomorphic, or coherently strained, when the film is stretched ($\varepsilon > 0$) or compressed ($\varepsilon < 0$) such that the substrate and the film in-plane lattice constants coincide. Lastly, there are the cases where elastic energy of mismatch is totally or partially relaxed by formation of islands and/or dislocations.

In mismatched heteroepitaxy, the stress-free and the pseudomorphic growths are in competition. In the pseudomorphic layer the elastic energy is accumulated and the cell strain along the directions parallel, $\varepsilon_{\parallel}$, and perpendicular, $\varepsilon_{\perp}$, to the surface are defined as

$$\varepsilon_{\parallel(\perp)} = \frac{|a_{f\parallel(\perp)} - a_f|}{a_f}, \tag{2}$$

where $a_{f\parallel}$ and $a_{f\perp}$ are the in-plane and out-of-plane lattice parameters of the deformed cell, whose tetragonal distortion, ν, called Poisson ratio, is

$$\nu = \frac{\varepsilon_{\perp}}{\varepsilon_{\parallel}}. \tag{3}$$

From the continuum elasticity theory of isotropic media, the elastic energy of a biaxially strained crystal is proportional to its volume; therefore, the elastic energy per unit area of an isotropic film with thickness d is

$$\Delta E_{\text{film}} = \frac{E}{1-\nu}\varepsilon^2 d, \tag{4}$$

where E is the Young modulus. Since this energy increases with d, the growth of the strained film remains coherent (2D-coh.) up to a critical thickness, above which the stored elastic energy is released, totally or in part, by formation of 3D islands (SK-coh.) or by misfit-dislocations (2D-MD), or both. The favoured growth morphology of the film is that of minimum Gibbs free energy. For the above-mentioned growth modes these energies (per unit area) are [4]

$$F_{\text{2D-coh.}} = \frac{E}{1-\nu}\varepsilon^2 d + \gamma_{\text{film}}, \tag{5}$$

$$F_{\text{SK-coh.}} = R\frac{E}{1-\nu}\varepsilon^2 d + \gamma_{\text{film}} + \Delta\gamma_{\text{3D}}, \tag{6}$$

$$F_{\text{2D-MD}} = \frac{2E_{\text{MD}}}{d_0} + \gamma_{\text{film}}, \tag{7}$$

where $\Delta\gamma_{\text{3D}} = n\Delta\gamma_{\text{is}}$ is the surface energy cost per unit area due to the formation of a density n of 3D islands relaxing a fraction $(1 - R)$ of the strain energy of the film of surface energy γ_{film}, while E_{MD} is the energy per unit length of $2/d_0$ noninteracting misfit dislocations (per unit area) at the equilibrium distance, d_0, fully relaxing the strain.

The critical thickness, d_c, for transition from 2D-coh. to SK-coh. growth and the critical thickness, d_{MD}, for transition from 2D-coh. to 2D-MD growth can be derived by equating (5) and (6), and (6) and (7), respectively. For each system film/substrate, the parameters $\Delta\gamma_{\text{3D}}$ and R depend on size and shape of the islands.

In most cases, such shapes are truncated pyramids with rectangular bases. Here, the energy cost $\Delta\gamma_{\text{is}}$ for the formation of an island "A" on the substrate "B" is [5]

$$\Delta\gamma_{\text{is}} = ST(\gamma_{\text{A}} + \gamma_{\text{AB}} - \gamma_{\text{B}}) + 2(S+T)[H\gamma_{\text{i}}\csc\vartheta - H\text{ctg}\,\vartheta(\gamma_{\text{A}} + \gamma_{\text{AB}} - \gamma_{\text{B}})/2], \tag{8}$$

where the parameters S, T, H and θ are defined in Fig. 1, and γ_i is the surface energy of the lateral face i of the island. For the particular case of coherent Stranski–Krastanov growth of a square-base pyramidal island on a wetted substrate (no intermixing), $\gamma_{\text{B}} = \gamma_{\text{A}}, \gamma_{\text{AB}} = 0$ and

$$\Delta\gamma_{\text{is}} = 4SH\gamma_i \csc\vartheta. \tag{9}$$

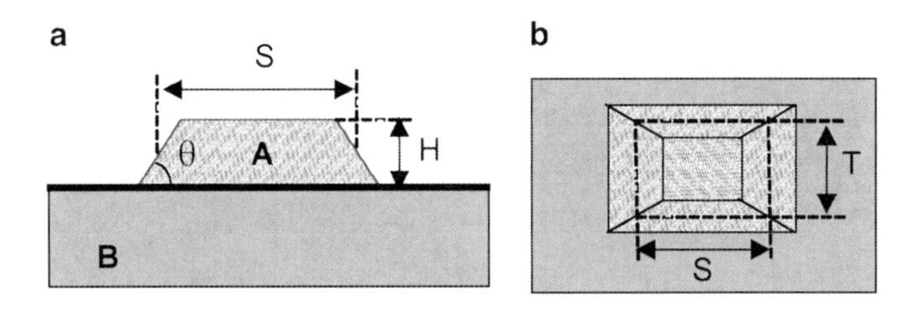

Fig. 1 The excess surface free energy due to the formation on the substrate B of a truncated-pyramid A of the specified sizes is given by the contribution, $[ST - (S + T)H\text{ctg}\ \theta]\gamma_A$, of the top face of surface energy γ_A, the contribution, $[2(S + T)H/\sin\theta]\gamma_i$, of the four lateral faces of surface energy γ_i, and the contribution, $[ST + (S + T)H\text{ctg}\ \theta](\gamma_{AB} - \gamma_B)$, of the interface with the substrate of surface energies γ_{AB} and γ_B, respectively

2.2 *Equilibrium Shape of Self-assembled Islands*

For an *isolated crystal*, "A," the equilibrium shape (ES) is found by minimising, at constant volume, the Gibbs surface free energy F_S given by

$$F_S = \sum_i \gamma_i S_i \tag{10}$$

where S_i is the area of the facet i of the crystal.

If γ is known for all directions $\mathbf{n}$ of the crystal, the equilibrium shape can be easily found from the $\gamma(\mathbf{n})$ plot with a geometrical construction based on the century-old Wulff's theorem,

$$\frac{\gamma_j}{h_j} = \lambda \tag{11}$$

stating the constant ratio, λ, between the surface energy γ_j of each facet j bounding the crystal of given volume and the distance h_j of that facet from a common point (Wulff's point) inside the crystal. It follows that, for a free-crystal growing in near-equilibrium conditions, its shape for different sizes (different λ) is self-similar around the common Wulff's point, as also results from the constancy of the shape ratios r_i of equilibrium facets i for increasing volume of the crystal (Fig. 2a):

$$r_i = \frac{h_A}{h_i} = \frac{\gamma_A}{\gamma_i}. \tag{12}$$

For a *supported crystal* "A" on a substrate "B" (Fig. 2b), the equilibrium shape *in the absence of misfit* (no strain) is again determined only by the minimisation of surface and interface energies. For a crystal "A" of volume V_A having free facets of

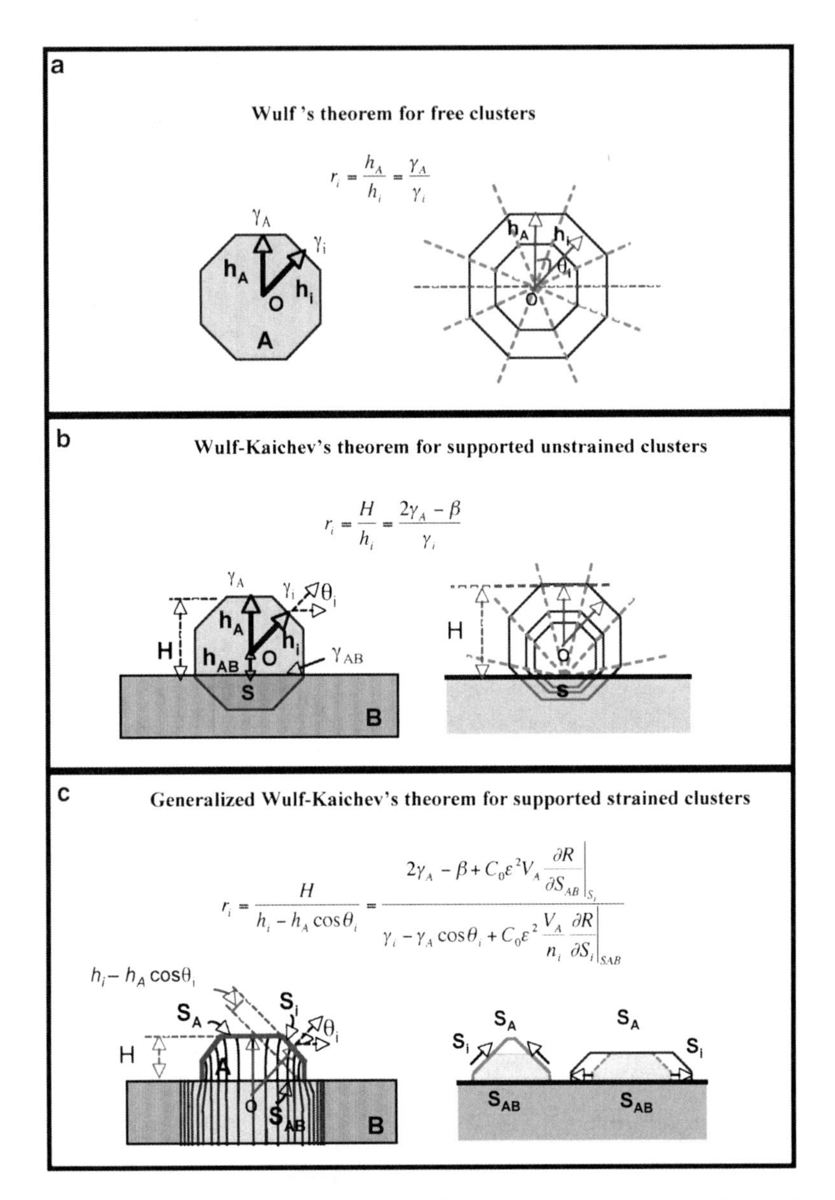

Fig. 2 Equilibrium shape of an unsupported crystal A (**a**); of an island A supported by a substrate B having the same lattice parameter (**b**); and of an island A supported by a substrate B that is lattice mismatched (**c**) (see text for details)

area S_i and a contact area S_{AB} with the substrate "B" of area S_{B}, the surface free energy, F_{S}, reads

$$F_{\mathrm{S}} = \sum_{i \neq \mathrm{AB}} \gamma_i S_i + S_{\mathrm{AB}}\gamma_{\mathrm{AB}} + (S_{\mathrm{B}} - S_{\mathrm{AB}})\gamma_{\mathrm{B}}, \tag{13}$$

being $\gamma_{AB} = \gamma_A + \gamma_B - \beta$ the interfacial energy, where β is the adhesion energy of A on B.

The volume V_A of the polyhedral crystal can be split into pyramids of height h_i and basis S_i, such that $V_A = (1/3)\sum_i h_i S_i$ and, to a first order, $dV_A = (1/2)\sum_i h_i\, dS_i$. Accordingly, to the emerging volume of the crystal (Fig. 2b) applies the variation

$$dV_A = (1/2)\sum_{i \neq AB} h_i dS_i + (1/2) h_{AB} dS_{AB}, \tag{14}$$

where the summation runs over the free facets $i \neq$ AB and h_{AB} is the common distance of the interface to all the pyramid apexes (Wulff's point) taken positive or negative according to whether the Wulff's point is outside or inside the substrate, respectively.

The ES is obtained by minimising F_S at constant volume V_A, i.e., equating to zero the first-order differential

$$dF_S + \alpha dV_A = \sum_i \gamma_i dS_i + (\gamma_A - \beta) dS_{AB} + \alpha\left(\frac{1}{2}\sum_{i \neq AB} h_i dS_i + \frac{1}{2} h_{AB} dS_{AB}\right) = 0, \tag{15}$$

where α is a Lagrange multiplier. This leads to the same set of equations defining the free facets (11), supplemented by the relation for the contact face AB (Wulff–Kaichew's theorem) [6],

$$\begin{cases} \lambda = \dfrac{\gamma_i}{h_i} \\ \lambda = \dfrac{2\gamma_{AB} - \gamma_B}{H} \end{cases}, \tag{16}$$

where $H = h_A + h_{AB}$ is the emerging height of the crystal. In this case, as shown in Fig. 2b, in absence of misfit the ES maintains self-similarity around the common point S on the substrate as a consequence of the size independence of the shape ratios

$$r_i = \frac{H}{h_i} = \frac{2\gamma_A - \beta}{\gamma_i}. \tag{17}$$

The last case considered here is the ES of an epitaxially *strained* polyhedral crystal A deposited on a rigid *lattice-mismatched* substrate B. Using 3D-Hooke's law, the bulk elastic energy, F_{el}, can be calculated for a cubic system [6], and, for the low index (001) orientation, it can be written as a function of the elastic constants c_{ij} as

$$F_{el} = \rho_{el} V_A = \frac{(c_{11} + 2c_{12})(c_{11} - c_{12})}{c_{11}} \varepsilon^2 V_A = C_0 \varepsilon^2 V_A, \tag{18}$$

where ρ_{el} is the elastic energy density. Equation (18) is identical to the result for an isotropic material, $F_{el} = [E/(1-\nu)]\varepsilon^2 V$.

For a noninfinitely rigid substrate and assuming coherence at the interface, the elastic strain accumulates during growth both in the crystal and in the nearby region of the substrate underneath. If the crystal and the substrate are allowed to relax, the stored elastic energy (18) is reduced by a *relaxation-energy factor*, R (Sect. 4.2), as [6]

$$F_{el,R} = C_0\varepsilon^2 V_A(R_A + R_B) = C_0\varepsilon^2 V_A R, \tag{19}$$

where $0<R<1$ accounts for the relaxation of both crystal R_A and substrate R_B, and depends, in complex way, upon the crystal shape.

The ES is, again, given by the minimisation, at constant volume V_A, of F_S, which now includes the elastic energy $F_{el,R}$ and reads [6]

$$\begin{aligned} dF_S + \alpha dV_A = &\sum_i \gamma_i dS_i + (\gamma_A - \beta)dS_{AB} + C_0\varepsilon^2 V_A dR + C_0\varepsilon^2 R dV_A \\ &+ \alpha\left(\frac{1}{2}\sum_{i\neq AB} h_i dS_i + \frac{1}{2}h_{AB}dS_{AB}\right) = 0. \end{aligned} \tag{20}$$

The solution is the set of equations, called *generalised Wulff–Kaishew theorem* [6],

$$\begin{cases} \lambda = \dfrac{\gamma_i - \gamma_A \cos\vartheta_i + C_0 v^2 \frac{V_A}{n_i}\frac{\partial R}{\partial S_i}\Big|_{AB}}{h_i - h_A \cos\vartheta_i} \\ \lambda = \dfrac{2\gamma_A - \beta + C_0 v^2 V_A \frac{\partial R}{\partial S_{AB}}\Big|_i}{H} \end{cases}, \tag{21}$$

where the first equation refers to the n_i equivalent crystallographic free facets i of area S_i, and the second equation refers to the interface AB. Now the shape ratios characterising the ES (Fig. 2c),

$$r_i = \frac{H}{h_i - h_A\cos\vartheta_i} = \frac{2\gamma_A - \beta + C_0\varepsilon^2 V_A \frac{\partial R}{\partial S_{AB}}\Big|_i}{\gamma_i - \gamma_A\cos\vartheta_i + C_0\varepsilon^2 \frac{V_A}{n_i}\frac{\partial R}{\partial S_i}\Big|_{AB}}, \tag{22}$$

are functions of the volume V_A and of the partial derivatives of the relaxation factor R; thus, there is no more self-similarity of the ES. As a general trend, since an extension of the interfacial area increases the elastic energy, then $dR/dS_{AB} > 0$ and since for a given interface area the extension of each facet increases relaxation through its free edges, $dR/dS_i < 0$.

To conclude, the ES of a supported strained island changes continuously during equilibrium growth, and precisely, facets change their relative extension while maintaining their crystallographic orientations as long as only bulk elasticity is considered and the strain-induced changes of the surface energy are disregarded. To the contrary, new crystallographic orientations can enter the ES of the island whenever strain-induced changes of surface energy occur, as it happens, in particular, for the InAs islands on GaAs(001) discussed in the following.

3 InAs Growth on GaAs(001): The 2D Phase

InAs on GaAs(001) is a prototypical example of highly mismatched growth where the 2D–3D transition takes place. Although of the SK-coh. type, such transition has a complex evolution since its initial stage. The 3D nucleation is completed within a narrow InAs coverage range, 0.2 ML, above the critical thickness by the formation of 10^{10}–10^{11} cm^{-2} islands with initial bimodal size distribution that evolves in a complex way as a function of deposition and temperature. Distinctive features differing from the conventional Stranski–Krastanov growth are as follows:

1. The In–Ga intermixing and In segregation that take place both at the QDs–substrate interface and in the 2D wetting layer (WL) that evolves toward an InGaAs alloy of precise surface composition and reconstruction.
2. The total volume of the QD that is larger than the total volume being deposited in the 2D–3D transition because of a sizable mass contribution from the substrate and the wetting layer [7, 8].

Other features of the InAs–GaAs heteroepitaxy concern the thermodynamic and kinetic aspects of the growth [9], the dependence of the critical thickness on substrate temperature and In flux [10], the coarsening process governing the evolution of the size distribution of islands as a function of the annealing temperature [11].

Figure 3 summarises the significant steps of the InAs growth on GaAs(001). The central graph reports the RHEED intensity monitored in situ during the growth process, where the sudden rise signals the onset of the 2D–3D transition. AFM/STM topographies and RHEED diffraction patterns are also shown which characterise the surface of the GaAs substrate before growth, the subcritical WL of InAs at 1.4 ML, the incipient 2D–3D transition at 1.6 ML and the well-developed 3D InAs islands at 3 ML.

The next sections will briefly review the experimental and theoretical work we have done on the InAs/GaAs system starting with the study of the 2D phase (the WL) and following with the study of the formation and evolution of the distribution of QDs.

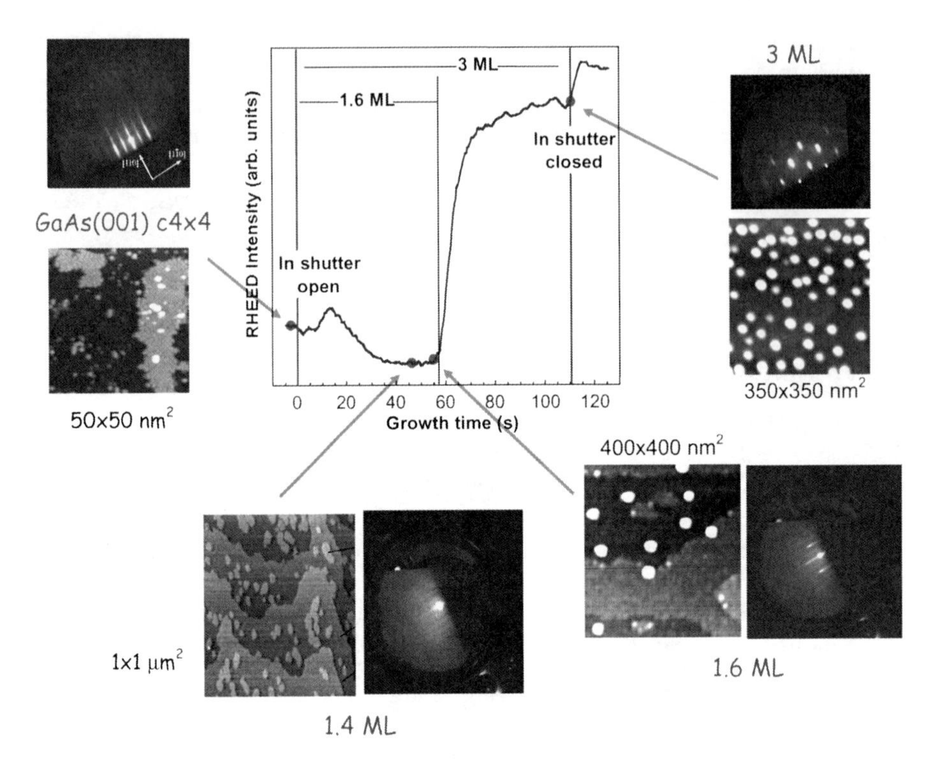

Fig. 3 The graph represents the RHEED intensity monitored during the deposition of 3 ML of InAs on GaAs(001). Rheed patterns, STM and AFM topographies are shown for coverages below, at and above the critical thickness for the transition (from [12])

Table 1 Parameter values of the lattice constant a, equilibrium distance d_0 and energy E_{MD} of a misfit dislocation for InAs on GaAs(001), surface energy γ, Poisson ratio v and Young modulus E

		InAs/GaAs				
	a (Å)	d_0 (Å)	E_{MD} (meV/Å)	$\gamma_{(001)}$ (meV/Å^2)	$v_{(001)}$	E (dyn/cm^2)
InAs	6.05	32	0.9	44	0.35	5.14×10^{11}
GaAs	5.65			45	0.31	8.59×10^{11}

3.1 *Wetting Layer and Critical Thickness*

Initially, due to the lower surface energy (Table 1) the InAs forms a strained 2D-coh. film (WL) wetting the GaAs substrate. On growing, the elastic energy accumulates in the film up to the point where further growth induces partial relaxation of the strain by islanding in the SK-coh. mode.

The critical thickness at which this occurs is

$$d_c = \frac{\Delta\gamma_{3D}}{(1-R)E\varepsilon^2/(1-v)}. \tag{23}$$

Table 2 Critical thickness d_c in MLs for the 2D-SK coh. transition in the InAs/GaAs(001) system evaluated using (9) and (23), experimental parameters measured in [13] for a distribution of pyramidal InAs island (volume V, basal area A, height H, aspect ratio r, number density n) and theoretical calculations of the relaxation factor R (Sect. 4.3)

	V (nm^3)	A (nm^2)	H (nm)	AR	θ (°)	n (cm^{-2})	$\Delta\gamma$ (erg/cm^2)	R	d_c (ML)
InAs QD	147.6	170.3	2.6	0.22	23.0	5.4×10^{10}	66.08	0.6	1.51

The evaluation of d_c, besides bulk parameters of the material reported in Table 1, requires the knowledge of the equilibrium density, size, shape and shape ratio of the islands, which, in turn, determine the strain relaxation factor R.

From our structural studies we derived that at 1.7 ML of InAs deposition on GaAs(001) at 500°C, the island shapes were approximately full pyramids bounded by {137} and {111} facets [13] with average aspect ratio $r = 0.22$, r being the height-to-base length ratio of the dot. Assuming this shape, we performed finite-element simulations of the elastic relaxation of InAs islands on GaAs and calculated a relaxation factor $R = 0.6$ (Sect. 4.3).

Using (9) and (23), we calculated the value $d_c = 1.51$ ML reported in Table 2 for the critical thickness of the 2D-SK coh. transition, in good agreement with the measured value, 1.59 ML, by RHEED [10]. Consistently with the simplified hypothesis of (7), the estimated critical thickness for 2D-coh. to 2D-MD growth mode, $d_{MD} \cong 8$ ML, was higher than d_c, confirming that the elastic misfit energy is first released by 3D island formation, rather than by bulk dislocations, in the InAs/GaAs system.

To the end of understanding the 2D–3D transition in detail, it is of major relevance to establish more precisely thickness and composition of the WL, since, as previously stated, the InAs epilayer is significantly alloyed with the GaAs substrate. To prove this point we have grow on the GaAs(001) by MBE, using the same experimental parameters (590°C for the 0.75 μm GaAs re-growth at a rate 1 μm/h and 500°C for the InAs deposition at a rate 0.028 ML/s), two subcritical films, 0.7 and 1.3 ML, of InAs on GaAs(001) and 45 ML of the ternary alloy $In_{0.2}Ga_{0.8}As$. After growth the samples were capped with As in order to preserve the surface for the STM analysis performed ex situ. Prior to the STM measurements, in ultra-high vacuum, the As cap was removed at ≈300°C and the crystalline surface was characterised by low-energy electron diffraction (LEED). Fig. 4a reports the topography of the WL at 1.3 ML at a large scale taken with AFM in air. The typical features of the mixed-growth regime of step flow and nucleation of two-dimensional islands result from the combined effect of the lower surface free energy of InAs and the reduced migration length of In cations due to strain [14, 15]. The large two-dimensional islands nucleated on terraces tend to coalesce at slightly higher InAs coverage, giving rise to a more uniform overlayer. Fig.4b–d report the atomically resolved STM images of the 0.7 and 1.3 ML WL and of the alloy, respectively. The formation of the alloyed interface can be clearly inferred by the comparison between the WL topographies (b) and (c) and that of the strained alloy (d) were domains of the same symmetry are observed.

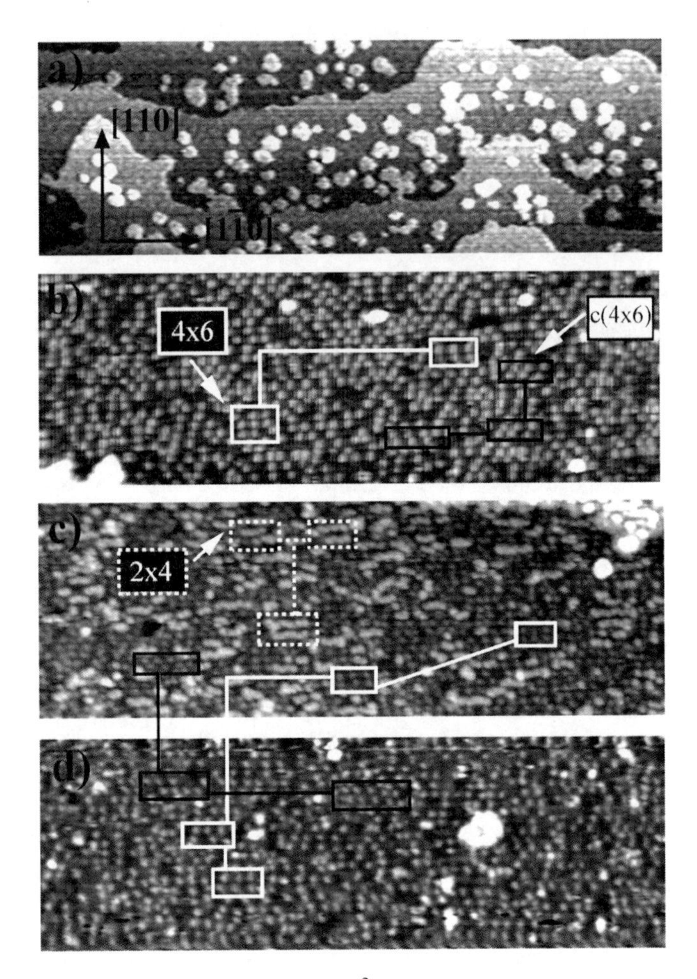

Fig. 4 (**a**) AFM topography, 1,500 × 1,500 nm^2, of the strained two-dimensional phase (WL) obtained for 1.3 ML of InAs on GaAs(001). Atomically resolved, 30 × 90 nm^2, STM images of (**b**) InAs WL at 0.7 ML, (**c**) InAs WL at 1.3 ML and (**d**) $In_{0.2}Ga_{0.8}As$ alloy, 450 ML, grown by MBE on GaAs(001). Domains of (4 × 3) and c(4 × 6) periodicity are highlighted on the WL and on the alloy. 2 × 4 InAs chains are detected only on the WL at 1.3 ML. The periodicity (n × m) in referred to the (001) surface cell (a_0 = 4 Å), n along the [1-10] and [110] directions (from [14])

The x3 translational symmetry of the top plane is the fingerprint for the In–Ga alloying [17]. On the atomic-scale STM images of Fig. 4, small domains of (4 × 3) and c(4 × 6) periodicity [4× along [1-10] and 3× (6×) along [110] directions] are identified both on the wetting layer at the two different thickness and on the alloy, as marked in panels (b), (c) and (d), respectively. Differently from our case where samples were analysed ex situ, more often, for in situ measurements, a 2 × 3 symmetry (with a short correlation for the 2× periodicity along the [1-10] direction) is reported by RHEED and X-ray diffraction data on InGaAs alloys [18] and by STM measurements of InAs films for depositions larger than 0.8 ML [17]. Indeed, a metastable 2 × 3 surface phase was observed upon several hours annealing at 300°C on the de-capped GaAs(001)c(4 × 4) surface [19]. However, as pointed out by Zhang et al. [20], the 2×3 unit cell of GaAs is not charge compensated and cannot be stable. As a matter of fact, high-resolution STM images [19] reveal that this symmetry consists of charge compensated (4×3) and $c\ (4 \times 6)$ domains, like those we detect on the WL and on the alloy surfaces. Height profiles taken on the topography of Fig. 4 show that the x3 (and x6) translational symmetry is generally maintained over large portions of the surface, while the 4× reconstruction along [1-10] has a correlation limited to two or three unit cells and is barely observed in LEED patterns. This reduced topological order, observed as well on in situ measurements, supports the idea of the existence of a "liquid" surface layer with a certain degree of order caused by the lowering of the melting point at the high hydrostatic pressure induced by the strain [21].

Many experimental evidences are reported in literature on the In segregation in epitaxial ternary III–V alloys leading to the formation of an In-rich surface [22]. Segregation of In to the very surface plane has been assumed to control the critical thickness of the 2D–3D transition in the growth of $In_xGa_{1-x}As$ on GaAs for $x \geq 0.25$ [23, 24]. Using the segregation model of ref [25], the saturation value $x \sim 0.85$ was derived at which strain in the alloy is released by islanding [23]. By applying the same model, we calculated the In content of the surface plane in the cases of 45 ML of $In_{0.2}Ga_{0.8}As$, and 1 ML and 2 ML of InAs on GaAs [16]. The values obtained were 0.83 for the top layer of the alloy, 0.82 for the top layer of the 1 ML film, and 0.99 and 0.83 for top and second layer of the 2 ML film of InAs. On account of the same detected atomic structure, we predict the same In fraction, close to the critical value 0.85, for the alloy surface, for the subsurface of the WL at 1.3 ML and for the surface domains at 0.7 ML coverage shown in Fig. 4.

An important difference exists between the alloy and the WL above 1 ML, i.e., the presence on the latter of zigzag chains, one atomic plane ($a_0/4 \sim 0.14$ ML) above the sub-surface, having a (2×4) periodicity different from the x3 (x6) periodicity of the substrate. One can speculate that, above 1 ML, the deposited In atoms form strained chains "floating" on top of the intermixed substrate, which are mainly pure InAs [26–29]. This is consistent with the segregation model that anticipates a surface In fraction 0.99 on completion of the second monolayer of InAs, and with the fact that the $In_{0.82}Ga_{0.18}As$ alloyed surface formed on 1 ML deposition is that with the minimum Gibbs free energy [30]. This model for the WL has important implications for the 2D–3D transition, at 1.59 ML, since the loosely bound In, estimated 0.5–0.6 ML from the image, can, in principle, participate in the surface mass transport responsible for the sudden volume increase of the 3D QDs at the transition, as will be discussed in Sect. 4.1.

3.2 In Diffusion on the Wetting Layer

From a theoretical point of view, the study of nucleation of self-assembled InGaAs quantum dots implies to know what happens during the mass transport that takes place at the 2D–3D transition. As a simplifying assumption, the coordinate motion

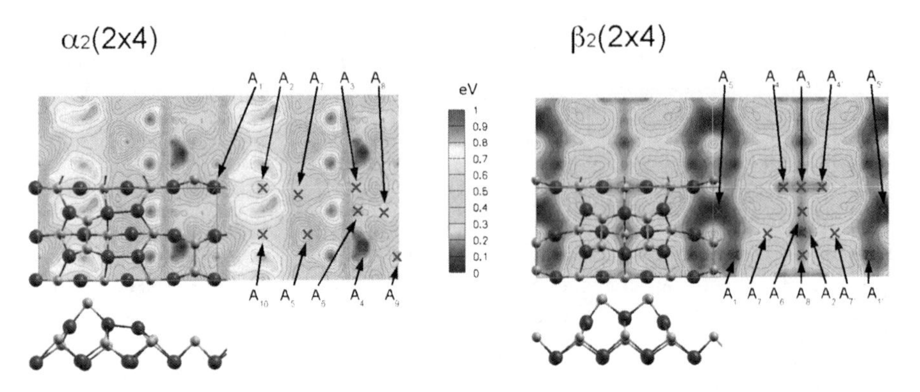

Fig. 5 PES for the In adatom on the $\alpha_2(2 \times 4)$ (*left*) and $\beta_2(2 \times 4)$ (*right*) surface reconstructions. The adsorption sites A_i are indicated. Note the effect of the symmetry lowering in α_2 with respect to β_2 (from [59])

of the adsorbates and surface atoms during mass transport is thought to be the sum of independent single surface-diffusion events.

We focussed our attention to In-rich WLs on GaAs(001), with a (2×4) surface reconstruction, which is that observed at the onset of quantum dot formation [17, 31–33]. For the (001) In-rich surface, the stable (2×4) reconstructions have been found to be α_2 and β_2 [34–36]. Both reconstructions consist of As dimers in the topmost, incomplete surface plane (ad-dimers), and As dimers formed on the lower complete As layer of the crystal (in-dimers) (Fig. 5). The β_2 reconstruction contains an additional As ad-dimer; thus, it is stabilised in a more As-rich atmosphere or/and at a lower growth temperature where As evaporation from the surface is reduced. The InGaAs WL grown on GaAs(001), which has a smaller lattice parameter, is subjected to compressive strain. The effect of strain on In diffusion can be evaluated by comparing the results of In diffusion on the In-covered $\beta_2(2 \times 4)$ surfaces with those obtained for the strain-free pure InAs $\beta_2(2 \times 4)$ surfaces reported in the literature [37].

It has been found [17] that as the In coverage increases, the (001) surface reconstruction changes from $c(4 \times 4)$ for pure GaAs, to (2×3) at a In $\theta \approx 0.7$ ML coverage, to finally a (2×4) reconstruction at a much higher In coverage. The characteristics of In adsorption have been previously studied for the $c(4 \times 4)$ GaAs (001) [36], the (1×3) and the (2×3) $In_{0.66}Ga_{0.33}As(001)$ [38] surfaces. In this contribution, we extend the above analysis to the (2×4) reconstruction and study the resulting In diffusion on the In-free $c(4 \times 4)$ GaAs(001), on the $(1 \times 3)/(2 \times 3)$ $In_{0.66}Ga_{0.33}As$, and on the $\alpha_2/\beta_2(2 \times 4)$ InAs WLs on GaAs(001). Notice that these WLs have a different reconstruction *and* a different In composition, implying increasing compressive strain conditions.

The description of In diffusion is based on the transition state theory [39–41]. In adatoms perform thermally activated jumps from an adsorption site on the WL to another one, overcoming energy barriers. The In adsorption sites on the reconstructed WLs and the corresponding diffusion barriers are calculated using a first-principles density functional theory (DFT) approach.

3.3 Surface Reconstructions and Potential Energy Surfaces

The first step of the theory requires the study of the potential energy surface (PES) perceived by the single In adatom deposited on the surface. From the PES the binding sites for the adatom and the energy barriers can be extracted. Obviously, the PES is a function of the particular surface reconstruction. We performed first-principles calculations for the α_2 and β_2 reconstructions, within the DFT in the generalised gradient approximation (DFT-GGA) with the PBE exchange and correlation functional [42] using the QuantumESPRESSO simulation package [43]. We used norm-conserving pseudopotentials treating the outermost s- and p-shells of Ga, In and As as valence electrons, and the electronic wave functions were expanded in plane waves, with a 15 Ry kinetic energy cutoff. The energy cutoff has been tested in order to reach convergence for the lattice bulk properties of GaAs and InAs. The core-corrected atomic pseudopotentials have been tested on bulk Ga, In and As, where the determined equilibrium configurations and elastic moduli compare well with the experimental data. The equilibrium lattice parameters obtained for the GaAs and InAs bulk phases are $a_0 = 5.779$ Å and $a_0 = 6.314$ Å, which are slightly higher than the experimental ones, that is, 5.653 and 6.058 Å, respectively. Thus, our calculated lattice mismatch is 9.3%, higher than the experimental value of 6.8%. This overestimates the effects due to the WL strain. We modelled the surface through a (001)-oriented supercell containing four layers of GaAs, covered with 1.75 layers of InAs arranged according to the $\alpha_2(2 \times 4)$ and $\beta_2(2 \times 4)$ surface reconstructions. The slab is repeated along the (001) direction with a periodicity $5a_0$ and a separation of about 10 Å of vacuum. We have calculated the optimised geometries and the formation energies of the $\alpha_2(2 \times 4)$ and $\beta_2(2 \times 4)$ reconstructions of the bare GaAs surface and of GaAs covered with a $\theta = 1.75$ ML InAs layer by relaxing the atomic positions. The lower layer of Ga atoms is kept fixed during the cell relaxation, in order to mimic the constraint due to the underlying semi-infinite bulk, and it is passivated with pseudo-hydrogen atoms of 1.25 electron charge. All the examined structures have been relaxed in order to find the equilibrium geometries, until all the forces acting on the atoms were less than 2.5 meV/Å. Brillouin zone (BZ) integration was carried out using a set of special k-points equivalent to 16 points in the 1×1 surface BZ. A smearing of 0.01 Ry has been used in order to deal with the possible metalisation of the surface electronic structure. These calculations provide the equilibrium atomic configuration and electronic properties of the WL covered GaAs substrate. Then the same calculations are repeated considering the presence of a single In adatom on the WL.

The PES for the adsorbate is defined as the two-dimensional function given by the minimum of the total energy of the WL *plus* adsorbate system, with respect to the distance Z of the adsorbate from the surface:

$$\mathrm{PES}(x, y) = \min_R \min_z U(R, (x, y, z)), \tag{24}$$

where R is the set of all the $3N$ degrees of freedom of the slab, and (x,y,z) is the position of the adsorbate, that is, for each adsorbate (x,y,z) position all the degrees of freedom of the entire system are included in the energy minimisation, but the only (x,y) coordinates of the adsorbate. In this way, the problem of the adsorbate interaction with the surface is reduced to a two-dimensional dynamics on the PES.

To calculate the PES, we have set up a grid of 64 points, corresponding to a mesh width of 1.4 Å, and for each point we have found the minimum energy configuration for the adsorbate plus surface system. Then, we have interpolated the grid data with a bi-cubic spline algorithm in order to find the positions of the minima and the saddle points (e.g., the energy barriers). The exact values of the minima have then been determined by relaxing also the (x,y) coordinates of the adsorbate. Since our calculation approach implies a three-dimensional periodicity, we have minimised the spurious interaction between the adatom and its periodic images, by doubling the surface unit cell along the $(\bar{1}10)$ direction.

These procedure is used to calculate the In potential energy surface *on top of* the WL. Additional potential minima, described in the literature [36, 44, 45], correspond to configurations where the In adatom breaks the As dimers and inserts itself into them. In these cases, in order to find the energy barriers to be overcome by the In adsorbate to enter in these binding sites, we have performed calculations of the saddle points using the nudged elastic band method (NEB) [46]. This method is able to find the In path formed by positions of minimum energy [and therefore called minimum energy path (MEP)] between two sites on the surface, by simulating a string of replicas of the system, where the different images are linked by springs. By minimising the energy associated to the path, an accurate description of the MEP and the corresponding saddle point is obtained.

The PESs for α_2 and β_2 surface reconstructions are reported in Fig. 5. The two surfaces are considerably different: the energy difference between the minimum and the maximum values of the PES is about 1 eV for α_2, and is much lower, 0.6 eV, for β_2. Thus, the β_2 PES is flatter than the α_2 one. Moreover, the PES on β_2 is symmetric with respect a [110] plane while α_2 is not. On both surfaces, low-energy trenches are present along the $[\bar{1}10]$ direction, at both sides of the in-dimer chain. These trenches include four adsorption sites, each separated by low confining barriers: this potential landscape should lead to a higher diffusion coefficient along this direction.

We found ten adsorption sites for In on the α_2 surface reconstruction. The deepest adsorption sites are A_4, A_3, A_8, A_9, and A_5: the first four sites are located in the low-energy trench along the $[\bar{1}10]$ direction; the latter adsorption site (A_5) sits on top of the uncovered In atoms at the topmost layer, and it is confined by high barriers.

On the β_2 reconstruction we found instead 12 adsorption sites. The ones with the lowest energies are $A_1, A_{1'}, A_5, A_{5'}$, and A_2. The binding site configuration is similar to that found on α_2, but their arrangement follows a more symmetric pattern, due to the higher symmetry of the β_2 reconstruction.

The transition networks derived from the PESs of α_2 and β_2 are shown in Fig. 6, where the adsorption sites described above are included. For the sites where the

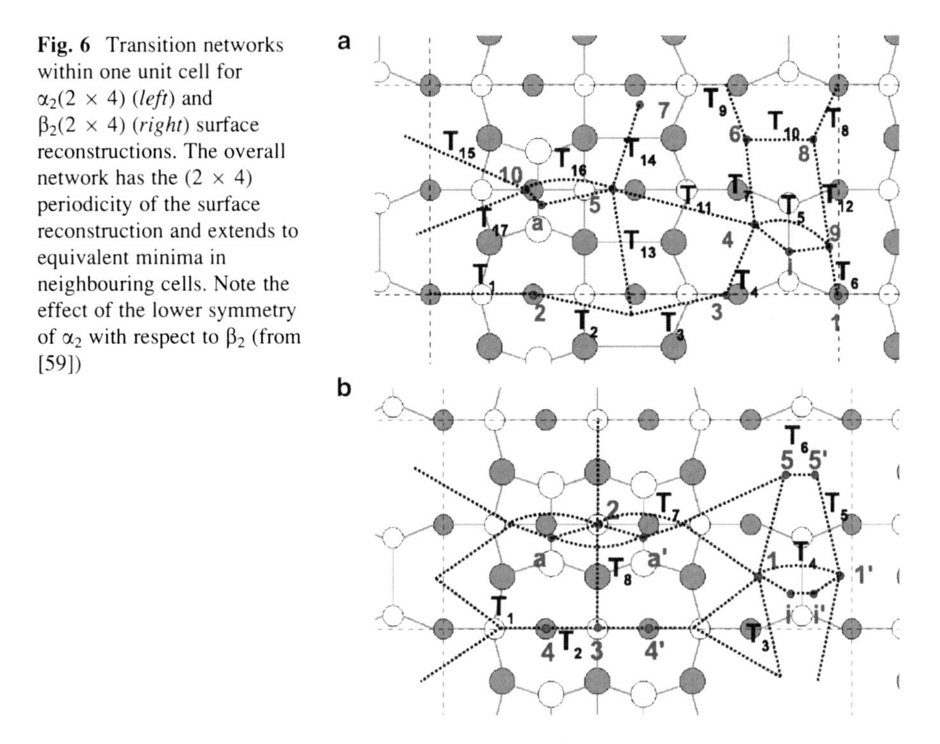

Fig. 6 Transition networks within one unit cell for $\alpha_2(2 \times 4)$ (*left*) and $\beta_2(2 \times 4)$ (*right*) surface reconstructions. The overall network has the (2×4) periodicity of the surface reconstruction and extends to equivalent minima in neighbouring cells. Note the effect of the lower symmetry of α_2 with respect to β_2 (from [59])

adatom is inserted into the ad-dimers we use the label "a" as indicated in the figure, while for the sites into the in-dimers we use the label "i." In the figure, the corresponding transitions to/from neighbouring sites are also indicated. Some shallow adsorption sites (e.g., A_6, A_7, $A_{7'}$, and A_8) have not been included in the β_2 transition network (Fig. 6), since the relative confining barriers are lower than 20 meV, and thus, at normal growth temperatures, they are "invisible" for the random walker.

3.4 *The General Problem of the Surface Diffusion*

The detailed knowledge of the In PES is used to calculate the properties of In diffusion on the surface described by that PES. The tracer diffusion tensor is defined as [47–49]

$$D^*_{\alpha,\beta} = \lim_{t\to\infty} \frac{1}{4t} \langle \Delta x(t) \Delta y(t) \rangle, \tag{25}$$

where $\Delta x(t)$, $\Delta y(t)$ is the adsorbate displacement with respect to the initial position. In the case of isotropic surface diffusion, it will be

$$D^*_{x,y} = D^* \delta_{x,y}, \tag{26}$$

where D^* is the tracer diffusion coefficient. The theory applies to the random walk of noninteracting adatoms that corresponds to an experimental situation of low adatom concentration and vanishing interaction among the adatoms.

The diffusion motion originates from the random walk of the adsorbate on the surface, consisting of a series of thermally activated jumps. The adsorbate in an initial adsorption state (j), after a given time τ, escapes to another adsorption site (k) with a transition probability per unit time $\Gamma_{kj} \equiv \Gamma_{k \leftarrow j} \equiv 1/\tau_{kj}$. It is clear that the tracer diffusion coefficient is a function of the whole set of Γ_{kj}.

In order to determine the coefficients Γ_{kj} , we apply the transition state theory [39–41]. Thus, the transition probability per unit time is expressed as

$$\Gamma_{kj} = \frac{k_\mathrm{B}T}{2\pi\hbar} \mathrm{e}^{-\frac{\Delta F}{k_\mathrm{B}T}}, \tag{27}$$

where ΔF is the activation free energy calculated as the difference between the Helmholtz free energies ($F = -k_\mathrm{B}T \log Z$, being Z the partition function) of the initial adsorption site j and that of the corresponding transition state at the top of the barrier connecting j to k. The last expression is usually rewritten as

$$\Gamma_{kj} = \Gamma_{kj}^{(0)} \mathrm{e}^{-\frac{\Delta U}{k_\mathrm{B}T}}, \tag{28}$$

where now ΔU is the difference between the $T = 0$ K energy of the adsorption site i and that of the transition state at the saddle point along the path $j \rightarrow k$ that is commonly referred to as diffusion barrier. $\Gamma_{kj}^{(0)}$ is interpreted as the attempt-to-escape frequency of the adatom in site i and it is related to the vibrational properties of the system and to the temperature. In this work, we consider $\Gamma_{kj}^{(0)} = \Gamma_0 = 10^{13}\,\mathrm{s}^{-1}$, for each j, k. This is a reasonable assumption since the prefactor $\Gamma_{kj}^{(0)}$ has been found to have a maximum variation of a factor 2 [50]. The determination of ΔU is more critical since it appears in the exponent of (28). The energy barriers ΔU that enter the process rates in (28) are those calculated from first principles.

3.5 *The Rate Equation*

The diffusion coefficient can be determined from the set of the Γ_{kj} defined above, through the solution of the rate equation for $P_j(\mathbf{n}, t)$, the probability of finding the random walker in the adsorption site j of the $\mathbf{n}$-th unit cell at time t. The rate equation is

$$\frac{\mathrm{d}}{\mathrm{d}t} P_j(\mathbf{n}, t) = \sum_{k=1, k \neq j}^{N_\mathrm{a}} \sum_{\mathbf{n}'} \Gamma_{jk}(\mathbf{n} - \mathbf{n}') P_k(\mathbf{n}', t) - P_j(\mathbf{n}, t) \sum_{k=1, k \neq j}^{N_\mathrm{a}} \sum_{\mathbf{n}'} \Gamma_{kj}(\mathbf{n} - \mathbf{n}'), \tag{29}$$

where $\mathbf{n} = (n_x, n_y)$ is the 2D lattice index along the directions x and y, $\Gamma_{kj}(\mathbf{n}' - \mathbf{n})$ is the transition probability per unit time from site $(j, \mathbf{n})$ to site $(k, \mathbf{n}')$ and N_a is the number of adsorption sites per surface unit cell. The conditions $P_j(\mathbf{n},t) \in [0,1]$ and $\sum_{j=1}^{N_a} \sum_{\mathbf{n}} P_j(\mathbf{n},t) = 1$ hold. The first term represents the adatom coming from site k to site j, the second one the adatom leaving j by jumping to all the other possible sites k.

By applying the Fourier transform

$$P_j(\mathbf{n},t) = \sum_{\mathbf{q}} e^{i\mathbf{qn}} P_j(\mathbf{q},t) \tag{30}$$

the master equation, (29), can be brought to the compact form:

$$\frac{d}{dt}\mathbf{P}(\mathbf{q},t) = \Gamma(\mathbf{q})\mathbf{P}(\mathbf{q},t), \tag{31}$$

where $\mathbf{P}$ is the array of N_a elements of the probability Fourier coefficients and Γ is a $N_a \times N_a$ matrix whose elements are

$$\Gamma_{jk}(\mathbf{q}) = \sum_{\mathbf{n}} e^{i\mathbf{qn}} \Gamma_{jk}(\mathbf{n}) - \delta_{jk} \sum_{l=1}^{N_a} \sum_{\mathbf{n}} \Gamma_{lj}(\mathbf{n}) \tag{32}$$

commonly referred to as the transition rate matrix [51].

We are now interested in the steady-state solutions of (29) and (31). Owing to the periodic boundary conditions and considering a very large system with an infinite number N of cells, the probability $P_j(\mathbf{n})$ should be constant over $\mathbf{n}$; thus, $P_j(\mathbf{q}) = P_j(0)\delta(\mathbf{q})$. The steady-state solution for P_j is calculated directly from

$$\Gamma(\mathbf{q})\mathbf{P}\delta(\mathbf{q}) = 0. \tag{33}$$

Another quantity useful for understanding diffusion is the mean life-time in each site, given by the sum of the probabilities of escaping from site j:

$$\frac{1}{\tau_j(\mathbf{n})} = \sum_{k=1}^{N_a} \sum_{\mathbf{n}'} \Gamma_{kj}(\mathbf{n}' - \mathbf{n}). \tag{34}$$

The tracer diffusion tensor can be obtained [52, 53] from the eigenvalues of the transition rate matrix (32). It can be demonstrated [54] that for this master equation there is one and only one eigenvalue, let it be $\gamma_1(\mathbf{q})$, such that

$$\gamma_1(0) = 0 \tag{35}$$

and the real part of all the other eigenvalues is negative. The tracer diffusion tensor is obtained as

$$\mathbf{D}^* = \mathbf{BHB}^{\mathrm{T}}, \tag{36}$$

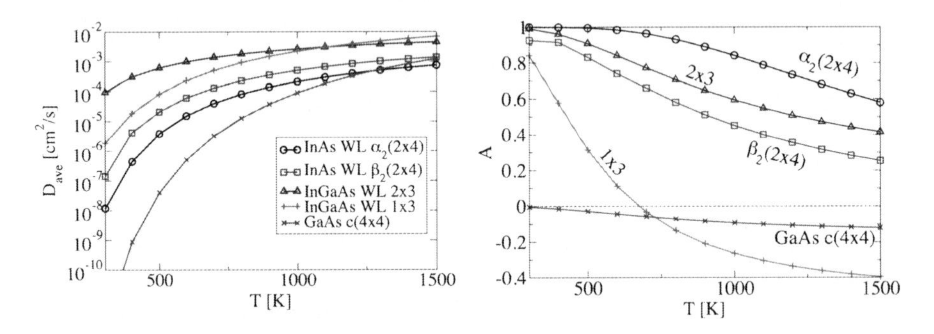

Fig. 7 Average diffusion coefficient (D_{ave}) and anisotropy (A) of the diffusion coefficient for the investigated systems, as a function of temperature (from [59])

where **B** is the transformation matrix from Cartesian coordinates to lattice indexes, and **H** is the Hessian of the above eigenvalue:

$$\mathbf{H} = -\frac{1}{2}\nabla_{\mathbf{q}}\nabla_{\mathbf{q}}\gamma_1(\mathbf{q})\bigg|_{\mathbf{q}=0}. \tag{37}$$

In order to illustrate the results of our calculation, we define the average In diffusion coefficient

$$D_{\mathrm{ave}} = \left(D^*_{[110]} + D^*_{[\bar{1}10]}\right)/2 \tag{38}$$

and the anisotropy

$$A = \frac{D[110] - D[\bar{1}10]}{D[110] + D[\bar{1}10]}. \tag{39}$$

The average diffusion coefficient and the anisotropy for In on each of the considered surface reconstructions are reported in Fig. 7, as a function of the temperature.

We know that α_2 is the WL reconstruction having the lowest symmetry among all the considered reconstructions. Comparison with the β_2 reconstruction which is, in many ways, similar but has a much higher symmetry can evidence the effects of symmetry. The first observation is that the In diffusion coefficient on β_2 is higher than on α_2: this is due to the flatness of the In PES on β_2, as shown in Fig. 5. In this case, the surface symmetry plays a dominant role in diminishing the PES corrugation, by limiting the atom bond deformation in the surface layer. The reconstruction α_2 is instead more capable to relieve the strain due to the lattice mismatch between the WL and the substrate, by strongly deforming the surface bonds. We also notice that there is a great difference between the diffusion along the two main surface directions: In diffusion is much larger along the $[\bar{1}10]$ direction than along the $[110]$ direction, for both surface reconstructions. This is a

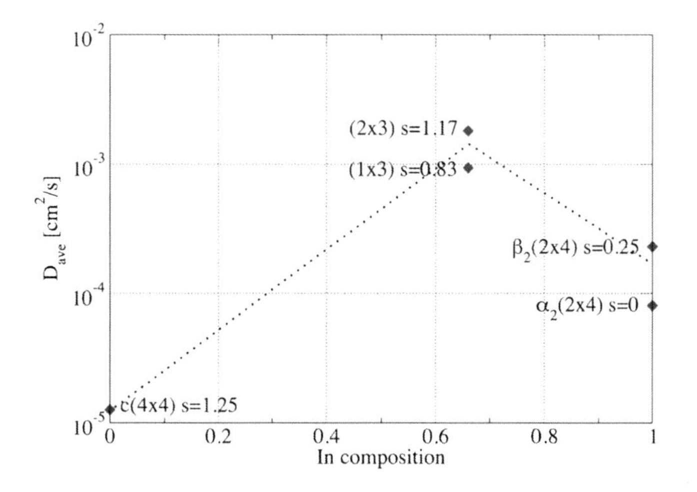

Fig. 8 Average diffusion coefficient D_{ave} for the investigated systems as a function of indium composition and As stoichiometry S, at $T = 800$ K (the *dotted line* is a guide for the eye) (from [59])

consequence of the PES shape [45]: low-energy regions, with weak energy barriers extending along the $[\bar{1}10]$ direction, act as deep channels for In motion, while along the orthogonal $[110]$ direction the adsorption sites are separated by higher barriers located at the As dimers (Fig. 5). Of course, the anisotropy is higher at low temperatures, since the probability to overcome the highest barriers is lower than at high temperatures. Therefore, at low T the diffusion occurs mainly along $[\bar{1}10]$. This means that material transport towards the quantum dots occurs preferentially along the $[\bar{1}10]$ directions. We observe that the most anisotropic diffusion tensors are found for In on the $(2 \times N)$ (2×3 and 2×4) surface reconstructions. This behaviour originates from the presence of the in-dimer trenches oriented along the $[\bar{1}10]$ directions having lower barriers for diffusion (Fig. 5).

In Fig. 8, D_{ave} is shown as a function of the In composition of the WL, which is directly related to the amount of surface strain present on the system. We notice that the trend of the In diffusion coefficient is not monotonic with the quantity of In incorporated in the WL. The lowest In diffusion coefficient is observed for the $c(4 \times 4)$ surface reconstruction, that is, the one which is not subjected to compressive strain. For 66% In composition, the In diffusion coefficient is maximum (two orders of magnitude higher than in the previous case), then it decreases again for the (2×4) WLs, corresponding to an In coverage of 1.75 ML, and 100% In composition. Since the WL reconstructions at 66% In composition are different from those studied at 100% In composition, it is not possible only on the basis of these results to discriminate the effects on diffusion due to the WL composition from those due to the surface reconstruction.

Finally, when comparing the β_2 reconstructed WL to the clean InAs $\beta_2(2 \times 4)$ surface [37], we see that the In adatom presents a higher diffusion coefficient on the WL than on the pure InAs surface ($D_{ave} = 1.07 \times 10^{-4}$ cm^2/s and $D_{ave} = 2.32 \times 10^{-4}$ cm^2/s, respectively). Moreover, the adsorption sites within the in-dimers were not considered in [37] (if considered, they would further increase the diffusion barriers). Thus, the compressive strain of the WL makes the adsorbate more mobile on the surface. To understand this point we observe that since the InAs WL is compressively strained on GaAs, the atom positions are much more displaced from their ideal sites than in the case of the unstrained but reconstructed InAs(001) β_2 surface. Their lateral distance is smaller in the WL than on the free surface, while the interplanar (001) distance is larger. This leads to the flattening of the PES and the In adatom moves more freely on the WL. The trend of the diffusion coefficient which decreases with an increasing compressive strain of the substrate agrees qualitatively with the trends reported in [36].

Our calculations agree with the general experimental behaviour. The anisotropy of diffusion is seen to strongly favour the $[\bar{1}10]$ direction. This is consistent with the experimentally observed elongation of self-assembled nanostructures, e.g., in the transformation of InAs quantum dots under slow overgrowth using a GaAs capping layer [55]. Other examples are the formation of mounds elongated along the $[\bar{1}10]$ direction [56] or the formation of elliptical nanorings [31], depending on the deposited GaAs thickness and the annealing time and temperature. The increase of diffusion with In coverage has been observed for In coverage increasing up to the critical thickness [7, 31].

The effect of the adsorption sites sitting in between the As dimers has been discussed in previous works [38, 44, 57]. In the cases of In on $c(4 \times 4)$ and on (1×3) and (2×3) reconstructions, the authors found these sites too shallow, with a low probability to be occupied at usual growth temperatures, and thus they do not need to be inserted in the calculation of diffusion. On the $\beta_2(2 \times 4)$ surface reconstructions, the in-dimer A_i adsorption sites are the ones with the lowest adsorption energy, while on the $\alpha_2(2 \times 4)$ the corresponding A_i adsorption site has an energy 0.5 eV higher than the PES bottom. Thus, on the β_2 the sites A_i are highly populated at the usual growth temperatures and have to be taken into account for a correct description of In diffusion. On the other hand, on the α_2 surface reconstruction at growth temperatures, the probability to find the adatom in those sites is very low, and their effect on diffusion is therefore negligible. We have verified that the diffusion coefficient is increased by a factor of 2 on β_2 if the adsorption sites A_i and A_a are not included in the transition network, while no remarkable difference is found on the α_2.

The probability of site population is connected with the mean life time τ [$P_j(\mathbf{n}, 0) = f_j(\mathbf{n})/\tau_j(\mathbf{n})$] being $f_j(\mathbf{n})$ the fraction of tracers in the particular $(j, \mathbf{n})$ configuration], also defined in (34). The analysis of the site life time τ_j reported in Fig. 9 reveals that the sites where the adsorbate spends most of its time are the ones where In is inserted in the As dimers. Other important high-permanence sites are A_1 and A_2 for the α_2 surface reconstruction, and A_4 and A_5 for the β_2 reconstruction [58, 59]. Let us notice that life time τ_j in the dimer sites is orders of magnitude

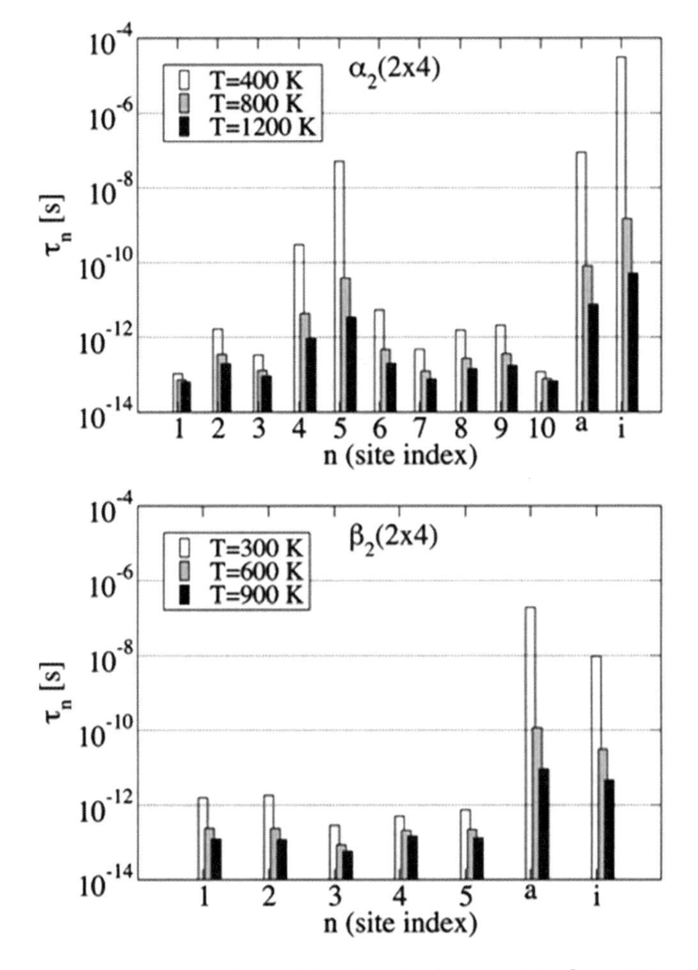

Fig. 9 Mean permanence time τ_n for each binding site, for α_2 and for β_2, at different substrate temperatures T. The additional sites where the In adatom is included in the As dimers are labelled by "a" and "i" (ad-dimer and in-dimer). The sites n' for the β_2 are not shown in the figure, since they are mirror symmetric to the n sites (from [59])

higher than that in the other sites, owing to the higher confining barriers. Thus, even if these sites, in the case of the α_2 reconstruction, do not play an important role on the diffusion coefficient, they are however essential if we want to address island nucleation. Indeed, sites with a high life time are the best candidates for nucleation. When the temperature is increased, the permanence times tend to become similar for all adsorption sites, because the thermal energy of the adsorbate is higher and the probability to overcome the barriers increases significantly everywhere.

4 InAs Growth on GaAs(001): The 2D–3D Transition

4.1 *The 2D–3D Transition and Size Distribution of Quantum Dots*

Approaching the critical coverage of 1.59 ML, the surface morphology becomes quite complex. This is exemplified by the AFM topography at 1.5 ML InAs coverage displayed in Fig. 10. One can distinguish on terraces three feature families: (1) large 2D islands 1 ML high (0.3 nm) corresponding to the as-yet incomplete WL; (2) small 3D islands, termed quasi-3D QDs, which are typically 2 ML high and sit preferentially at the upper edge of large 2D-islands and terraces (see Fig. 10, the inset and the left line-profile); and (3) well-developed 3D QDs with 3–4 nm height and ~40 nm base size.

With increasing coverage, the family of 3D QDs undergoes an explosive nucleation with one order of magnitude change in the density within 0.2 ML above the critical coverage. Their total volume is larger than the deposited volume because of a sizable surface-mass transport during growth. The WL is also involved and its thickness is reduced progressively by the supply of adatoms to QDs until the barrier for nucleation increases and the nucleation process ends [12].

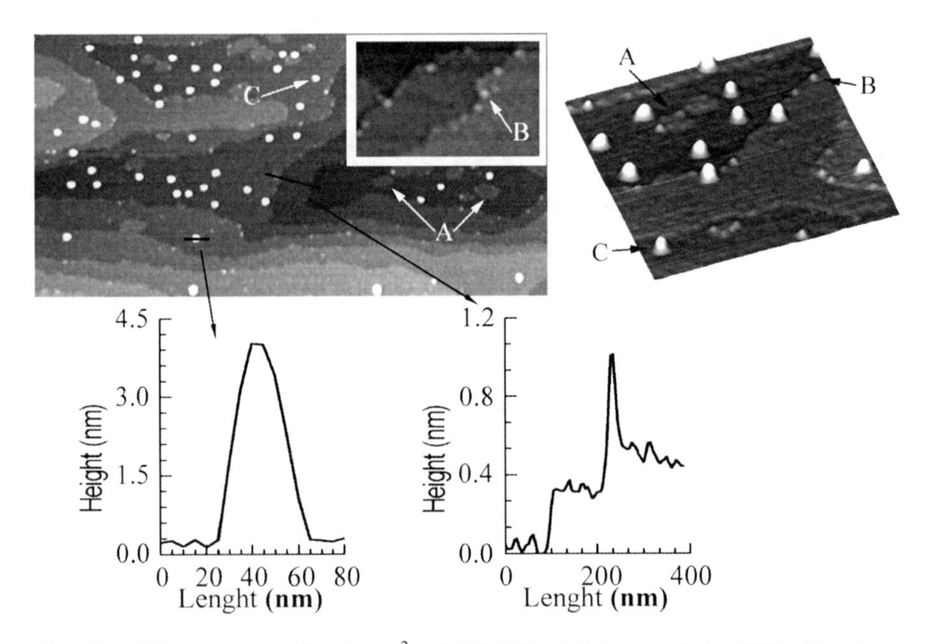

Fig. 10 AFM topography, $2 \times 1\ \mu m^2$, of 1.5 ML of InAs on GaAs (001). The *inset*, $400 \times 250\ nm^2$, evidences the nucleation of small 3D dots on the upper edge of steps. The labelled features are: (**A**) large and small 2D islands one monolayer high; (**B**) small 3D dots (quasi-3D QD) of height <2 nm and base size ~20 nm; (**C**) 3D quantum dots (3D QD) of height 3–4 nm and base size ~40 nm. Height profiles are shown of a 3D QD (*left*), and of a sequence of two steps with a *quasi*-3D QD nucleated at the upper step-edge (*right*) (from [14])

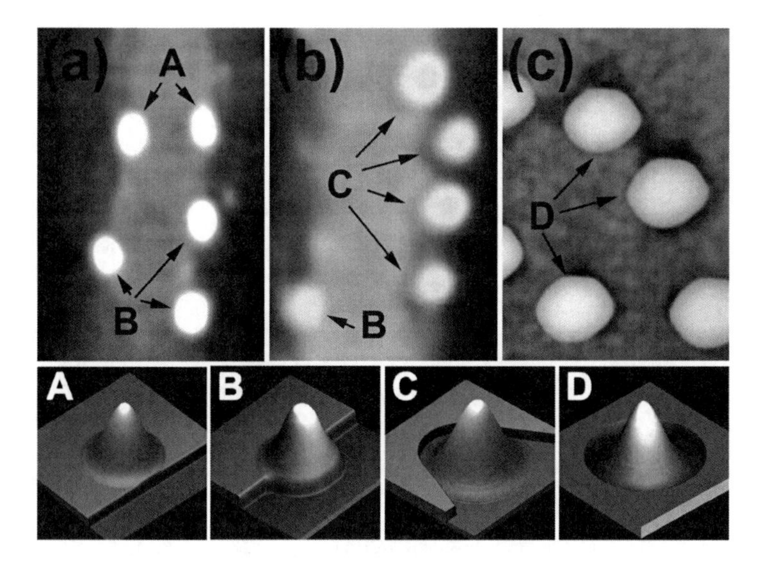

Fig. 11 120 × 160 nm^2 AFM images for coverage of 1.54 ML (**a**) and 1.61 ML (**b**). "A" are QDs nucleated over the step-edge, "B" are QDs that have partially eroded the step-edge and "C" are QDs that have completely eroded it. (**c**) 80 × 120 nm^2 AFM image showing plane erosion occurring for 2.4 ML of InAs grown at 522°C ("D" are QDs after plane erosion). Bottom panels show schematic 3D views of QDs before and after the process of step and plane erosion (from [8])

About the origin of this surface mass transport, we stress that, besides the possible contribution of the "floating" In on the WL (see Sect. 3.1), there is convincing experimental evidence that a substantial surface mass contribution comes from the mechanism of step erosion by dots nucleated at step edges.

This is clearly shown in Fig. 11 that compares two regions of the surface of InAs coverage 1.54 ML (panel a) and 1.61 ML (panel b) where the average volume of the 3D-QDs is ≈220 nm^3 and ≈340 nm^3, respectively. QDs are observed nucleating preferentially at the upper side of step edges, as shown in panel a. Comparing images for successive coverage, one observes that QDs nucleated close to the step edge (A in panel a) progressively erode the step edge (B) until they appear detached (C in panel b), as modelled in pictures A–C below the topographies. Such erosion is self-evident for dots at the step edges but is unclear for dots nucleated on flat regions of the surface. Nevertheless, signs of erosion of the plane (D in panel c) were observed for dots grown at higher temperature (522°C) and coverage (2.4 ML). In the latter case, the larger size of the dot (average volume 1,500 nm^3) and, consequently, the larger strain energy at its base could explain the erosion. As a matter of fact, around the QD a compressive strained area forms where the elastic energy differs from that of the WL far from the island by an amount that is large and positive. This is due to the strain redistribution between island and substrate that will be discussed in detail in Sect. 4.3 where the elastic energy map is shown (Fig. 19) for the limit case of an InAs dot on a GaAs substrate in absence of the WL.

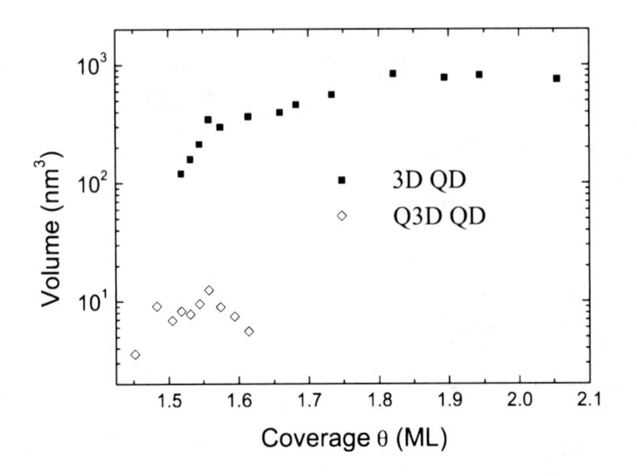

Fig. 12 Average volume measured from AFM topographies of quasi-3D QDs (Q3D QD) and 3D QDs (3D QD) as a function of InAs coverage for a sample grown at 500°C with growth interruptions, i.e., by cycling the InAs deposition in 5 s of evaporation followed by 25 s of growth interruption

We speculate that, because of the strain profile encircling dots, the detachment rate of adatoms from steps is higher than the attachment rate favouring the consumption of the steps around dots.

4.2 *Kinetics of the 2D–3D Transition*

At the very beginning of the 2D–3D transition, the size distribution of the islands is bimodal comprising two families of islands: the quasi-3D QDs and the 3D QDs.

As shown in Fig. 12, the family of quasi-3D QDs (precursors) tends to disappear with increasing coverage and to convert to 3D QDs. The kinetics of this process is strongly dependent on the parameters and procedures of the growth [61].

We analysed the volume density of InAs QDs on GaAs(001) substrates obtained by scanning atomic force microscopy (AFM) with a growth model based on kinetic rate equations. In order to describe the time evolution of the amounts of adatoms, n_1, precursors, n_2, and quantum dots, n_3, on the surface, we write the kinetic rate equations in the mean-field approximation [62, 63]:

$$\begin{cases} \dot{n}_1 = F - \beta(n_1 + n_3) - n_1(\kappa_2 n_2 + \kappa_3 n_3) \\ \dot{n}_2 = \kappa_2 n_1 n_2 - \gamma n_2 \\ \dot{n}_3 = \beta n_3 + \kappa_3 n_1 n_3 + \gamma n_2 \end{cases}, \quad (40)$$

where F is the InAs flux (ML/s), $k_j = \sigma_j D$ $(j = 2, 3)$ are the rate coefficients for adatom attachment to precursors and QDs, respectively, σ_j are the size-normalised diffusion capture factors and D is the surface diffusion coefficient of adatoms. The detachment rate of adatoms from precursors is small and can be disregarded, whereas the direct capture rate of adatoms, β, by the WL and QDs is unity at the growth temperature of 500°C [64]. In models of strained epitaxy based on the classical theory of nucleation [65, 66], the mass transfer is a thermally activated process described by the nucleation rate, γ, which, in turn, depends exponentially on the superstress $\zeta = (\theta - \theta_c)/\theta_c$, where θ and θ_c are the coverage and critical coverage, respectively. ζ measures the instability (2D–3D transition) of the WL for thickness larger than the equilibrium thickness [67]. Recently, Song et al. [61] used $\gamma = \gamma_0 \exp[\zeta E/kT]$, where γ_0 and E are constant parameters to fit the high-energy electron-diffraction intensity from the WL of InAs deposited on GaAs for $\theta < \theta_c$. They found $\gamma_0 = 0.09 \pm 0.02\ \mathrm{s}^{-1}$ and $E = 2.0 \pm 0.3$ eV and a direct dependence of E on the deposition flux at small growth rates. $E\zeta$ represents the formation energy of QDs.

In Fig. 13, we compare the volumes (in unit of ML) of the InAs nanostructures on GaAs(001) substrates as a function of θ, calculated by solving the system of (40) with the corresponding experimental distributions [67]. As initial conditions we assume that adatoms are dynamically in equilibrium with the WL under the flux F before the appearance of precursors and QDs and that precursors, in the absence of an exact criterion to fix their critical size [62], have the measured density [67] lower than $4 \times 10^{-7}\mathrm{nm}^{-2}$ around $\theta = 1.4$ ML. Using the optimised set of parameters given by Song et al. [61], we find good agreement between our calculations and the experimental data of Ramachandran et al. [67]. It is worth noting that the model supports strongly the binomial size distribution of self-assembled QDs of InAs on GaAs. In detail, the transition from the 2D-island morphology of the WL to the 3D morphology of QDs occurs in the narrow range 1.45–1.74 ML in which the WL increases consuming the adatoms and the precursors grow up to $\theta \sim 1.61$ ML. Below θ_c, ζ is negative and the metastable WL [69] continues to grow layer by layer under the InAs flux. The relief of the strain energy takes place through the progressive increase of the height of the 2D islands to form Q3D clusters with small aspect ratio r (height-to-base length ratio). Beyond θ_c, ζ turns to positive and the nucleation rate rises exponentially with θ starting from the value $\gamma \sim 0.09\ \mathrm{s}^{-1}$. At $\theta = 1.61$ ML, the superstress is large enough that nucleation of 3D QDs with higher aspect ratio starts dominating because taller clusters are more relaxed than flatter ones [69]. Concurrently, the precursors drop abruptly down to 1.74 ML and convert to QDs at high rate. We find that at $\theta \sim 2.2$ ML the contribution from the impinging flux after the 2D–3D transition to the final QD volume of 0.89 ML is 0.69 ML, and the rest, 0.1 ML, comes from precursors and 0.1 ML from the adatoms directly attaching to QDs. Therefore, part of the mass transfer to QDs comes from the precursors in agreement with [67]. For $\theta \geq 1.8$ ML the decrease of the 2D island volume with increasing InAs delivery signals the attachment (at a rate β) of adatoms to QDs (Fig. 13c). As remarked by Ramachandran et al. [67], this is consistent with the disappearance beyond 1.74 ML of the large 2D islands in the WL as well as of precursors, as evident from Fig. 13a. Moreover, post-growth ripening of the QDs must be excluded since the surface morphology was frozen very quickly after growth [63].

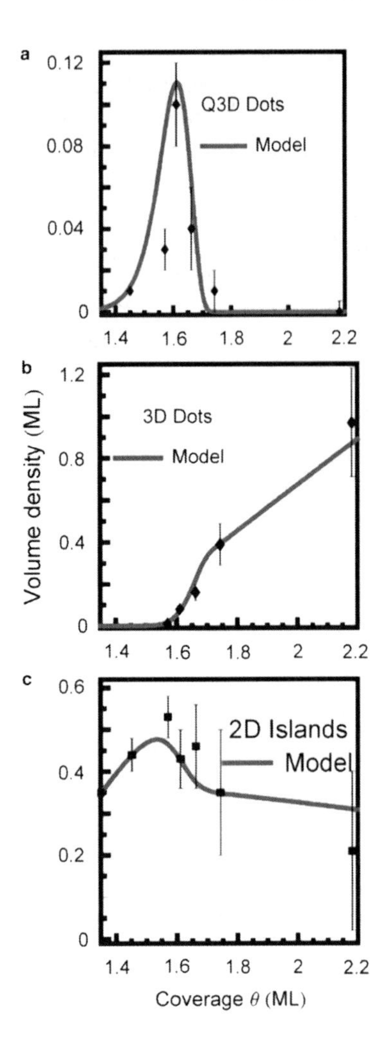

Fig. 13 Coverage dependence of the volume density (in ML) of InAs: (**a**) quasi-3D quantum dots n_2, (**b**) 3D quantum dots n_3, and (**c**) 2D islands $\theta - 1 - (n_2 + n_3)$, growing over a 1-ML wetting layer of GaAs(001). Scattered symbols with error bars are the experimental data from [60] measured at 500°C using a growth rate $F = 0.220 \pm 0.022$ ML/s and continuous growth mode. *Solid curves* are calculated according to the model (24). The parameters' values are $E = 3$ eV, $n_1(t_0) = 0.22$ ML, $n_2(t_0) = 6 \times 10^{-4}$ ML, $n_3(t_0) = 0$, $t_0 = 5.6$ s, and the size-normalised capture rates $\kappa_2 = 20$, $\kappa_3 = 40$ (from [61])

The above quantitative analysis enables us to understand the main modifications that occur when deposition of InAs is performed with growth interruptions, as in our experiment (Fig. 12). Once the In flux is stopped after 5 s of deposition, precursors are allowed to convert into 3D-QDs without refilling from the adatoms. The rate equation for n_2 can approximately be written as

$$\dot{n}_2 = -\gamma n_2, \tag{41}$$

because the adatom volume density n_1 is small at the early stages of growth and γ is nearly constant and equal to γ_0 for $F = 0$. Thus, the volume of quasi-3D dots decreases exponentially with time with a time constant $\tau = \gamma^{-1}$ during the 25 s of growth interruption. The amount of InAs contained in the small quasi-3D clusters, calculated from (41), taking as initial values for n_2 at $t_0 = 5$ s, the ones

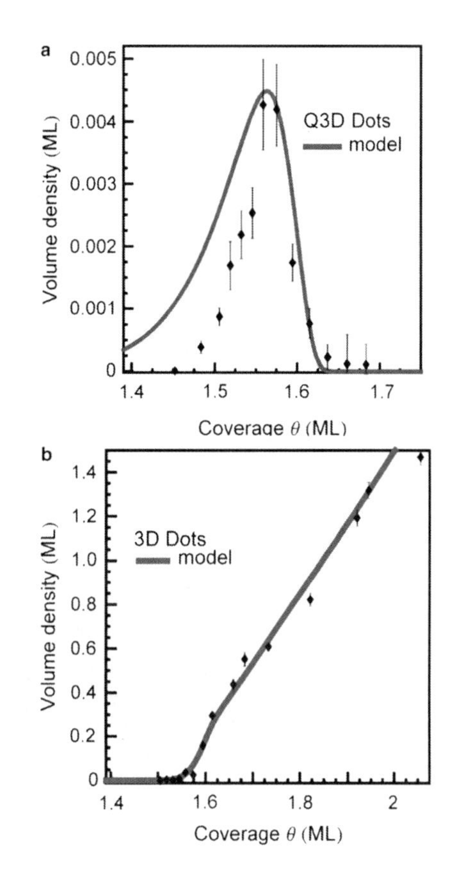

Fig. 14 (**a**) Coverage dependence of the volume density (in ML) of InAs/GaAs (001) quasi-3D quantum dots (Q3D) calculated from (41) after 25 s of growth interruption and (**b**) 3D quantum dots volumes n_3 scaled by a constant factor of ~2. The scattered symbols with error bars are the experimental data measured at 500°C using a growth rate $F = 0.029$ ML/s and interrupted growth mode. The parameters values are $E = 6$ eV, $n_1(t_0) = 0.029$ ML, $n_2(t_0) = 6 \times 10^{-4}$ ML, $n_3(t_0) = 0$, $t_0 = 45$ s, and the size-normalised capture rates $\kappa_2 = 20$, $\kappa_3 = 40$ (from [60])

obtained by solving the system (40), is shown in Fig. 14a. It is found that the quasi-3D volume density is reduced by more than an order of magnitude compared to that of continuous growth mode, in close agreement with the experimental data.

Above θ_c the calculated n_3 values must be scaled up by a factor of about 2 to agree with the experiment, as shown in Fig. 14b.

4.3 *Equilibrium Shape of InAs Islands*

The size and shape of self-organised QDs are affected both during the first stage of the growth, depending on the kinetics of the process, as well as during the second stage, depending on capping of the islands or annealing processes. Moreover, intermixing between the QD material and the substrate occurs to a varying degree according to temperature and depositions rates of the growth. Despite theoretical predictions [70, 71] based on equilibrium properties, most experiments [8, 72, 73]

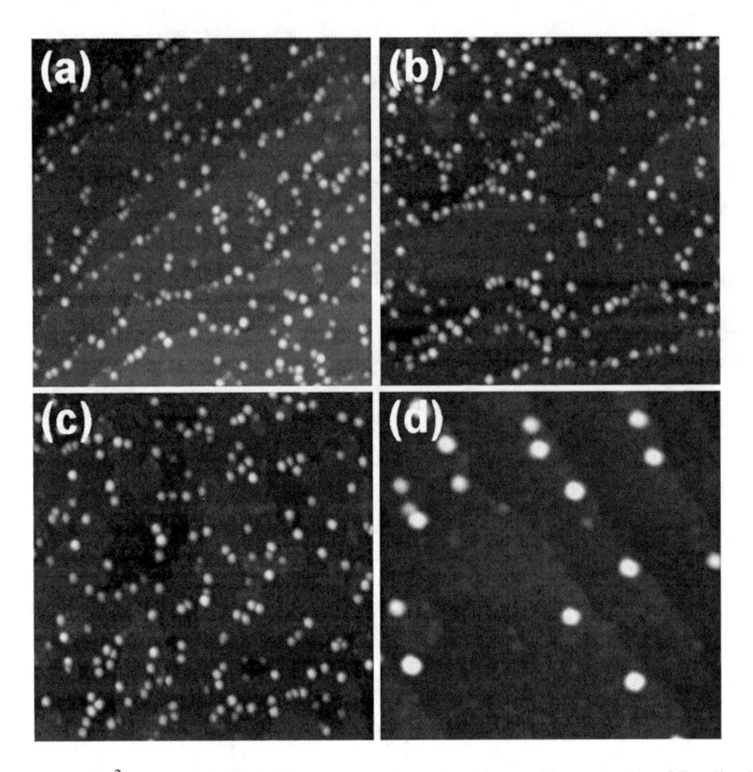

Fig. 15 $1 \times 1\ \mu m^2$ AFM topographies for 1.7 ML InAs on GaAs(001) showing QDs distribution immediately quenched (**a**), and with 30 min post-growth annealing at 420°C (**b**), 460°C (**c**), 500°C (**d**) (from [13])

report an aspect ratio (AR) r of the InAs islands on GaAs(001) considerably lower (r between 0.2 and 0.3) than expected (between 0.3 and 0.4). In fact, InAs QDs are found bounded predominantly by {137} facets with $r = 0.2$ (height-to-base, along the [110] direction).

We have performed a combined experimental and theoretical study of r and faceting of QDs immediately after the growth and of their evolution upon annealing at several temperatures. This has been done by means of AFM and RHEED measurements and by simulations with the finite element method (FEM) of the elastic relaxation of InAs islands of appropriate shape on GaAs(001). Our aim was to understand how kinetics drives the formation and shape of QDs under extremely different growth conditions.

The InAs/GaAs(001) QDs were grown at 500°C by depositing 1.7 ML of InAs with a rate of 0.033 ML/s. Immediately after the InAs growth, the samples were annealed at several temperatures: 420, 460 and 500°C. For comparison, samples without post-growth annealing and immediately quenched after the InAs deposition were grown as well.

Figure 15 shows the AFM topographies of 1.7 ML of InAs deposited on GaAs (001) and immediately quenched (a), after 30 min annealing at 420°C (b), 460°C (c)

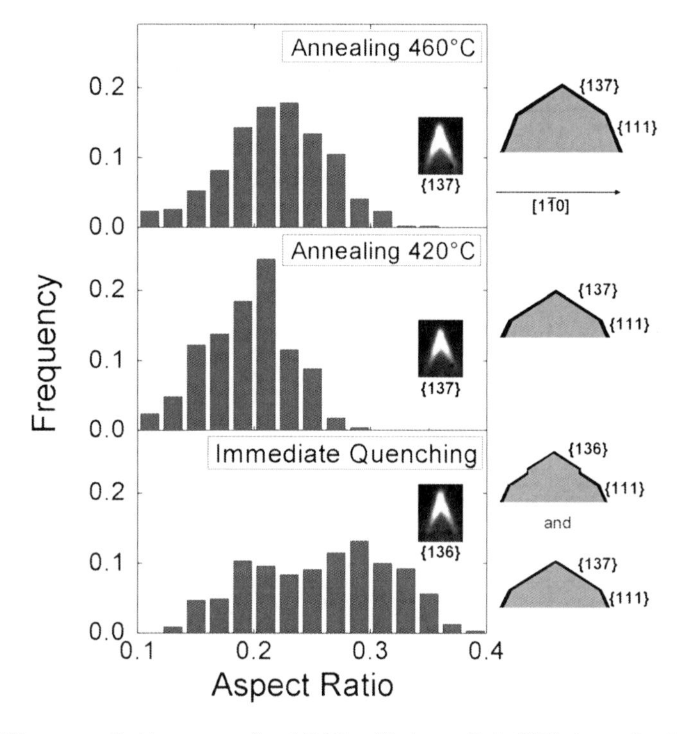

Fig. 16 QD aspect ratio histograms after 1.7 ML of InAs on GaAs(001): immediately quenched growth (lower panel); after 30 min post-growth annealing at 420°C (medium panel) and 460°C (upper panel). On the *right* of the image the schematic shape of the QDs is shown. In the insets RHEED chevrons and facets determined by RHEED and AFM analyses are reported (from [13])

and 500°C (d). The images clearly evidence coarsening effects on the sample annealed at 500°C while the slight effects present at 420 and 460°C are not clearly visible from the images.

Figure 16 shows the r histograms for all the samples examined. The r value was determined by measuring the facets angle. Unlike the size distribution of islands is unimodal at 1.7 ML of InAs deposition (Sect. 4.1), it is clearly evident that in the sample without annealing the AR distribution is bimodal, centred at $\approx$0.22 and $\approx$0.28. To the contrary, when annealing is applied, a single distribution (more or less wide) is centred at about 0.22. The value $r \approx 0.22$ matches with islands mainly composed by {137} facets (pyramids with $r = 0.24$), often reported in literature [72, 74, 75]. RHEED patterns along [1$\bar{1}$0] azimuth reveal a chevron angle that confirms the presence of these facets. For all the samples that underwent post-annealing, the AFM gradient-image analysis suggests a structure fully compatible with the structure indicated by Jacobi et al. [76], with about 80% of the whole island surface terminating with {137} facets, as labelled in Fig. 16, with the smaller {110} and {111} facets along the [110] direction. For the immediately quenched sample, the AFM analysis is more complex because of bimodality of AR distribution. For $r \approx 0.22$, the AFM analysis confirms the presence of {137} facets. On the other

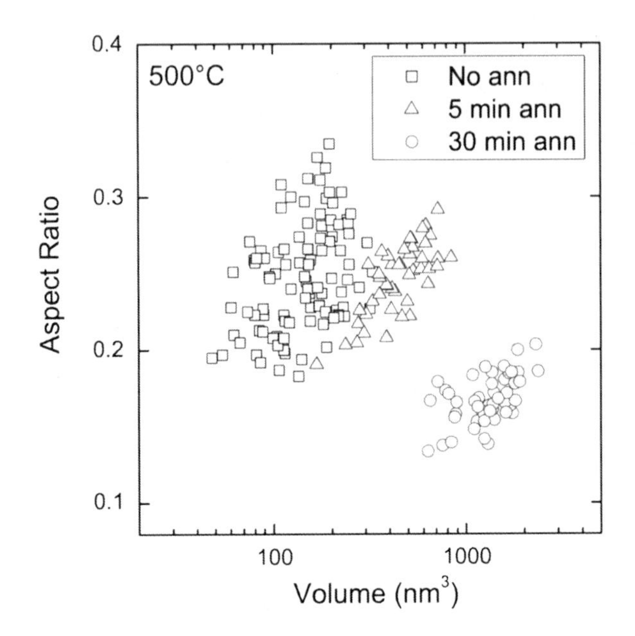

Fig. 17 QD aspect ratio versus island volume at 500°C as function of the annealing time: 0 min (*squares*), 5 min (*triangles*), and 30 min (*circles*) (from [13])

hand, understanding of $r \approx 0.28$ is possible by a comparison with RHEED patterns along $[1\bar{1}0]$ azimuth (insets in Fig. 16): the chevron angle clearly supports the presence of steeper facets, precisely {136}, as reported in literature [77, 78]. Although an $r \approx 0.26$ is expected for {136} facet, the small difference with respect to RHEED observations is likely due to the presence of transitional QDs which are observed when atoms have less time to diffuse and arrange the island equilibrium shape. The most likely situation in this case is described in [73] and schematically shown on the right of AR histogram, where the growth proceeds mainly layer by layer on {136} facets and the topmost layers do not make contact with the (001) substrate. This refers to the occurrence of faceted islands with atomic layers only present on the top. With such a shape, the QDs exhibit a higher AR from that expected. Differently from [73], we do not see successive formation of domes with steeper facets, due to the lower growth temperature in our case. The presence of {136} and {137} facets is then related to a competitive kinetic–thermodynamic growth conditions. Indeed, in comparison to experimental data in the literature, {136} facets are measured when the InAs growth rate is higher than 0.1 ML/s [77, 78], while {137} facets are found when growth rate is ≈0.01 ML/s or less [72, 74, 79]. The absence of QDs with $r = 0.3$ upon annealing is thus related to the kinetics of growth: When the system has time enough to evolve, the thermodynamically more stable {137} facet is found.

Annealing at higher temperatures or for a longer period increases the mean volume of QDs, changing the AR and faceting. The evolution of QD volume with post-growth annealing is related to QDs coarsening (Fig. 15). In Fig. 17, it can be noted an increase in QDs' volume in the first 5-min annealing at 500°C. The AFM analysis reveals the same {137} faceting and pyramidal shape of samples undergoing a 30-min annealing at 420 and 460°C. Besides a considerable increase in QD volume, 30-min annealing at 500°C, the QD shape changes from pyramidal to dome and the AR decreases. This does not disagree with experimental findings in literature [74, 79], which report an AR increase due to transition from pyramids to domes. In fact, because of the long-period annealing at 500°C, a massive island coarsening and a significant In–Ga intermixing may occur, thus increasing the Ga concentration in QDs and flattening the QD shape as expected by theory [80] and predicted by total-energy simulations for increasing volume of islands of Fig. 21.

From AFM profiles we find that large islands have a structure formed of facets forming an average 18.8° angle with the (001) plane, compatible with {114} facets, sometimes found on these systems [81]. Moreover, AFM profiles evidence that the dome's central area is formed of gentler facets: the measured average angle with the (001) plane is 11.6°, in good agreement with {117} facets. The faceting evolution during 500°C annealing gives an estimate on the time scale of In–Ga intermixing in QDs around 10 min.

The equilibrium shape of the strained coherent epitaxial island is derived by minimising its total energy with respect to its shape, at constant volume.

As described in Sect. 2.2, the excess free energy, F_{tot}, associated with the formation of 3D islands is divided into the elastic contribution originating from bulk strain relaxation, $F_{\mathrm{el},R}$ (19), and surface terms accounting for the additional formation of island facets, F_{S}, and edges, F_{edge},

$$F_{\mathrm{tot}} = F_{\mathrm{el},R} + F_{\mathrm{S}} + F_{\mathrm{edge}}, \tag{42}$$

where each term refers to the energy difference between a 3D island and a homogeneous planar InAs film of the same volume.

We have applied the finite element method (FEM) to the treatment of the first term, the elastic relaxation, for simulations of the stress and strain distribution in nanometric InAs islands on GaAs(001) surface. In the following, we discuss briefly both the effectiveness of the strain redistribution and the thermodynamic stability for islands with different r.

The calculation of the strain distributions was performed within the elasticity continuum theory. Thus, the structure under consideration is considered as an elastic continuum and the physical parameters under study are obtained by solving the elasticity equilibrium equations. In particular, the elastic free energy density ρ_{el} in linear approximation is given by the elastic constants c_{klmn} and the strain tensor ε:

$$\rho_{\mathrm{el}} = \frac{F_{\mathrm{el}}(\varepsilon)}{V} = \frac{1}{2}\sum_{klmn} c_{klmn}\varepsilon_{kl}\varepsilon_{mn}. \tag{43}$$

Table 3 Anisotropic material properties of GaAs and InAs

	c_{11} (GPa)	c_{12} (GPa)	c_{44} (GPa)
GaAs	118.8	53.8	59.4
InAs	83.4	45.4	39.5

By specifying the strain tensor in the canonical coordinate system of the crystal, i.e., for $\mathbf{e}_x||[100]$, $\mathbf{e}_y||[010]$ and $\mathbf{e}_z||[001]$, for structures with cubic symmetry the elastic energy reduces to

$$\rho_{\text{el}}(\varepsilon) = \frac{c_{11}}{2}\left(\varepsilon_{xx}^2 + \varepsilon_{yy}^2 + \varepsilon_{zz}^2\right) + 2c_{44}\left(\varepsilon_{xy}^2 + \varepsilon_{yz}^2 + \varepsilon_{xz}^2\right) + 2c_{12}\left(\varepsilon_{xx}\varepsilon_{yy} + \varepsilon_{xx}\varepsilon_{zz} + \varepsilon_{yy}\varepsilon_{zz}\right), \tag{44}$$

where the moduli of elasticity c_{11}, c_{12} and c_{44} of InAs and GaAs are given in Table 3 [82].

In FEM, both the island and the thick slab representing the substrate are divided into mesh elements on the vertices of which the displacement field is tabulated [83]. Periodic boundary conditions were applied to the side planes of the material in the simulation. The distribution of the finite elements was such that the number of finite elements was increased in the parts of the island where the elastic energy density was large. Within each element of this partitioning, the displacement field is uniquely determined by the linear interpolation of the values at the corners. The elastic energy density of the system after minimisation is evaluated as follows:

$$\rho_{\text{el}} = \frac{1}{V_{\text{is}}}\left[\int_{\text{is}} \rho(\mathbf{r})\mathrm{d}\mathbf{r} + \int_{\text{sub}} \rho(\mathbf{r})\mathrm{d}\mathbf{r}\right]. \tag{45}$$

The above expression, in which the two integrals are extended over the island and the substrate, is normalised to the island volume, V_{is}, in order to guarantee the self-similarity of the calculation.

For all the shapes calculated, we evaluated the energy release by QD formation, i.e., the elastic energy gain which stabilises the islands by strain relief compared to the homogeneously strained InAs film, for which the elastic energy per unit volume, from (18), can be given in analytical form as

$$\rho_{\text{film}} = \left(c_{11} + c_{12} - 2\frac{c_{12}^2}{c_{11}}\right)\varepsilon^2, \tag{46}$$

where ε is the mismatch between InAs and GaAs. The energy relaxation accompanying the 3D island formation is expressed by the relaxation factor [see (19)]

$$R = \frac{F_{\text{el},R}}{\rho_{\text{film}}}, \tag{47}$$

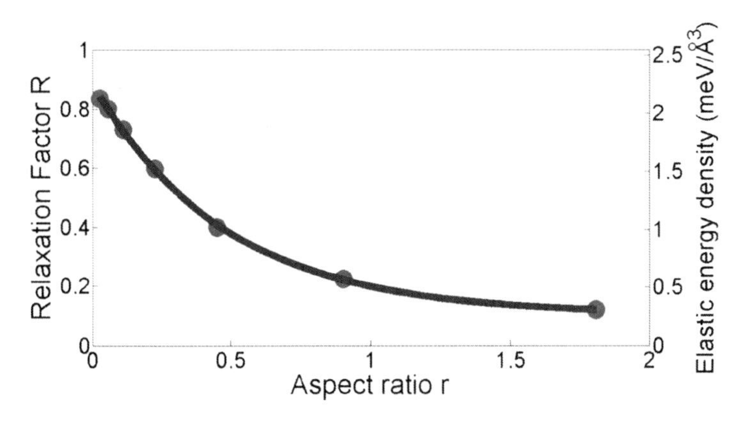

Fig. 18 Relaxation factor as a function of island aspect ratio. Calculated values (*points*) are fitted to the curve of (48) (*solid line*)

which ranges between 0 (fully relaxed configuration) and 1 (unrelaxed configuration). In Fig. 18, R is plotted as a function of r. The vertical scale on the right reports the elastic energy density ρ_{el} corresponding to each r.

R shows a monotonic behaviour, which can be fitted to the model function (solid line in Fig. 18):

$$R(r) = 0.80\mathrm{e}^{-2.0r} + 0.090\mathrm{e}^{0.055r^2}. \tag{48}$$

The dominant first term in (48) is a decreasing exponential function, which represents the more effective strain relaxation in steeper islands. In particular, the strain release is located at the island facets and close to the island apex (Fig. 19).

The second exponential term in (48), which is modulated by a smaller pre-factor, increases monotonically as the aspect ratio increases. This is due to the larger strain redistribution from island to the substrate, which causes a compressive load close to the island base and the substrate below, and is as effective as the steeper is the facet inclination (Fig. 20).

The elastic term in (42) scales linearly with the volume of the QD, being a volume effect, and ultimately becomes dominant for the larger islands. On the other hand, the second term, which is due to the creation of side facets, is an energetic cost for all calculated shapes. Being a surface term, it scales like $V^{2/3}$ to leading order, i.e., disregarding the renormalisation of surface energies due to strain. The latter term in (42) is the contribution from the edges, which scales as $V^{1/3}$, is quite small and is disregarded in the present calculation. The influence of the edge energies is small. Hence, the total energy gain in forming islands can be rewritten as follows:

$$F_{\mathrm{tot}} = (\rho_{\mathrm{el}} - \rho_{\mathrm{film}})V + \gamma(C_{\mathrm{i}} - C_{\mathrm{B}})V^{2/3}, \tag{49}$$

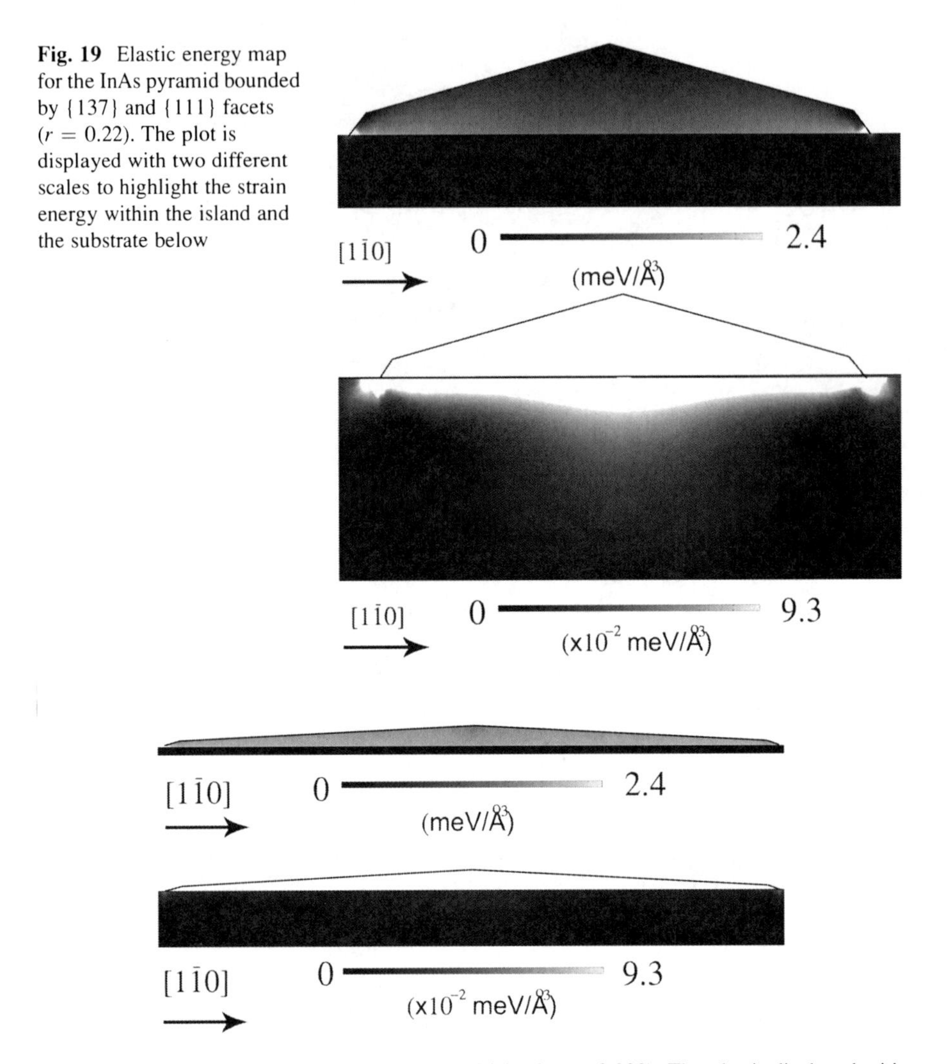

Fig. 19 Elastic energy map for the InAs pyramid bounded by {137} and {111} facets ($r = 0.22$). The plot is displayed with two different scales to highlight the strain energy within the island and the substrate below

Fig. 20 Elastic energy map for a mound-shaped island ($r = 0.032$). The plot is displayed with two different scales to highlight the strain energy within the island and in the substrate below

where C_i and C_B are the shape-dependent ratios between the exposed and the base surface area of the island and $V^{2/3}$, respectively. The surface energy value γ was taken 44 meV/Å^2 all over the exposed InAs facets. In Fig. 21, (49) is plotted as a function of the island volume for different aspect ratios r.

We note that for volumes between ~80 and ~280 nm^3 the most stable islands predicted by the calculation are InAs pyramids bounded by {137} and {111} facets (r ~ 0.22). This agrees with the experimental observation (Fig. 17) that the 3D QD volume, for immediately quenched samples grown at 500°C, is ~300 nm^3 on average.

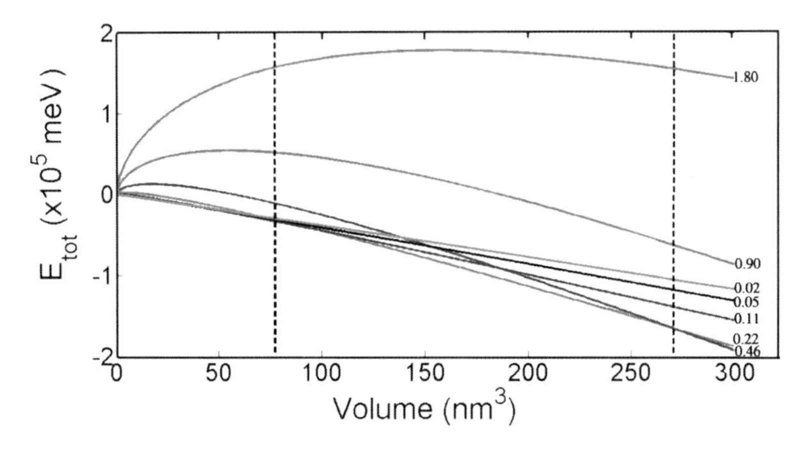

Fig. 21 Total energy of InAs islands on GaAs(001) surface for different island aspect ratio. The region where pyramids bounded by {137} and {111} facets ($r \sim 0.22$) are the more stable is marked by vertical *dashed lines*

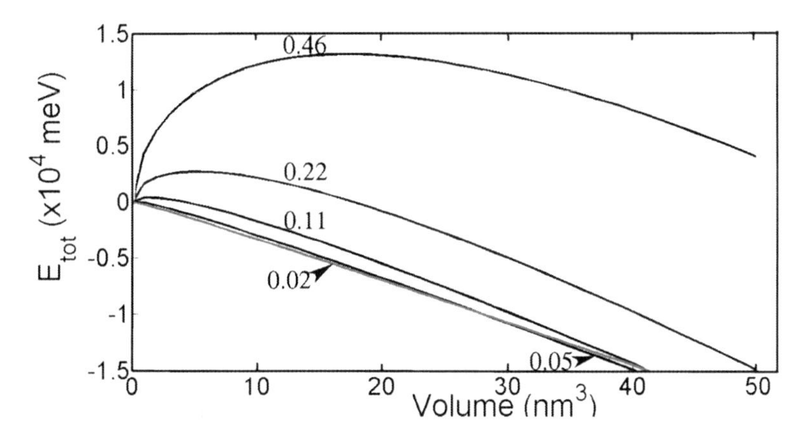

Fig. 22 Total energy of flat mound-shaped islands at low island volumes

For volumes smaller than 80 nm^3 the equilibrium shapes are flat mounds, with r ranging between 0.02 and 0.11 (Fig. 22). This result, again, is consistent with the finding, at the beginning of the 2D–3D transition, of the small quasi-3D QDs, discussed in Sect. 4.1.

It is noteworthy that this result seems to be not influenced by the neglected edge term, F_{edge}, in (42). The energy gain between the $r = 0.22$ curve and the $r = 0.11$ curve at $V = 10$ nm^3 is $\approx$4 eV. Accurate values of the edge energies are unknown. Assuming that the edge energies are comparable to the step energies ($\approx$25 meV/Å) on the GaAs (100), the edge contribution for an island of 10 nm^3 is about 2 eV.

4.4 Ripening of Quantum Dots

The size and shape of InAs Qds are deeply related to their emission wavelength; consequently, understanding the evolution of these quantities is of primary importance for the possibility of tuning the energy spectrum of QDs.

It is well known that post-growth annealing of QDs is responsible for several processes which influence the size distribution and shape of the islands, i.e., coarsening, where the larger islands grow at the expenses of the smaller ones that shrink and disappear (Ostwald ripening), and intermixing with the substrate atoms. The relative importance of these processes depends on annealing conditions and initial configuration after deposition. The kinetics of the Ostwald ripening (OR) is usually discussed in two limiting cases depending on two rate-determining processes: diffusion limited (DL) [84] and attachment-detachment limited (ADL) [84, 85]. The former is the coarsening process where islands are immobile but adatoms move to and from islands. The latter, in which processes at the island edges limit the kinetics, is characterised by a chemical potential constant among islands and discontinuous at the island edge.

Experimental studies suggest that, for annealing temperatures ranging between 440 and 475°C, coarsening is an Oswald ripening of the ADL type [86, 87].

By measuring the size distribution function of QDs as a function of the annealing temperature and time, we have identified the different regimes of coarsening and the mechanism that governs the evolution of the size distribution [11]. The analysis was performed on a set of samples grown as follows: after a 500-nm re-growth of GaAs on the GaAs(001) substrates, a 30-min annealing at 660°C under As_4 flux was performed to flatten the surface. 1.63 ML of InAs was deposited at a rate of 0.032 ML/s at 500°C when a clear $c(4 \times 4)$ reconstruction of the GaAs surface was observed. Finally, the samples were annealed at 420, 460 and 500°C. For purposes of comparison, samples without post-growth annealing and immediately quenched were also grown.

Figure 23 reports the size distribution of InAs islands for different annealing temperature and time. In order to describe our results within the theoretical framework of Ostwald ripening, a set of standardised histograms were created as a function of $V^{1/3}$, with V being the island volume (Fig. 23a–e). The histograms shown in Fig. 23f–j are the theoretical standardised distribution functions for DL and ADL Ostwald ripening [84, 85] to be compared with the experimental histograms.

The bimodal distribution of dots shown in Fig. 23a, f (no annealing) has been reported several times [9, 12, 67] (see Sect. 4.1). The size distributions after annealing, shown in Fig. 23b–d, g–i, are unimodal. In particular, Fig. 23b–d display a mean value which increases with increasing temperature, a trend suggesting that coarsening must have occurred. In Fig. 23e we observe the unexpected reappearance of a bimodal distribution. In correspondence, AFM images [11] (not shown) evidence the presence of small islands in shape of flat platelets 1 ML high. A likely interpretation is that either they are remnants of the coarsening process, where QDs have shrunk in favour of the larger ones or they simply have evaporated. For this

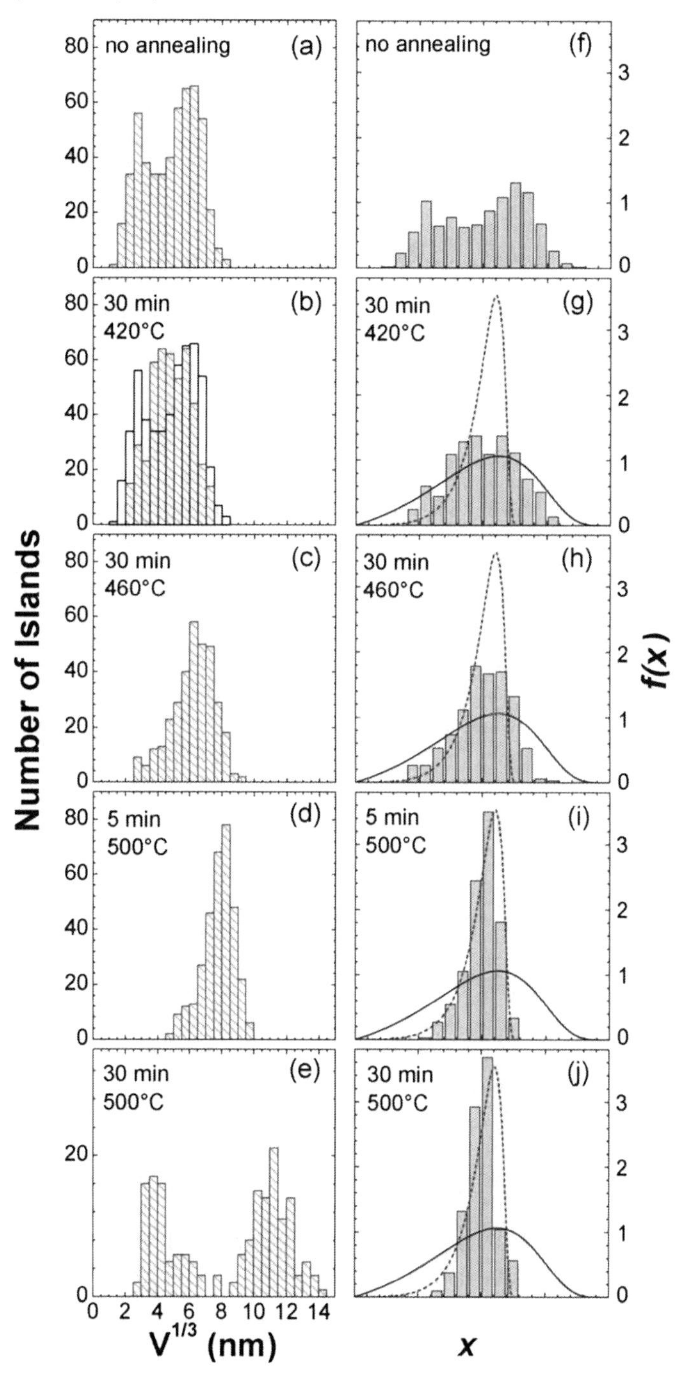

Fig. 23 (**a**)–(**e**) $V^{1/3}$-histograms as a function of annealing temperature and time. (**a**) is superimposed to (**b**) for comparison. (**f**)–(**j**) Corresponding experimental standardised distribution of (**a**)–(**e**) plotted as a function of the scaling variable $x = V^{1/3}/<V>^{1/3}$. The theoretical scaling functions $f_{ADL}(x)$ (*continuous line*) and $f_{DL}(x)$ (*dashed line*) are reported for comparison. In (**e**) the distribution at low volume refers to flat platelets (see text) (from [11])

reason in Fig. 23j, only the distribution of large-size islands has been taken into account and the platelets have been excluded from the statistical analysis.

Although the change from a bimodal (Fig. 23a, f) to unimodal (Fig. 23b–e, g–j) QD size distribution demonstrates that a coarsening process is underway, it is not the only process even at the lowest annealing temperature of 420°C. In Fig. 23b, we have superimposed, for a direct comparison, the distribution of Fig. 23a (empty histogram) to the distribution obtained after annealing at 420°C. It is evident that not only do the smaller QDs dissolve in favour of the larger ones (as expected for a standard coarsening process) but, because of the elastic barrier due to strain at the island edges, also the largest QDs lose atoms making the distribution more symmetrical in a kind of self-sizing process [88]. This latter statement is corroborated by the fact that, after annealing, the transition from {136} to {137} facets (see Sect. 2.4.3) is accompanied by a reduction of the respective volume of the QDs [13]. In Fig. 23g the sample annealed at $T = 420°$C exhibits a broad and unimodal standardised distribution which does not differ significantly from the theoretical ADL-curve (continuous line) [84, 85]. The small change observed is ascribed to the self-sizing process cited above, which makes the distribution quite symmetrical [11].

After annealing at 460°C, the size distribution function (Fig. 23c) is modified. The volume mean value increases, the distribution shows a clear reduction of the standard deviation, becomes more asymmetric, and the QDs density decreases of about 10%. This is evidence of a coarsening process occurring on the surface. On the other hand, in Fig. 23h the experimental distribution is plotted along with the theoretical ADL and DL functions (continuous line and dashed line, respectively). Neither theoretical functions describe the experiment. We remark that both theories are correct for an infinitely dilute distribution of islands, and do not take into account correlation effects in island growth leading to broadening of the distribution and to a shift of its maximum to higher values, in contrast to our experimental observations [11] (see evolution in Fig. 23h–j). More important, standard OR models are based on the assumption that the total volume of the islands is constant, while, for the system under study, an important mass transfer to the QDs from both the wetting layer and the substrate occurs for temperature larger than 450°C, as discussed in Sect. 4.1.

At 500°C, the annealing time becomes crucial. Within the first 5 min, we observe an increase of the mean volume of the QDs accompanied by a reduction of their number density such that the total volume remains constant. This fact indicates that coarsening occurs on a time scale faster than that of mass transport [89]. This fact also justifies the comparison between theoretical and experimental distributions: the inspection of Fig. 23i suggests that the DL theory explains the experimental distribution better than the ADL theory. A change from ADL to DL kinetics implies higher-energy barriers for monomer diffusion among the islands. As a matter of fact, the calculated diffusion coefficient of In is much lower on GaAs surface than on strained InAs/GaAs surface [38, 58]. So, the interdiffusion process, which increases the surface Ga concentration at 500°C reducing the strain, is likely responsible for such increase of the barrier energies.

Annealing the sample for 30 min at 500°C determines a reduction of the total volume contained in QDs [11] but does not change further the standard deviation of the distribution that becomes more symmetrical (see Fig. 23j). By considering the reduction of the total volume contained in QDs [11], the OR theory is not applicable, and the distribution is probably the result of the competition between coarsening, WL/substrate erosion and desorption.

5 Selective Area Epitaxy of InAs on GaAs(001)

5.1 Nanoscale Epitaxy of Quantum Dots

Advances in new generation devices for nanophotonics based on InAs/GaAs compounds rely on the capability to accurately control size and position of quantum dots on the substrate surface, well beyond the limit achievable in standard self-assembled growth by MBE. This requirement has given rise to several attempts to spatially confine the growth of InAs or InGaAs QDs to restricted areas of the GaAs surface with a variety of methods, including lithographic patterning, natural or artificial nanostructuring of the surface or exploiting facet-dependent migration [90–98]. Among these methods, electron-beam lithography is of particular interest by making the patterning of the substrate accessible at dimensions approaching the lateral size of dots, and thus opening the way for the control of the nucleation of the single dot [99, 100]. The most used procedure for spatially confining dots is the "top-down" approach where first dots self-aggregate freely on the surface and after nano-mesas are defined by e-beam lithography. The post-growth lithographic processing, however, albeit selective and flexible, introduces defects that may severely decrease the emission efficiency of QDs and the performance of the device. A viable alternative is the "bottom-up" approach where first confined substrate areas are lithographed and after, QDs are selectively self-assembled on them.

An interesting "bottom-up" approach for a large number of applications is to mask the GaAs(001) substrate with a SiO_2 film and open by e-beam lithography holes of the desired size and shape to expose areas of the GaAs substrate where QDs are subsequently grown [101], Crucial to this procedure is to achieve selective growth of both GaAs and InAs on the SiO_2, i.e., a negligible deposition on the oxide that must be maintained throughout the growth of the heterostructure. In terms of growth parameters this implies that negligible sticking coefficient and/or desorption-time conditions must hold throughout the deposition sequence.

In the following, we show that InAs dots can be grown on nanoscale holes of SiO_2-masked GaAs substrate while maintaining emission performances close to those of dots on the free GaAs surface. The sample was processed as follows: A SiO_2 film, 160 nm thick, was deposited on the semi-insulating epi-ready GaAs substrate by electron-cyclotron-resonance-activated O_2 plasma and SiH_4 (ECR-PECVD). The film was patterned by e-beam lithography followed by reactive ion etching. The SiO_2 mask consisted of $80 \times 80\ \mu m^2$ matrices of circular-shaped holes of diameter ranging from 200 to 1,000 nm, and spacing among holes ranging from 100 nm to 5 μm. The mask-patterned GaAs substrate was mounted in a MBE chamber where a

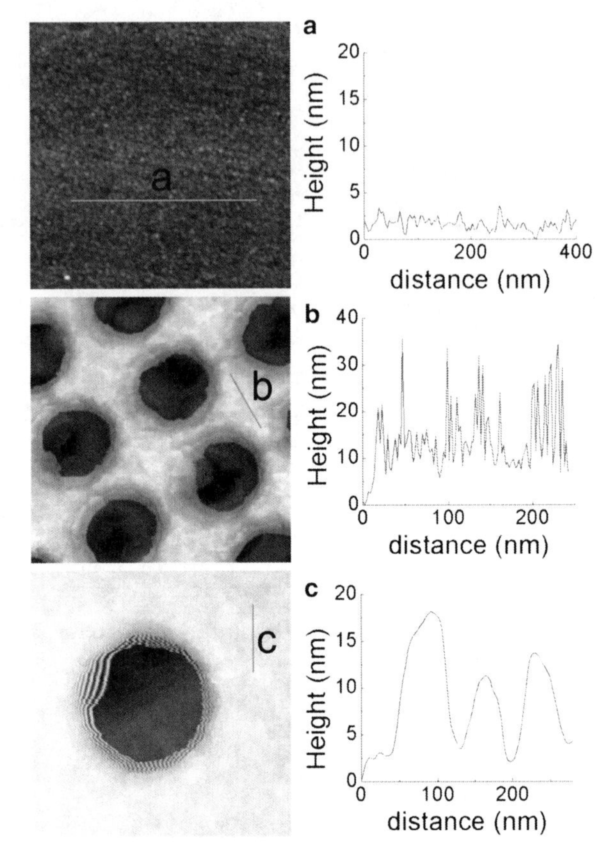

Fig. 24 AFM images, on the *left*, and the corresponding height profiles, on the *right*, taken: (**a**) on the SiO_2 film before the lithographic processing; (**b**) and (**c**) on the SiO_2 film after deposition 2 ML InAs followed by 40 nm of $In_{0.25}Ga_{0.85}As$ cap; in (**b**) holes have 200 nm diameter and are spaced by 100 nm, while in (**c**) holes have 500 nm diameter and are spaced by 5 μm (from [101])

6-nm GaAs buffer layer was deposited at 592°C with a flux rate of 0.167 ML/s, followed by the deposition of InAs at 500°C and at a rate of 0.034 ML/s. These conditions are compatible with the selective growth of the GaAs buffer and the InAs dots. The growth proceeded by covering the dots with a 40 nm capping layer of $In_{0.18}Ga_{0.82}As$ deposited at 500°C and at a rate of 0.21 ML/s, in order to perform photoluminescence experiments. This last step is crucial since the low temperature, 500°C, required to minimise segregation of In from dots to cap layer lies in the range where both GaAs and InGaAs growths are nonselective on SiO_2 [102].

Moreover, when Ga (In) atoms diffusing on SiO_2 cover distances of the order of or larger than their migration length, they can be captured by the defects and damages of the oxide introduced by the lithographic processing that act as seeds for nucleation of Ga(In)As clusters. Nevertheless, insofar as the dimension of the oxide surface available for diffusion is of the same order as the migration length of the In–Ga adatoms, the growth can effectively be considered selective. The above statements are confirmed by images in Fig. 24.

The top image *a*, on the left, and the corresponding height profile, on the right, were taken on the SiO_2 film *before* the lithographic processing; here the height profile shows roughening of the surface in the 0–3 nm scale. The topographies *b* and *c* below, and the corresponding height profiles, refer to two differently patterned regions of the same sample *at the end* of the growth process, i.e., after 2 ML, InAs followed by 40 nm of $In_{0.25}Ga_{0.85}As$ cap has been deposited. After capping, the SiO_2 surface in between the 100-nm close-spaced holes with 200-nm diameter (Fig. 24b) appears highly roughened, the corrugation being in the range 0–30 nm. Although such corrugation is larger than that of the unprocessed oxide surface (Fig. 24, line profile *a*), no clearly defined clusters are seen. By contrast, in the case of 5-μm large-spaced holes with 500-nm diameter (Fig. 24c), the line profile evidences formation of clusters on the oxide surface with base sizes larger than 60 nm and heights larger than 8–10 nm. At such large inter-hole distances, clusters nucleate on the SiO_2 even during GaAs and InAs depositions, in spite of the selective growth conditions.

An issue that deserves to be highlighted is the reduced surface mass transport occurring in the 2D–3D transition when island nucleation is restricted to nano-sized areas of the substrate. It has already been shown in Sect. 4.1 that, in the 2D–3D transition, the total volume of the InAs QDs is larger than the InAs volume being deposited above the critical thickness [7], due to surface mass transport. The same phenomenon occurs for the confined growth. For instance, for 2 ML InAs deposition, the excess volume of dots nucleated inside 500 nm diameter holes was 0.28 ML.

Figure 25 shows, on the left, the line profiles of dots nucleated inside a 100×100 nm^2 hole and of dots inside a 500 nm diameter hole. The comparison evidences the smaller size of dots inside the smaller area hole. The results discussed so far indicate that, upon a suitable choice of growth parameters and of an appropriate patterning of the masked GaAs substrate, it is possible to achieve a limited lateral control of the nucleation process by localising a few InAs dots on selected nano-areas of the surface.

In Fig. 26 (central part), we show AFM topographies of two 100×100 nm^2 holes on the SiO_2-masked GaAs substrate having three to five InAs QDS nucleated at their bottom.

This achievement is a significant step towards the lateral control of the single QD, which is the target required for last generation optoelectronic devices. To this end, however, it is essential that single QDs, at the end of processing, maintain good luminescence properties.

5.2 Micro-photoluminescence of Single Quantum Dots

The micro-photoluminescence (μPL) technique is particularly suitable to the investigation of "single" nanostructures, since the experimental setup can combine traditional photoluminescence apparatuses for time-integrated or time-resolved measurements with a spatial resolution which is diffraction limited. The possibility of detecting a single nano-object is fundamental for the comprehension of their

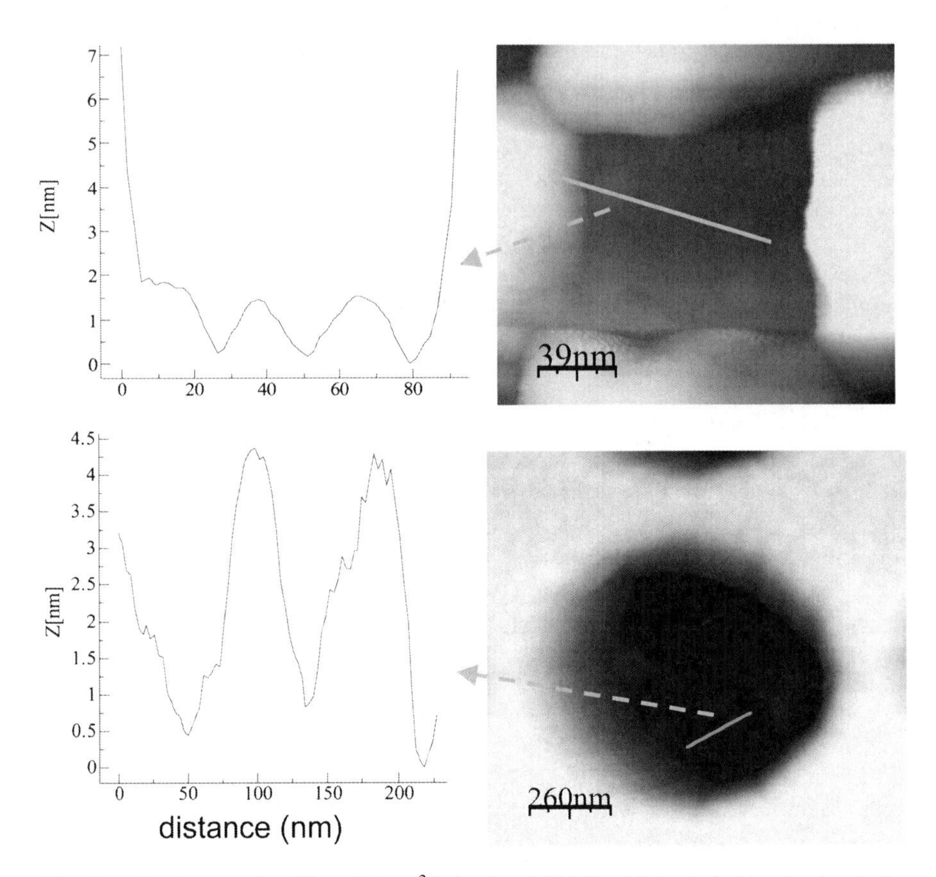

Fig. 25 AFM images of a 100 × 100 nm^2 hole after 1.97 ML of InAs deposition (*top*) and of a 500 nm diameter hole after 1.87 ML of InAs deposition (*bottom*). The line profiles of dots nucleated inside are shown on the *left*. Note the smaller size of dots nucleated in the smaller area hole (*top*)

physical properties. It is well known, in fact, that the electronic properties of nanoscale materials strongly depend on their size, structural disorder and their environment; therefore, the investigation on an ensemble can provide only an average information. Moreover, the continuous improvement on the growth and post-growth processes, a reduced size dispersion and a controlled dot nucleation without a significant loss of radiative efficiency, as shown for the samples here presented, makes μPL spectroscopy a powerful tool for the investigation of a wide range of semiconductor QDs.

In Fig. 27 our experimental setup for μPL measurements is drawn. The signal detection is done in the far field, using a confocal geometry. The sample is mounted on the cold finger of a continuous flow helium cryostat designed for microscopy experiments, which allows a tuning of the sample temperature between 6 K and room temperature. The cryostat is characterised by a short working distance (few millimetres) and by a very high mechanical stability. The excitation beam can be

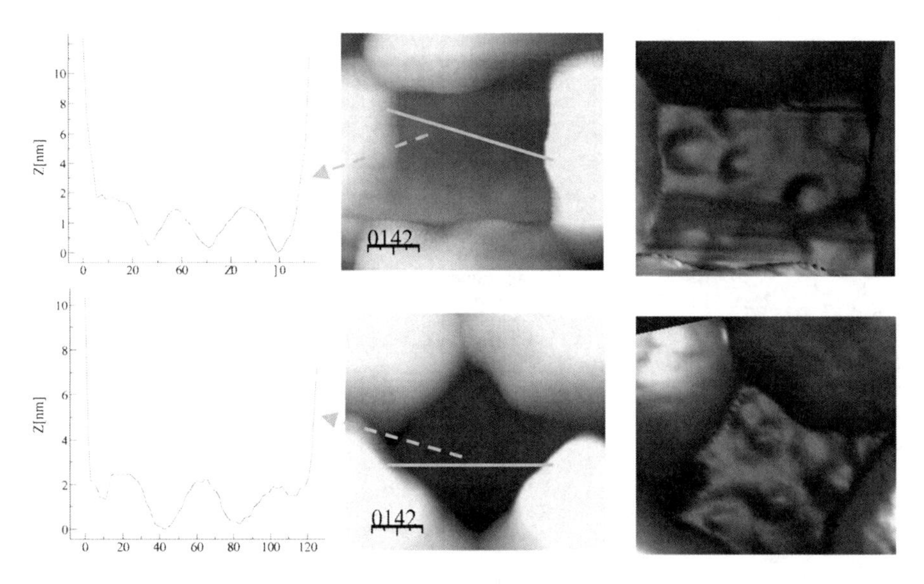

Fig. 26 AFM topographies of two 100 × 100 holes of a pattern where holes are spaced out by 100 nm. Three to five InAs dots are nucleated at their bottom, as better evident from 3D projections on the *right* part of the figure. On the *left*, height profiles are shown along the corresponding lines marked on the topographies

focussed on the sample using the objective O1 (path *a* in figure). The laser beam is aligned on the PL optical path by using a dichroic mirror that allows the separation of the collected PL and the excitation beam. In this configuration, the spot size is set by the focal length of O1 and it is usually of 1–2 μm. Alternatively, the sample excitation can be realised by focussing the laser beam on the sample with a lens (path *b* in figure). In this case, the spot diameter is about 50–100 μm, much larger than the microscope resolution. This assures an almost uniform excitation over a large portion of the sample. The PL signal is collected by the high numerical aperture (NA) microscopy objective (O1, focal length f_1), working in confocal configuration with a second microscopy objective (O2, focal length f_2). A pinhole set at the focal plane of O2 allows for the spatial resolution R. The value of the lateral spatial resolution is determined by the focal lengths f_1, f_2, the objectives NA and the pinhole diameter d. If M is the magnification of the microscope (given by the ratio f_2/f_1), an estimate of the resolution is given by $R = d/M$; then, for example, if $M = 50$ and $d = 100$ μm, the resolution is ≈2 μm. Since the signal is detected in the far-field region, spatial resolution cannot overcome the limit by light diffraction that results $R = 0.61$ (λ/NA), with λ the light wavelength, by the Rayleigh criterion. In the visible–near infrared region ($\lambda \approx 800$ nm), the best-achieved spatial resolution is, in our case, of the order of 0.7 μm (NA = 0.55).

To realise a more flexible confocal setup, instead of using the pinhole shown in Fig. 27, we align on the focal plane of O1 a single-mode optical fibre, which acts as the pinhole, making easier the delivery of light to the spectrometer.

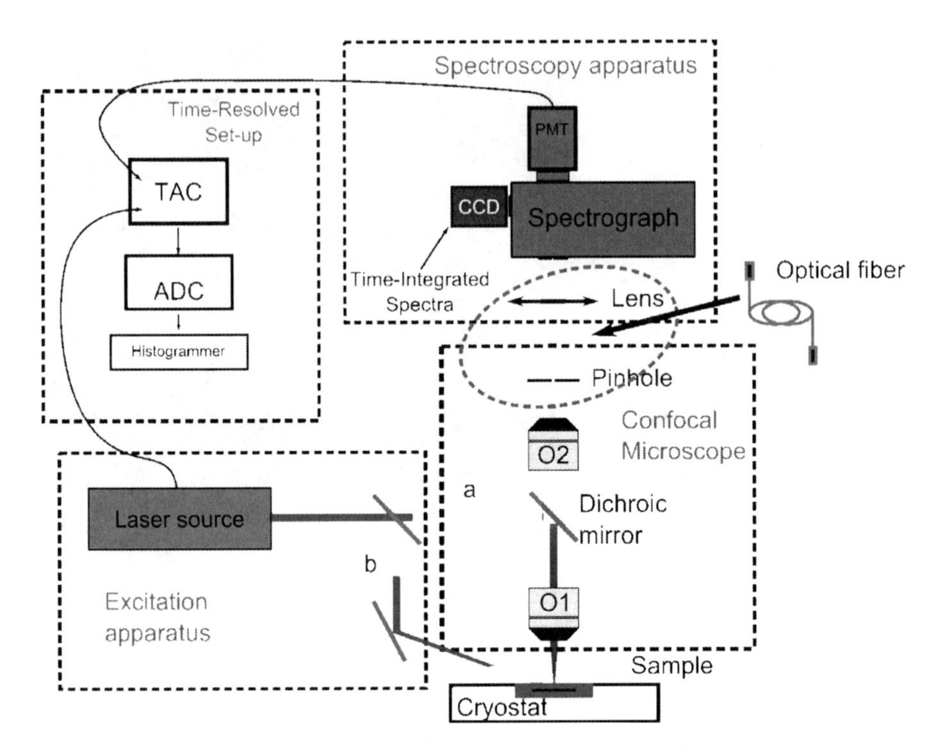

Fig. 27 Outline of the micro-photoluminescence apparatus described in the text

Moreover, changing the magnification and the fibre core size modify the experimental lateral resolution. It is worth noticing that such a spatial resolution cannot be enough to select a single emitter when their density is greater than 10^4 cm^{-2}. For example for typical self-assembling of QDs on the extended substrate [7], the high density ($\approx 4 \times 10^{10}$ cm^{-2}) makes impossible the study of a single dot; therefore, in unprocessed samples, μPL experiments collect roughly a few thousands of emitters and the PL spectra are dominated by the inhomogeneous broadening. Single nano-objects can be isolated exploiting post-growth techniques with a top-down approach to nano-patterning for fabricating on sub-micron length [103, 104]. Differently, as presented in this paper, a bottom-up approach can provide a controlled site nucleation of single emitters in nano-holes [99, 103].

As already mentioned, the flexibility of μPL setups gives the possibility of combining spatial resolution, achieved with confocal microscopy detection, with several spectroscopy apparatuses. As shown in Fig. 27, the collected PL can be dispersed by a spectrograph and then focussed on a detector. By using a CCD camera or a photomultiplier, it is possible to perform CW and time-integrated PL measurements. In Fig. 28, we compare the macro-emission spectrum (spatial resolution 100 μm) of the QDs sample described in Sect. 5.1 with the corresponding μPL spectrum of a single nano-hole at low temperature. In macro-PL, we observe

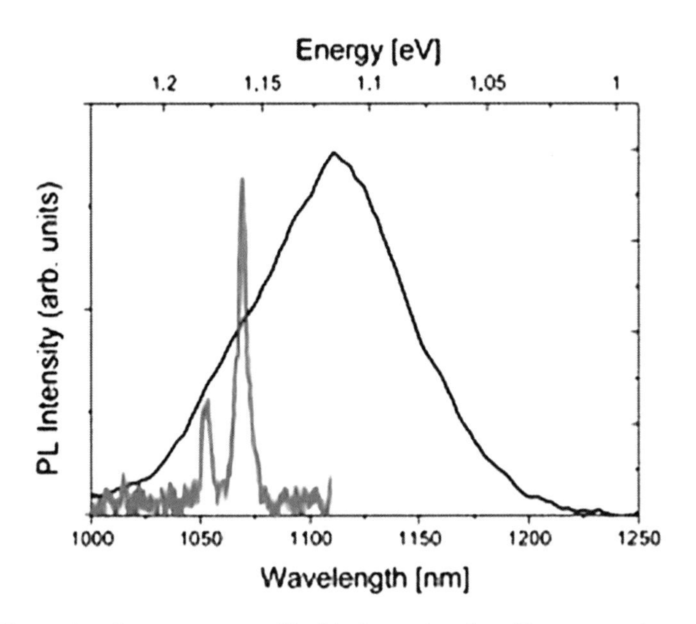

Fig. 28 Comparison between a macro-PL (*black curve*) and a μPL spectrum (*gray curve*), at $T = 10$ K, of a single 500 nm diameter hole. The sharp emission lines in the μPL spectrum appear when few QDs are selected

an inhomogeneous band due to the size dispersion of the QD ensemble. Sharp emission lines, characteristic of atom-like behaviour of a single QD, appear when few QDs are selected.

The microscope can also be coupled to time-resolved setups, such as a time-correlated single photon counting (TCSPC) apparatus (see figure setup). With this configuration, the time evolution of the single emitter PL can be reconstructed by the statistics of photon time arrivals, requiring a long time acquisition. Therefore, the high stability of the microscope is mandatory for the micro-PL time-resolved experiment.

Figure 29 (upper panel) reports the low-temperature PL spectra measured in different regions of the patterned sample depicted in the inset, where 2 ML InAs dots were buried with 40 nm of $In_{0.25}Ga_{0.85}As$. A broad peak centred at 1.064 eV and assigned to QDs emission [99] is measured for the *a* region (extended GaAs surface). The peak shifts to higher energy (1.104 eV) when the signal is measured from the *b* region where dots were grown in 200 nm holes of the SiO_2-masked GaAs. This blue shift is in agreement with the reduced size, discussed in Sect. 5.1, of dots nucleated within the nano-scale holes compared to dots on the extended GaAs surface or on micro-scale holes (Fig. 25). Note that the spectrum *c* taken on the oxide surface does not show significant PL emission in the QD region. For a better comparison, the intensity of spectrum *c* was rescaled to spectra *a* and *b* in the wavelength region 900–950 nm where PL emission is due to defects of the GaAs substrate underneath. PL decays at the QD peak emission, reported in Fig. 29 (lower

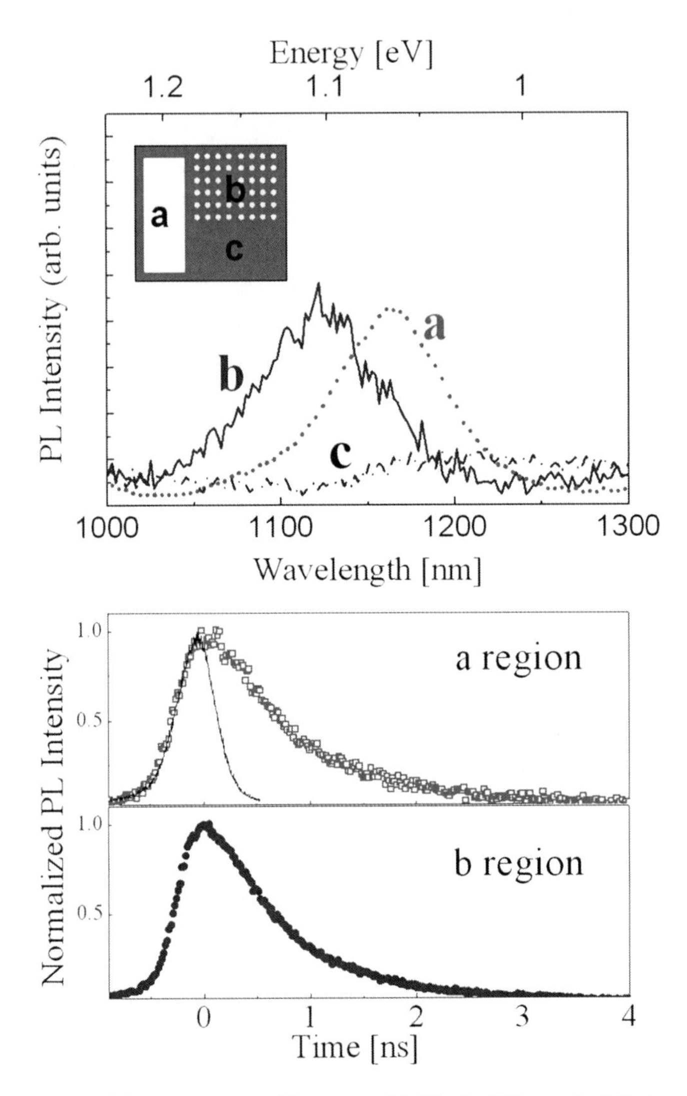

Fig. 29 *Upper panel*: low temperature PL spectra (10 K) of a SiO_2-masked GaAs sample after deposition of 2 ML InAs followed by 40 nm of $In_{0.25}Ga_{0.85}As$ cap. The inset shows the different regions of the sample where the spectra were taken: i.e., (*a*) on the extended SiO_2-free GaAs surface; (*b*) on the SiO_2 pattern of holes masking the GaAs surface; (*c*) on the unpatterned SiO_2 film. *Lower panel*: PL decays at the QD peak emission showing that the recombination kinetics is almost identical in regions *a* and *b*. A decay time of 650 ps can be extracted from a fitting procedure when the finite experimental resolution (*black line* – a region) is taken into account (from [101])

panels), show that the recombination kinetics is almost identical to the recombination from regions *a* and *b*. A decay time of 650 ps can be extracted from a fitting procedure when the finite experimental resolution (black line) is taken into account. This value, if compared with results for Stranki–Krastanov InAs QDs on GaAs(001), denotes that the sample processing does not introduce efficient nonradiative channels. Such a result is confirmed by a comparison of the quantum yield with an un-patterned sample. The radiative recombination efficiency decreases by $\approx 10\%$ when we consider the processed sample.

Acknowledgement We acknowledge the support of the MIUR PRIN 2007 under contract No. 2007S4FAA4.

References

1. Michler, P., Kiraz, A., Becher, C., Schoenfeld, W.V., Petrof, P.M., Zhang, L., Hu, E., Imamoglu, A.: A quantum dot single-photon turnstile device. Science **290**, 2282 (2000)
2. Santori, C., Pelton, M., Salomon, G., Dale, Y., Yamamoto, Y.: Triggered single photons from a quantum dot. Phys. Rev. Lett. **86**, 1502 (2001)
3. Zwiller, V., Blom, H., Jonsson, P., Panev, N., Jeppesen, S., Tsegaye, T., Goobar, E., Pistol, M., Samuelson, L., Björk, G.: Single quantum dots emit single photons at a time: antibunching experiments. Appl. Phys. Lett. **78**, 2476 (2001)
4. Mariette, H.: Formation of self-assembled quantum dots induced by the Stranski–Krastanow transition: a comparison of various semiconductor systems. C. R. Phys. **6**, 23 (2005)
5. Tersoff, J., Tromp, R.M.: Shape transition in growth of strained islands: spontaneous formation of quantum wires. Phys. Rev. Lett. **70**, 2782 (1993)
6. Müller, P., Kern, R.: Equilibrium nano-shape changes induced by epitaxial stress (generalised Wulf–Kaishew theorem). Surf. Sci. **457**, 229 (2000)
7. Patella, F., Arciprete, F., Placidi, E., Fanfoni, M., Sessi, V., Balzarotti, A.: Reflection high energy electron diffraction observation of surface mass transport at the two- to three-dimensional growth transition of InAs on GaAs(001). Appl. Phys. Lett. **87**, 252101 (2005)
8. Placidi, E., Arciprete, F., Sessi, V., Fanfoni, M., Patella, F., Balzarotti, A.: Step erosion during nucleation of InAs/GaAs(001) quantum dots. Appl. Phys. Lett. **86**, 241913 (2005)
9. Arciprete, F., Placidi, E., Sessi, V., Fanfoni, M., Patella, F., Balzarotti, A.: How kinetics drives the two- to three-dimensional transition in semiconductor strained heterostructures: the case of InAs/GaAs(001). Appl. Phys. Lett. **89**, 041904 (2006)
10. Patella, F., Arciprete, F., Placidi, E., Fanfoni, M., Balzarotti, A.: Apparent critical thickness versus temperature for InAs quantum dot growth on GaAs(001). Appl. Phys. Lett. **88**, 161903 (2006)
11. Arciprete, F., Fanfoni, M., Patella, F., Della Pia, A., Balzarotti, A., Placidi, E.: Temperature dependence of the size distribution function of InAs quantum dots on GaAs(001). Phys. Rev. B **81**, 165306 (2010)
12. Placidi, E., Arciprete, F., Fanfoni, M., Patella, F., Balzarotti, A.: InAs/GaAs(001) epitaxy: kinetic effects in the two-dimensional to three-dimensional transition. J Phys. Cond. Mat. **19**, 225006 (2007)
13. Placidi, E., Della Pia, A., Arciprete, F.: Annealing effects on faceting of InAs/GaAs(001) quantum dots. Appl. Phys. Lett. **94**, 021901 (2009)
14. Patella, F., Nufris, S., Arciprete, F., Fanfoni, M., Placidi, E., Sgarlata, A., Balzarotti, A.: Tracing the two- to three-dimensional transition in the InAs/GaAs(001) heteroepitaxial growth. Phys. Rev. B **67**, 205308 (2003)
15. Leem, J.W., Lee, C.R., Noh, S.K., Son, J.S.: RHEED oscillation studies of pseudomorphic InGaAs strained layers on GaAs substrate. J. Cryst. Growth **197**, 84 (1999)
16. Krzyzewski, T.J., Joyce, P.B., Bell, G.R., Jones, T.S.: Wetting layer evolution in InAs/GaAs (0 0 1) heteroepitaxy: effects of surface reconstruction and strain. Surf. Sci. **517**, 8 (2002)

17. Belk, J.G., McConville, C.F., Sudijono, J.L., Jones, T.S., Joyce, B.A.: Surface alloying at InAs–GaAs interfaces grown on (001) surfaces by molecular beam epitaxy. Surf. Sci. **387**, 213 (1997)
18. Sauvage-Simkin, M., Garreau, Y., Pinchaux, R., Véron, M.B., Landesman, J.P., Nagle, J.: Commensurate and incommensurate phases at reconstructed (In,Ga)As(001) surfaces: X-ray diffraction evidence for a composition lock-in. Phys. Rev. Lett. **75**, 3485 (1995)
19. Chizhov, I., Lee, G., Willis, R.F., Lubyshev, D., Miller, D.L.: GaAs(001)-"2×3" surface studied by scanning tunneling microscopy. Phys. Rev. B **56**, 1013 (1997)
20. Zhang, S.B., Zunger, A.: Method of linear combination of structural motifs for surface and step energy calculations: application to GaAs(001). Phys. Rev. B **53**, 1343 (1996)
21. Bottomley, D.J.: The physical origin of InAs quantum dots on GaAs(001). Appl. Phys. Lett. **72**, 783 (1998)
22. Moison, J.M., Guille, C., Houzay, F., Barthe, F., Van Rompay, M.: Surface segregation of third-column atoms in group III-V arsenide compounds: ternary alloys and heterostructures. Phys. Rev. B **40**, 6149 (1989)
23. Walther, T., Cullis, A.G., Norris, D.J., Hopkinson, M.: Stranski–Krastanow transition and epitaxial island growth. Phys. Rev. Lett. **86**, 2381 (2001)
24. Cullis, A.G., Norris, D.J., Migliorato, M.A., Hopkinson, M.: Nature of the Stranski-Krastanow transition during epitaxy of InGaAs on GaAs. Phys. Rev. B **66**, 081305 (R) (2002)
25. Dehaese, O., Wallart, X., Mollot, F.: Kinetic model of element III segregation during molecular beam epitaxy of III-III′-V semiconductor compounds. Appl. Phys. Lett. **66**, 52 (1995)
26. Garcia, J.M., Silveira, J.P., Briones, F.: Strain relaxation and segregation effects during self-assembled InAs quantum dots formation on GaAs(001). Appl. Phys. Lett. **77**, 409 (2000)
27. Grandjean, N., Massies, J., Tottereau, O.: Surface segregation in (Ga,in)as/GaAs quantum boxes. Phys. Rev. B **55**, R10189 (1997)
28. Kaspi, R., Evans, K.R.: Improved compositional abruptness at the InGaAs on GaAs interface by presaturation with In during molecular-beam epitaxy. Appl. Phys. Lett. **67**, 819 (1995)
29. Brandt, O., Ploog, K., Tapfer, L., Hohenstein, M., Bierwolf, R., Phillipp, F.: Formation and morphology of InAs/GaAs heterointerfaces. Phys. Rev. B **45**, 8443 (1992)
30. Bottomley, D.J.: Formation and shape of InAs nanoparticles on GaAs surfaces: fundamental thermodynamics. Jpn. J. Appl. Phys. **39**, 4604 (2000)
31. Patella, F., Arciprete, F., Placidi, E., Nufris, S., Fanfoni, M., Sgarlata, A., Schiumarini, D., Balzarotti, A.: Morphological instabilities of the InAs/GaAs(001) interface and their effect on the self-assembling of InAs quantum-dot arrays. Appl. Phys. Lett. **81**, 2270 (2002)
32. Grandjean, N., Massies, J.: Epitaxial growth of highly strained InxGa1−xAs on GaAs(001): the role of surface diffusion length. J. Cryst. Growth **134**, 51 (1993)
33. Bone, P.A., Ripalda, J.M., Bell, G.R., Jones, T.S.: Surface reconstructions of InGaAs alloys. Surf. Sci. **600**, 973 (2006)
34. Ratsch, C.: Strain-induced change of surface reconstructions for InAs(001). Phys. Rev. B **63**, 161306(R) (2001)
35. Kratzer, P., Penev, E., Scheffler, M.: Understanding the growth mechanisms of GaAs and InGaAs thin films by employing first-principles calculations. Appl. Surf. Sci. **216**, 436 (2003)
36. Penev, E., Kratzer, P., Scheffler, M.: Effect of strain on surface diffusion in semiconductor heteroepitaxy. Phys. Rev. B **64**, 085401 (2001)
37. Rosini, M., Kratzer, P., Magri, R.: In adatom diffusion on InxGa1−xAs/GaAs(001): effects of strain, reconstruction and composition. J. Phys. Condens. Mat. **21**, 355007 (2009)
38. Fujiwara, K., Ishii, A., Aisaka, T.: First principles calculation of indium migration barrier energy on an InAs(001) surface. Thin Solid Films **464–465**, 35 (2004)
39. Penev, E., Stojkovic, S., Kratzer, P., Scheffler, M.: Anisotropic diffusion of In adatoms on pseudomorphic InxGa1-xAs films: first-principles total energy calculations. Phys. Rev. B **69**, 115335 (2004)

40. Pechukas, P.: In: Miller, W.H. (ed.) Dynamics of Molecular Collisions, Part B. Plenum, New York (1976)
41. Truhlar, D.G., Hase, W.L., Hynes, J.T.: Current status of transition-state theory. J. Phys. Chem. **87**, 2664 (1983)
42. Truhlar, D.G., Garret, B.C., Klipenstein, S.J.: Current status of transition-state theory. J. Phys. Chem. **100**, 12771 (1996)
43. Perdew, J.P., Burke, K., Ernzerhof, M.: Generalized gradient approximation made simple. Phys. Rev. Lett. **77**, 3865 (1996)
44. Giannozzi, P., et al.: http://www.pwscf.org
45. Kley, A., Ruggerone, P., Scheffler, M.: Novel diffusion mechanism on the GaAs(001) surface: the role of adatom-dimer interaction. Phys. Rev. Lett. **79**, 5278 (1997)
46. Rosini, M., Magri, R., Kratzer, P.: Adsorption of indium on an InAs wetting layer deposited on the GaAs(001) surface. Phys. Rev. B **77**, 165323 (2008)
47. Mills, G., Jónsson, H., Schenter, G.: Quantum and thermal effects in H2 dissociative adsorption: evaluation of free energy barriers in multidimensional quantum systems. Phys. Rev. Lett. **72**, 1124 (1994)
48. Gomer, R.: Diffusion of adsorbates on metal surfaces. Rep. Prog. Phys **53**, 917 (2002)
49. Bortolani, V., March, N.H., Tosi, P. (eds.): Interaction of Atoms and Molecules with Solid Surfaces. Plenum, New York (1990)
50. Binh, V.T. (ed.): Surface Mobility on Solid Materials, vol. 86 of NATO ASI B. Plenum, New York (1983)
51. Ratsch, C., Scheffler, M.: Density-functional theory calculations of hopping rates of surface diffusion. Phys. Rev. B **58**, 13163 (1998)
52. Haus, J.W., Kehr, K.W.: Diffusion in regular and disordered lattices. Phys. Rep. **150**, 263 (1987)
53. Festa, R., Galleani, E.: Diffusion coefficient for a Brownian particle in a periodic field of force: I. Large friction limit. Physica A **90**, 229 (1978)
54. Smoluchowski, M.V.: Über brownsche molekularbewegung unter einwirkung äußerer kräfte und den zusammenhang mit der verallgemeinerten diffusionsgleichung. Ann. Phys. Leipzig **48**, 1103 (1915)
55. Schnakenberg, J.: Network theory of microscopic and macroscopic behavior of master equation systems. Rev. Mod. Phys. **48**, 571 (1976)
56. Costantini, G., Rastelli, A., Manzano, C., Acosta-Diaz, P., Songmuang, R., Katsaros, G., Schmidt, O.G., Kern, K.: Interplay between thermodynamics and kinetics in the capping of InAs/GaAs(001) quantum dots. Phys. Rev. Lett. **96**, 226106 (2006)
57. Sztucki, M., Metzger, T.H., Chamard, V., Hesse, A., Holy, V.: Investigation of shape, strain, and interdiffusion in InGaAs quantum rings using grazing incidence x-ray diffraction. J. Appl. Phys. **99**, 033519 (2006)
58. Penev, E., Kratzer, P.: In: Joyce, B.A., et al. (eds.) Quantum Dots: Fundamentals, Application and Frontiers. Springer, New York (2005)
59. Rosini, M., Righi, M.C., Kratzer, P., Magri, R.: Indium surface diffusion on InAs (2 $\times$ 4) reconstructed wetting layers on GaAs(001). Phys. Rev. B **79**, 075302 (2009)
60. Balzarotti, A.: The evolution of self-assembled InAs/GaAs(001) quantum dots grown by growth-interrupted molecular beam epitaxy. Nanotechnology **19**, 505701 (2008)
61. Song, H.Z., Usuki, T., Nakata, Y., Yokoyama, N., Sasakura, H., Muto, S.: Formation of InAs/GaAs quantum dots from a subcritical InAs wetting layer: a reflection high-energy electron diffraction and theoretical study. Phys. Rev. B **73**, 115327 (2006)
62. Dobbs, H.T., Vedensky, D., Zangwill, A., Johansson, J., Carlsson, N., Seifert, W.: Mean-field theory of quantum dot formation. Phys. Rev. Lett. **79**, 897 (1997)
63. Kobayashi, N.P., Ramachandran, T.R., Chen, P., Madhukar, A.: In situ, atomic force microscope studies of the evolution of InAs three-dimensional islands on GaAs(001). Appl. Phys. Lett. **68**, 3299 (1996)

64. Osipov, A.V., Schmitt, F., Kukushkin, S.A., Hess, P.: Stress-driven nucleation of coherent islands: theory and experiment. Appl. Surf. Sci. **188**, 156 (2002)
65. Dubrovskii, V.G., Cirlin, G.E., Ustinov, V.M.: Kinetics of the initial stage of coherent island formation in heteroepitaxial systems. Phys. Rev. B **68**, 075409 (2003)
66. Muller, P., Kern, R.: The physical origin of the two-dimensional towards three-dimensional coherent epitaxial Stranski-Krastanov transition. Appl. Surf. Sci. **102**, 6 (1996)
67. Ramachandran, T.R., Heitz, R., Chen, P., Madhukar, A.: Mass transfer in Stranski–Krastanow growth of InAs on GaAs. Appl. Phys. Lett. **70**, 640 (1997)
68. Eisenberg, H.R., Kandel, D.: Wetting layer thickness and early evolution of epitaxially strained thin films. Phys. Rev. Lett. **85**, 1286 (2000)
69. Shchukin, V.A., Ledentsov, N.N., Kopev, P.S., Bimberg, D.: Spontaneous ordering of arrays of coherent strained islands. Phys. Rev. Lett. **75**, 2968 (1995)
70. Moll, N., Scheffler, M., Pehlke, E.: Influence of surface stress on the equilibrium shape of strained quantum dots. Phys. Rev. B **58**, 4566 (1998)
71. Wang, L.G., Kratzer, P., Moll, N., Scheffler, M.: Size, shape, and stability of InAs quantum dots on the GaAs(001) substrate. Phys. Rev. B **62**, 1897 (2000)
72. Márquez, J., Geelhaar, L., Jacobi, K.: Atomically resolved structure of InAs quantum dots. Appl. Phys. Lett. **78**, 2309 (2001)
73. Kratzer, P., Liu, Q.K.K., Acosta-Diaz, P., Manzano, C., Costantini, G., Songmuang, R., Rastelli, A., Schmidt, O.G., Kern, K.: Shape transition during epitaxial growth of InAs quantum dots on GaAs(001): theory and experiment. Phys. Rev. B **73**, 205347 (2006)
74. Costantini, G., Rastelli, A., Manzano, C., Songmuang, R., Schmidt, O.G., Kern, K., von Kanel, H.: Universal shapes of self-organized semiconductor quantum dots: striking similarities between InAs/GaAs(001) and Ge/Si(001). Appl. Phys. Lett. **85**, 5673 (2004)
75. Hasegawa, H., Kiyama, H., Xue, Q.K., Sakurai, T.: Atomic structure of faceted planes of three-dimensional InAs islands on GaAs(001) studied by scanning tunneling microscope. Appl. Phys. Lett. **72**, 2265 (1998)
76. Jacobi, K., Geelhaar, L., Màrquez, J.: Structure of high-index GaAs surfaces – the discovery of the stable GaAs (2 5 11) surface. Appl. Phys. A **75**, 113 (2002)
77. Saito, H., Nishi, K., Sugou, S.: Shape transition of InAs quantum dots by growth at high temperature. Appl. Phys. Lett. **74**, 1224 (1999)
78. Lee, H., Lowe-Webb, R., Yang, W., Sarcel, P.C.: Determination of the shape of self-organized InAs/GaAs quantum dots by reflection high energy electron diffraction. Appl. Phys. Lett. **72**, 812 (1998)
79. Kudo, T., Inuoe, T., Kita, T., Wada, O.: Real time analysis of self-assembled InAs/GaAs quantum dot growth by probing reflection high-energy electron diffraction chevron image. J. Appl. Phys. **104**, 074305 (2008)
80. Spencer, B.J., Tersoff, J.: Equilibrium shapes and properties of epitaxially strained islands. Phys. Rev. Lett. **79**, 4858 (1997)
81. Jacobi, K.: Atomic structure of InAs quantum dots on GaAs. Prog. Surf. Sci. **71**, 185 (2003)
82. Burenkov, Y.A., Davydov, S.Y., Nikanorov, S.P.: Elastic properties of indium-arsenide. Sov. Phys. Solid State **17**, 1446 (1975)
83. Liu, G.R., Quek Jerry, S.S.: A finite element study of the stress and strain fields of InAs quantum dots embedded in GaAs. Semicond. Sci. Technol. **17**, 630 (2002)
84. Chakraverty, B.K.: Grain size distribution in thin films–1. Conservative systems. J. Phys. Chem. Solids **28**, 2401 (1967)
85. Wagner, C.: Theorie der alterung von niederschlagen durch umlosen (Ostwald-reifung). Z. Elektrochem. **65**, 581 (1961)
86. Shaadt, D.M., Hu, D.Z., Ploog, K.H.: Stress evolution during ripening of self-assembled InAs/GaAs quantum dots. J. Vac. Sci. Technol. B **24**, 2069 (2006)
87. Kremzow, R., Pristovsek, M., Kneissl, M.: Ripening of InAs quantum dots on GaAs (0 0 1) investigated with in situ scanning tunneling microscopy in metal–organic vapor phase epitaxy. J. Cryst. Growth **310**, 4751 (2008)

88. Kamins, T.I., Stanley Williams, R.: A model for size evolution of pyramidal Ge islands on Si(001) during annealing. Surf. Sci. **405**, L580 (1998)
89. Krzyzewski, T.J., Jones, T.S.: Ripening and annealing effects in InAs/GaAs(001) quantum dot formation. J. Appl. Phys. **96**, 668 (2004)
90. Marzin, J.Y., Gérard, J.M., Izraël, A., Barrier, D., Bastard, G.: Photoluminescence of single InAs quantum dots obtained by self-organized growth on GaAs. Phys. Rev. Lett. **73**, 716 (1994)
91. Chang, W.H., Chen, W.Y., Chang, H.S., Hsieh, T.P., Chyi, J.I., Hsu, T.M.: Efficient single-photon sources based on low-density quantum dots in photonic-crystal nanocavities. Phys. Rev. Lett. **96**, 117401 (2006)
92. Monat, C., Alloing, B., Zinoni, C., Li, L.H., Fiore, A.: Nanostructured current-confined single quantum dot light-emitting diode at 1300 nm. Nano Lett. **6**, 1464 (2006)
93. Lachab, M., Sakaki, H.: Optical probe of InAs/GaAs self-assembled quantum dots grown using low growth rate and growth interruptions. Appl. Surf. Sci. **254**, 3385 (2008)
94. Grundmann, M., Christen, J., Ledentsov, N.N., Böhrer, J., Bimberg, D., Ruvimov, S.S., Werner, P., Richter, U., Gösele, U., Heydenreich, J., Ustinov, V.M., Yu Egorov, A., Zhukov, A.E., Kopev, P.S., Alferov, Zh.I.: Ultranarrow luminescence lines from single quantum dots. Phys. Rev. Lett. **74**, 4043 (1995)
95. Atkinson, P., Ward, M.B., Bremner, S.P., Anderson, D., Farrow, T., Jones, G.A.C., Shields, A.J., Ritchie, D.A.: Site control of InAs quantum dot nucleation by ex situ electron-beam lithographic patterning of GaAs substrates. Physica E **32**, 21 (2006)
96. Rastelli, A., Ulhaq, A., Kiravittaya, S., Wang, L., Zrenner, A., Schmidt, O.G.: In situ laser microprocessing of single self-assembled quantum dots and optical microcavities. Appl. Phys. Lett. **90**, 073120 (2007)
97. Schramboeck, M., Screnkt, W., Roch, T., Andrews, A.M., Austerer, M., Strasser, G.: Nano-patterning and growth of self-assembled quantum dots. Microelectron. Eng. **37**, 1532 (2006)
98. Wang, Z.M., Lee, J.H., Liang, B.L., Black, W.T., Kunets, V.P., Mazur, Y.I., Salamo, G.J.: Localized formation of InAs quantum dots on shallow-patterned GaAs(100). Appl. Phys. Lett. **88**, 233102 (2006)
99. Patella, F., Arciprete, F., Placidi, E., Fanfoni, M., Balzarotti, A., Vinattieri, A., Cavigli, L., Abbarchi, M., Gurioli, M., Lunghi, L., Gerardino, A.: Selective growth of InAs quantum dots on SiO_2-masked GaAs. Appl. Phys. Lett. **93**, 231904 (2008)
100. Lee, S.C., Stintz, A., Brueck, S.R.: Nanoscale limited area growth of InAs islands on GaAs (001) by molecular beam epitaxy. J. Appl. Phys. **91**, 3282 (2002)
101. Arciprete, F., Placidi, E., Patella, F., Fanfoni, M., Balzarotti, A., Vinattieri, A., Cavigli, L., Abbarchi, M., Gurioli, M., Lunghi, L., Gerardino, A.: Single quantum dot emission by nanoscale selective growth of InAs on GaAs: a bottom-up approach. J. Nanophotonics **3**, 031995 (2009)
102. Lee, S.C., Dawson, L.R., Brueck, S.R.J.: Heteroepitaxial selective growth of $In_xGa_{1-x}As$ on SiO_2-patterned GaAs(001) by molecular beam epitaxy. J. Appl. Phys. **96**, 4856 (2004)
103. Kalliakos, S., Garcia, C.P., Pellegrini, V., Zamfirescu, M., Cavigli, L., Gurioli, M., Vinattieri, A., Pinczuk, A., Dennis, B.S., Pfeifer, L.N., West, K.W.: Photoluminescence of individual doped GaAs/AlGaAs nanofabricated quantum dots. Appl. Phys. Lett. **90**, 181902 (2007)
104. Michler, P. (ed.): Single Quantum Dots, Topics of Applied Physics, vol. 90. Springer, Berlin (2003)

Self-assembled Quantum Dots: From Stranski–Krastanov to Droplet Epitaxy

Yu. G. Galitsyn, A.A. Lyamkina, S.P. Moshchenko, T.S. Shamirzaev, K.S. Zhuravlev, and A.I. Toropov

Abstract The results of investigation of InAs QDs in Al(Ga)As matrix grown by Stranski–Krastanov method and droplet epitaxy are presented. The atomic and energy structure of InAs/AlAs QDs was investigated in different growth conditions for Stranski–Krastanov method, and the coexistence of direct and indirect band structures is revealed. However, the lack of carrier transfer due to the low quality of a heterointerface and high concentration of nonradiative recombination centers is challenging. To overcome these problems, we used droplet epitaxy. As QD density in droplet epitaxy is determined by nucleation, we studied the initial stage of homoepitaxy in model system of GaAs to analyze nucleation processes. A proposed statistical approach is also very effective to describe InAs/GaAs QD formation in Stranski–Krastanov mode. The array of In metal droplets on the GaAs surface is studied as an initial stage of droplet epitaxy, and a model of droplet evolution is proposed. Indium dose dependence of QD properties reveals a critical phenomenon of a growth mode transition. Finally high-quality InAs/$Al_{0.9}Ga_{0.1}As$ QDs structures with a perfect heterointerface and high efficiency of the carrier capture from wetting layer to QDs were grown by droplet epitaxy.

Yu.G. Galitsyn • S.P. Moshchenko • T.S. Shamirzaev • K.S. Zhuravlev • A.I. Toropov (✉)
A.V. Rzhanov Institute of Semiconductor Physics SB RAS, Pr. Lavrentieva, 13,
630090 Novosibirsk, Russia
e-mail: sergem@isp.nsc.ru; toropov@thermo.isp.nsc.ru

A.A. Lyamkina
A.V. Rzhanov Institute of Semiconductor Physics SB RAS, Pr. Lavrentieva, 13,
630090 Novosibirsk, Russia

Novosibirsk State University, Pirogova 2, 630090 Novosibirsk, Russia
e-mail: lyamkina@thermo.isp.nsc.ru

S. Bellucci (ed.), *Self-Assembly of Nanostructures: The INFN Lectures, Vol. III*,
Lecture Notes in Nanoscale Science and Technology 12,
DOI 10.1007/978-1-4614-0742-3_3,

1 Introduction

Self-assembled quantum dots have been the focus of attention in the last years due to the promising perspectives of their application for advanced devices. Thanks to significant efforts of many research groups, diverse nanostructures can be reproducibly obtained [1–7]. However, some important aspects of QDs formation still remain unclear. From our point of view, only a systematic approach to QD investigation can lead to a breakthrough in this area of knowledge and give a better insight into the self-assemblance processes. This chapter contains the main results of InAs/GaAs and InAs/AlAs QDs investigation by our group with special attention to origin and formation of quantum dots, as we assume them to be key points for understanding the physics of the processes. All structures studied in this work were grown by molecular beam epitaxy using a Riber-32P system with the valve arsenic source on semi-insulating (001)-oriented GaAs substrate. Starting with fundamental studies like the statistical approach to QD formation, we then present the results of investigation of the features of droplet epitaxy. We utilize this method instead of Stranski–Krastanov one to overcome growth problems.

The chapter is organized as follows. In the first part, we investigate initial stage of QD formation. The GaAs surface relaxation in homoepitaxy under ultra-low coverage is studied to understand nuclei formation process. A fundamental thermodynamic description with phase transition is applied to 2D–3D transition in InAs/GaAs system. The efficiency of general physical approach for QD initial stages is shown.

The second part is devoted to the features of the droplet epitaxy. Here, we start with the initial stage for this technique – indium metal droplets on GaAs surface which are centers of subsequent QD formation. Another research line presented in next subsection is the indium dose dependence, which is of great interest for QD properties control. It reveals a critical phenomenon of growth mode transition.

In the next section, the investigation of atomic and energy structure of grown InAs/AlAs QD structures is presented. The studies reveal a lack of carrier transfer caused by low quality of samples grown by Stranski–Krastanov. The final section demonstrates high quality of structures fabricated by droplet epitaxy.

2 Initial Stage of MBE Growth

Homoepitaxy of gallium arsenide on (001) surface is a classical model system to study kinetics and mechanisms of 2D growth of A_3B_3 semiconductors and subsequent formation of nanostructures. Below, we investigate GaAs homoepitaxy with a special attention to the nucleation process and surface evolution, considering both being key points for subsequent quantum dots (QDs) formation. We demonstrate RHEED investigation of growing surface at the initial stage. The formation of 2D island is a crucial moment for both the MBE growth and the nanostructure fabrication. A theoretical description of this process is provided, based on a general thermodynamic approach and therefore suitable for both initial stage of the growth, including reconstruction transitions, and QD density evolution.

2.1 Dynamics and Kinetics in GaAs Homoepitaxy on (001)-β(2 × 4) Surface Under Ultra Low Coverage

Homoepitaxial growth of GaAs on (001)-β(2 × 4) surface starts with fluxes of Ga and As_4 or As_2 and goes through 2D nuclei formation with their subsequent growth until a complete monolayer of new phase appears on the surface. This process causes intensity oscillations of both specular and fractional spots in reflection high-energy electron diffraction (RHEED) patterns during the growth. In spite of many efforts, a lot of problems are still to be investigated [8].

Processes of the formation of monolayer-height 2D islands of GaAs new phase are usually simulated by kinetic Monte Carlo method (KMC) [9–16]. In this method, statistical sums for random configuration appearing on the surface are estimated. It requires a significant number of physical parameters, such as kinetic barriers for anisotropic diffusion of gallium and arsenic atoms, bounding energies of absorbed atoms, interaction energies, and so on. Thus, in the work [16], 30 parameters were used to describe a sequence of processes resulting in 2D island nucleation. In fact, in KMC, a choice of numeric parameter values is rather arbitrary. For example, in [14–16], 0.7 eV was used as the activation energy of Ga diffusion, whereas in [10], it was assumed to be 1.4 eV and in [12] – 1.0 eV. The other parameters vary in a wide range as well [8]. The diversity of processes on (001) GaAs surface during the growth results from a complexity of β(2 × 4) surface reconstruction. Every elementary cell (2 × 4) consists of two arsenic dimers and two dimer vacancies in the upper layer, i.e., the upper layer has an arsenic coverage of 0.5. The second layer of gallium atoms has the coverage of 0.75, as two atoms of gallium are absent. Then the third layer of arsenic atoms is completed, but at uncovered part being a trench, arsenic atoms form a dimer. This reconstruction has the lowest surface energy for substrate temperatures between 540 and 600°C and arsenic equivalent pressures between 10^{-5} and 10^{-6} torr.

However, a more significant disadvantage of the approach used in some works [9, 11] should be mentioned. They are mostly limited by a low coverage of the initial surface by a new monolayer: $\Theta < 0.1$. During epitaxy, the filling of a monolayer always happens before the nucleation of following one starts. It is unclear if such theoretical approach results in complete filling of a monolayer before the next one starts to be filled.

The main point is that GaAs homoepitaxy, going through the nucleation and evolution of 2D islands of a new phase on the surface, is actually a first-order 2D phase transition (PT) from a low-density gas phase (lattice gas) to a condensed phase with the high density (2D crystalline layer). For PT, the rate of new phase nucleation (spontaneous process) is determined by a state of a media where a PT takes place. For our case, this is a morphological state of the growing surface. The growth of the new phase is determined by kinetic parameters. In [8–16], a strict predefined scheme of subsequent processes that result in the formation of new phase 2D islands was assumed. First, the processes of the nucleation and growth are not separated there. Second, the nucleation is simulated on an ideal surface that does not correspond to a real system. Therefore, it is not surprising that some authors obtained strange results contradicting the classical nucleation theory. In [15, 16], the calculated density of 2D nuclei turned out to be nonmonotonic function of the temperature with a maximum of 2D island density at $T = 800$ K. Moreover, the shape of nuclei and their surface reconstruction obtained by KMC are not in a good agreement with the ones that are experimentally observed in homoepitaxy [8, 14].

Anyway, it is worth to mention that all these works are a great contribution to the description of initial stage kinetics in GaAs homoepitaxy. They used a standard technique of statistical mechanics combined with surface energy calculations for a different configuration by a density functional theory. In principle, this is a good approach to explain macroscopic laws of the growth with microscopic physics and chemistry of initial stages. But it is unclear, if this approach being extended to large coverages results in experimentally observed layer-by-layer 2D growth and doesn't lead to 3D nuclei formation. In others words, this microscopic approach is to be agreed with macroscopic one. The macroscopic approach implies to obtain the state equation of the growing surface with a first-order phase transition.

In the review [17] by B. Joyce and D. Vvedensky, the sequence of elementary stages of a critical 2D nucleus formation is stated on the base of KMC ideas. But the authors themselves consider it to be an abstract scheme with no experimental proof of any of its stage. In spite of the very title "Self-assembled growth," the authors do not determine a mechanism and driving force of homoepitaxial self-assembly and underline elementary local growth processes. From our point of view, only a statistical approach to the homoepitaxy can reveal reasons of the self-assembled growth and driving forces of the 2D island formation of the new crystalline phase. Two-dimensional island nucleation is a consequence of the first-order phase transition in the lattice gas that forms on the surface with the adsorption of growth components. The 2D phase transition itself results from the lateral attraction between lattice gas cells. We found the type of this interaction in an earlier work [18]. The driving force of the formation of a new crystalline layer is an effective interaction that can be expressed in the framework of mean field theory as follows:

$$E = E_{\mathrm{st}}\,\Theta\Theta_{\mathrm{GaAs}} = -F_{\mathrm{st}}(\Theta)\Theta_{\mathrm{GaAs}}^{2}, \qquad (1)$$

where Θ denotes a surface coverage with an intermediate lattice gas, and Θ_{GaAs} is a surface coverage with new elementary cells of $\beta2$-(2×4) structure. Cells of $\beta2$-(2×4) play a role of the intermediate lattice gas. Equation (1) contains already the self-organization mechanism: the larger the value of Θ_{GaAs}, the larger is an energy gain.

In Fig. 1, phase transition isotherms are presented, i.e., the state equation of growing surface at usual growth temperatures for GaAs homoepitaxial growth: 540–580°C. The coverage at which the gas branch transforms to the crystalline one is a key characteristic of the phase transition. This point (Θ_{p}) is the beginning of a vertical line in Fig. 1, and it denotes the appearance of a stable 2D

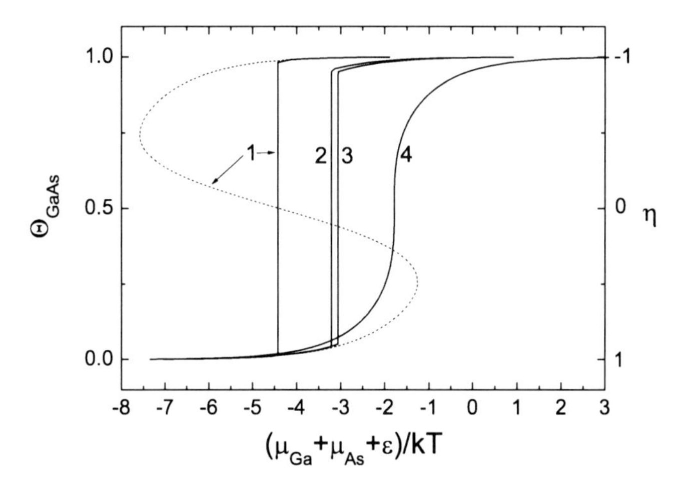

Fig. 1 Phase transition isotherms for different parameter values: *1* – $E_{st} = 0.8$ eV, $\Delta E = 0.4$ eV, $T = 580°C$ ($\theta_{trans} = 0.01$); *2* – $E_{st} = 0.6$ eV, $\Delta E = 0.3$ eV, $T = 540°C$; *3* – $E_{st} = 0.6$ eV, $\Delta E = 0.3$ eV, $T = 580°C$ ($\theta_{trans} = 0.05$); *4* – critical isotherm $T_c = 1840°C$. Metastable and unstable PT branches of isotherm 1 are dotted

nuclei of the new phase on the surface. The state of growing surface before this point is complex and includes different metastable structures. It is appropriate to recall that to form an elementary β2-(2 × 4) cell, eight gallium atoms and eight arsenic atoms are needed to be embedded into lattice sites. Within our simplified epitaxy description in [18], only one type of intermediate state was considered, that is, the lattice gas of β2-(2 × 4) cells. This intermediate state was supposed in [16] and also in theoretical analysis of homoepitaxial process sequence on β2-(2 × 4) (001)GaAs. It should be mentioned that in spite of a long period of the homoepitaxial processes investigation on (001)GaAs, there is still a lack of experimental data in situ.

Below we investigated experimentally surface states that appear in GaAs homoepitaxy, mainly for a coverage range below Θ_p. We used RHEED to obtain in situ the surface evolution during the deposition of growth components. An ultra low dose of GaAs in the range of 0.03–0.12 monolayer (Ml) was deposited to the substrate and time dependence of diffraction pattern was studied. All the experiments were conducted in UHV chamber of the Riber-32P setup. The sample size was 3 × 3 mm^2, providing sufficient homogeneity of both morphological state and temperature of the sample and therefore ensures the reproducibility of the experiments.

Various intermediate metastable centers appear on the surface before the formation of stable 2D nuclei of a GaAs new phase. If the growth is interrupted under surface coverage of $\Theta < \Theta_p$, these centers are destroyed, resulting in the relaxation of a RHEED spot intensity up to an initial value. If stable 2D nuclei form, the intensity does not relax to the initial one due to additional electron scattering by these stable nuclei. This clear idea was implemented in the present work. The intensities of both specular and fractional spots were obtained from the beginning of deposition to the relaxation after surface material redistribution.

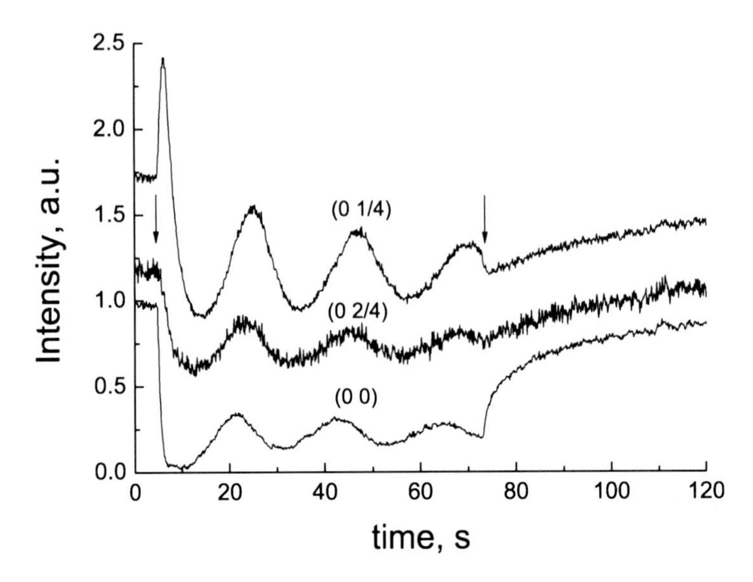

Fig. 2 The intensity of (0 0), (0 1/4), (0 2/4) diffraction spots during the growth at $T = 540°C$ and $P_{As} = 1.7 \times 10^{-6}$ torr. The beginning and end of gallium deposition are marked with *arrows*

In Fig. 2, typical kinetic curves corresponding to different diffraction spots in $[1\bar{1}0]$ azimuth at the beginning of GaAs monolayer are shown. It is seen that the initial intensities of (0 0) and (0 2/4) spots decrease, while the intensity of (0 1/4) increases. For a coverage $\Theta > 0.06$, the intensities of all spots decrease. This results from the formation of scattering centers on the surface under the coverage larger than 0.06. For a coverage below 0.03, the following process takes place. Gallium atoms are adsorbed in trenches of the initial β2-(2 × 4) reconstruction and domains of a new phase nonrelaxed α-(2 × 4)-n appear. Such phase was considered in [19]. Relative intensities of RHEED spots were calculated theoretically, and it was found that the intensities of (00) and (0 1/4) are almost equal, while the intensity of (0 2/4) spot is very low. Therefore, we suppose that the intensity evolution obtained at the growth beginning indicates the formation of the metastable phase α-(2 × 4)-n.

The behavior of this phase at different temperatures was investigated. In Fig. 3a, the dependence of (0 1/4) spot intensity in the temperature range of 550–570°C is shown, i.e., the intensity relaxation after the growth interruption after a surface coverage of 0.03. At this coverage, the formation of scattering centers can be neglected, then the intensity behavior in a maximum I reflexes the behavior of long-range order $I = |\eta|^2$ of the phase α-(2 × 4)-n. As it can be clearly observed, the maximum intensity decreases for an increasing temperature, but the half width of a peak increases in such a way that the area under the curve remains constant. Such behavior is characteristic for an order–disorder transition in the α-(2 × 4)-n phase [19].

Thus, gallium atoms embedding into lattice sites in trenches of β2-(2 × 4) reconstruction results in the formation of the metastable ordered phase α-(2 × 4)-n at relatively low temperatures. Typical order–disorder transition takes place.

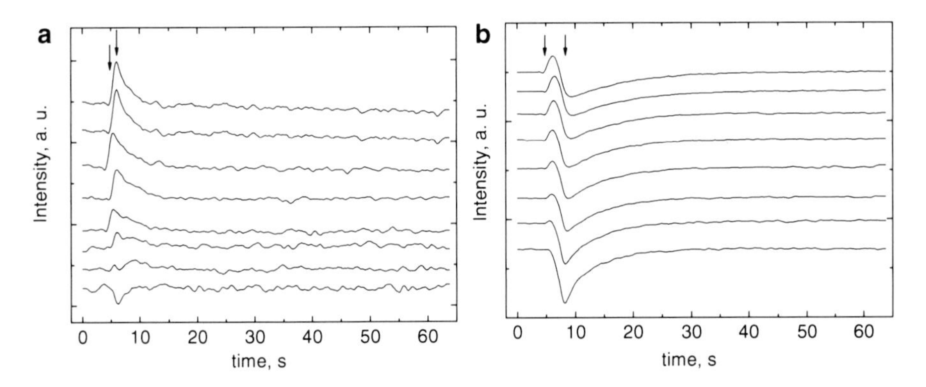

Fig. 3 Kinetic curves for $\theta = 0.03$ Ml (**a**) and $\theta = 0.09$ Ml (**b**). The *curves* are shifted vertically for the clarity. The moments of opening and closing Ga source are marked with the *arrows*. The temperatures from *up to down* are 550, 552, 555, 558, 561, 564, 567, 570°C, respectively

At the coverage range 0.06–0.09, the intensity of all the spots decreases. Therefore, we can assume that isolated scattering centers are formed. Under the growth interruption, the spot intensities are restored to the initial values, so these centers are metastable. In Fig. 3b, the dependence of (0 1/4) spot intensity on the temperature is shown under the growth interruption at the coverage of 0.09. It is seen that complete recovery up to the initial value takes place. With the increase in the amount of gallium adsorbed on the surface, the arsenic adsorption coefficient increases abruptly. The most possible way to change the state is the transformation of α-(2 × 4)-n local centers to β1-(2 × 4) ones after arsenic adsorption. The formation of such scattering centers is an activation process, because with the temperature increase their concentration increases exponentially (Fig. 4). The decay of centers is also an activation process. On the basis of our analysis, we obtain the decay constant $K(T) = 1/\tau_{1/2} = 3 \times 10^{10} \exp(-E_a/kT)\,\mathrm{s}^{-1}$, where $\tau_{1/2}$ is a time of β1-(2 × 4) cell half decay. An activation dependence Arrhenius plot is shown in Fig. 5, and the activation energy E_a is found to be 1.84 eV. Indeed, under the conditions of the β2-(2 × 4) stable phase existence, the phase β1-(2 × 4) is metastable and, consequently, under the growth interruption the surface relaxes to the initial state. It is interesting to note that under common conditions the lifetime of β1-(2 × 4) centers is about 1 s. Therefore, for a growth rate over one monolayer per second, local centers of β1-(2 × 4) can be a transition state for the formation of stable 2D nuclei, as it was proposed by Ishii et al. [10, 11].

The exponential decay of β1-(2 × 4) centers is similar to a process of monomolecular desorption from the surface. Then the assembly of such centers is the intermediate lattice gas for the statistical approach, like it was supposed in [11], i.e., $(\Theta; \Theta_{\mathrm{GaAs}})^{\mathrm{gas}} \rightarrow \Theta_{\mathrm{GaAs}}^{\mathrm{cryst.}}$. The phase transition in the lattice gas of Θ_{GaAs} with the interaction (1) results in the appearance of 2D crystalline phase.

The investigation of the relaxation after the growth interruption with the coverage over 0.12 Ml revealed that the spot intensities do not restore completely. It means

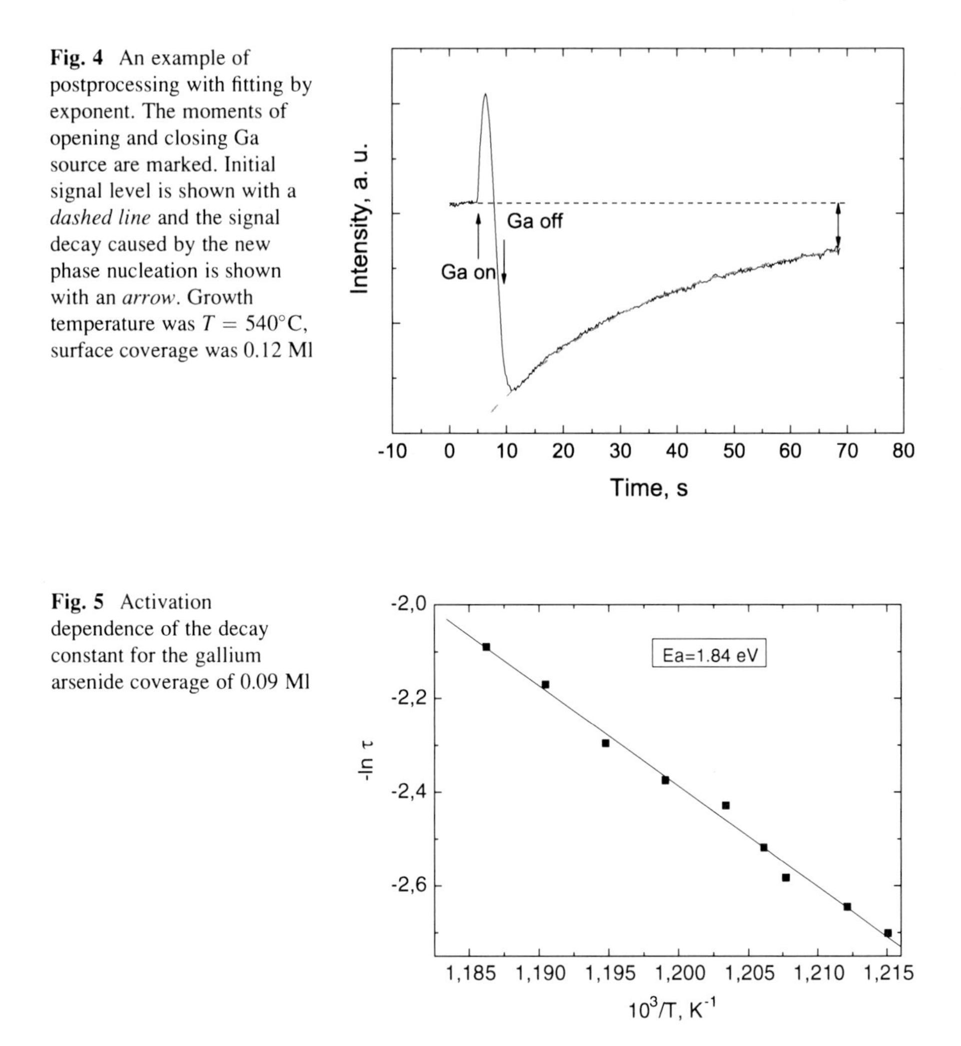

Fig. 4 An example of postprocessing with fitting by exponent. The moments of opening and closing Ga source are marked. Initial signal level is shown with a *dashed line* and the signal decay caused by the new phase nucleation is shown with an *arrow*. Growth temperature was $T = 540°C$, surface coverage was 0.12 Ml

Fig. 5 Activation dependence of the decay constant for the gallium arsenide coverage of 0.09 Ml

that stable 2D nuclei of new phase with β2(2 × 4) structure are already formed. To restore the initial structure, the annealing of a sample with the temperature increase up to 610°C is needed.

Therefore, our method of RHEED intensities relaxation process appeared to be extremely fruitful for the homoepitaxy investigation. The experiments confirmed the scheme that we proposed earlier in [18]: $L(P,T) \rightarrow (\Theta; \Theta_{\mathrm{GaAs}})^{\mathrm{gas}} \rightarrow \Theta_{\mathrm{GaAs}}^{\mathrm{cryst.}}$. Under low surface coverages (below 0.03 Ml,) the metastable phase α-(2 × 4)-n forms, which exists in an ordered state at low temperatures. At the temperature above 570°C, this phase transforms to a common Langmuir adsorbed phase. Under coverages of 0.03–0.09, the transition centers with the β1(2 × 4) local structure form. For statistical analysis, an ensemble of such centers with the same size can be considered as an intermediate lattice gas. Finally, under coverage above 0.09 Ml, stable crystalline nuclei of a new phase form.

2.2 *Statistical Approach to the Strain-Driven Formation of InAs Quantum Dots on GaAs(001)*

Heteroepitaxial growth system InAs on GaAs has a large lattice mismatch (7%); therefore, Frank-van der Merve growth mode cannot be realized. Nevertheless, three-dimensional islands or QDs arise only after the formation of indium arsenide two-dimensional film with a thickness of about 1.6 monolayers. Then three-dimensional coherent (dislocation-free) islands with a narrow size distribution form on the surface due to elastic relaxation in the film.

This 2D→3D growth mode transition known as Stranski–Krastanov (SK) transition is very abrupt. Over the critical coverage of 1.6 Ml, InAs three-dimensional islands with a density of 10^9-10^{10} cm^{-2} arise almost immediately. Their volume exceeds an incident flux of the growth components greatly, which means the formation of these islands involves both indium and gallium surface fluxes and therefore the QD composition is $Ga_xIn_{1-x}As$ ($x = 0.2$–0.5). Currently, there is no sufficient explanation of the transition abruptness. An attempt to describe the island density evolution in the framework of mean field kinetic equations can be found in [20]. According to [19], sharp and rapid growth of the islands can be explained and agreed with the experimental results within the assumption of a strong elastic repulsive interaction between the islands that value exceeds the typical one by three orders of magnitude. Similar results were obtained in [22, 23], where experimental results and an abrupt jump of islands density from 10^8 to 10^9 cm^{-2} were also considered on the basis of mean field kinetic equations.

Obviously in these works, the kinetics predicts a rapid density increase from zero to certain values, but recently a slight change in QD density from 1.5 to 2×10^9 cm^{-2} was obtained, when a subcritical amount of indium arsenide from 1.4 to 1.6 Ml was applied to the surface [24]. There was a QD density jump up to 10^{10} cm^{-2} for a dose of 1.6 Ml, and then in the dose range up to 2.4 Ml further smooth increase in QD density up to the limiting value was obtained. These detailed experiments have revealed a new property of 2D→3D transition. It was considered before that QDs arise only after the wetting layer reaches a critical thickness. But in [24, 25], it was demonstrated experimentally that QDs with a small density formed below the critical thickness and a QD density jump by an order of magnitude occurred only after the critical thickness had been reached. Actually there is a typical first-order phase transition (PT) in the ensemble of QDs on the surface. Studying QDs by STM, it is very difficult to detect experimentally a low density of QDs and investigate accurately its dependence on the incident flux. We can therefore suggest that in [20, 22] a weak dependence of QD density for subcritical thickness was not found experimentally and only in [24] its detailed study was made successfully. More precise experimental results obtained in [24, 25] allow us to revise the suggestions about the formation of InAs QDs on (001) GaAs. Principally the formation of QDs should be considered as a first-order PT in QD ensemble with a density jump, so there is a complete analogy with van der Waals gas–liquid transition. Considering that the described transition occurs on the surface, we can state that two-dimensional first-order PT takes place.

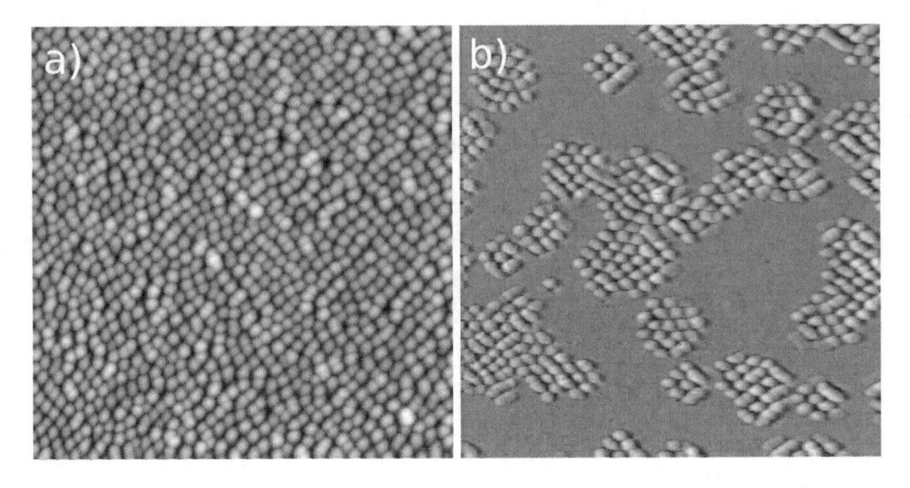

Fig. 6 AFM image of InAs/GaAs QD arrays. (**a**) Maximum density array corresponding to the QD liquid. (**b**) The droplets of the QD liquid

Two-dimensional PT is usually considered in the framework of so-called lattice gas model (LG) [26]. Indeed, QDs have similar sizes; hence, the whole surface can be divided into N identical square cells and then the process of QD formation is just the filling of these cells. Toward the filling of a cell, both surface flow and external incident InAs flux can be considered as a general external flux applied to the cell. We define a local filling variable in such a way that $c_i = 1$ if a cell is filled, i.e., a quantum dot has formed, otherwise $c_i = 0$. The surface filling by QDs α equals to the filling averaged by all the cells for a given temperature:

$$\alpha = \frac{1}{N}\Sigma\langle c_i\rangle. \tag{2}$$

Therefore, in QD ensemble, the PT from the low-density lattice gas to the condensed state of high-density QD ensemble occurs. We can state that during gas–liquid PT, the droplets of quantum dots form; complete filling of LG cells corresponds to the formation of two-dimensional QD film. It is demonstrated with AFM images of InAs/GaAs QDs fabricated by Stranski–Krastanov technique in different conditions that are presented in Fig. 6. For two-dimensional LG PT, lateral interactions between quantum dots have a key role. Since PT has a first order, these interactions should have attraction potential; however, pair interactions between QDs are characterized by the repulsion [27]. Consequently, taking into account only this interaction, a first-order PT cannot occur, and there should be a second-order PT that is usually described by one-parameter Fouler–Guggenheim isotherms.

Evidently LG lateral interactions have a more complex nature and indirect interactions can be considered that results in an effective attraction between QDs. In this case, PT must be described by a more universal three-parameter isotherm;

we have used such isotherms already in the analysis of reconstruction PT on surface (001) GaAs [18, 28, 29].

Using the conventional statistical analysis, we express the free energy F corresponding to the lattice gas α in the form:

$$F = -\varepsilon\alpha + kT[\alpha \ln \alpha + (1 - \alpha)\ln(1 - \alpha)] + \frac{1}{2}E_i\alpha, \tag{3}$$

where ε is the so-called vertical binding energy, E_i is the pairwise repulsive energy between neighboring sites, $\alpha = \rho/\rho_0$ is a relative QD density by a surface unit (ρ_0 – the maximum QD density for a specified size L, $\rho_0 = 1/L^2$), ω denotes an intermediate lattice gas increasing surface energy, which transforms then to the lattice gas of filled cells α. The free energy corresponding to the formation of QD through the cell of intermediate lattice gas ω is expressed as

$$F = V\omega + kT[\omega\ln\omega + (1 - \omega)\ln(1 - \omega)] - U\alpha\omega, \tag{4}$$

where V is an energy loss for the formation of intermediate lattice gas cell, and U is a stabilization energy (energy gain with QD formation). Minimizing the expression (3) with respect to the concentration ω and substituting the result into the expression (2), we obtain the subsequent state equation:

$$\mu = \frac{\partial F}{\partial \alpha} \tag{5}$$

or

$$\frac{(\mu + \varepsilon)}{kT} = \ln\left[\frac{\alpha}{1 - \alpha}\right] + \left[\frac{E_i\alpha}{kT} - \frac{U/kT}{1 + \exp[(V - U\alpha)/kT]}\right], \tag{6}$$

where μ is the QD chemical potential, ε is a vertical energy of QD surface bound [18, 28, 29]. Equation (5) is similar to the van der Waals equation of state; i.e., at rather low temperatures ($T < T_c$) the first-order PT occurs. The isotherms of type (5) describe well both first-order and second-order PTs and, moreover, the universality of such isotherms allows to apply them both for symmetric and for asymmetric PTs [18, 28, 29]. First-order transition is symmetric relatively to the transition point if the isotherm parameters are related as: $E_i < U$ and $|U| = 2V$, otherwise the PT is asymmetric.

Analyzing the results of the QD formation precise measurements as a function of InAs external flux to (001) GaAs surface presented in [24], we conclude that the isotherm describing their formation should be asymmetric. The isotherm with the parameters $U = 0.8$ eV, $V = 0.16$ eV, $E_i = 1.1$ eV is in a good agreement with the QD density measured experimentally (Fig. 7). The QD density jump by an order of magnitude from 2×10^9 to 2.5×10^{10} cm^{-2} is caused by two-dimensional

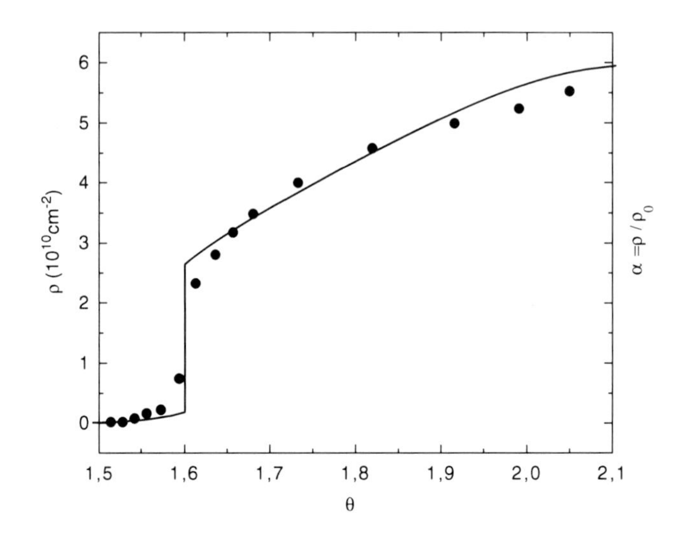

Fig. 7 Three-parameter isotherm of 2D phase transition (*line*) describing experimental data from [24] (*dots*). The values of isotherm parameters are $U = 0.8$ eV, $V = 0.16$ eV, $E_{\rm i} = 1.1$ eV

PT in QD LG from the gas to the condensed phase (QD droplets). This jump cannot be described with kinetic methods in principle. Statistical laws of the lattice gas determine the behavior of QD ensemble as a two-dimensional surface ensemble, where effective lateral interactions between QDs have a key role.

The interaction between QDs is carried out through elastic stress field in the InGaAs film that forms on GaAs surface when 1.5 Ml of InAs is applied. The QD density jump and the droplet formation are the consequences of the formation of condensed QD phase that is more stable than LG.

Some comments should be performed. Although the values of μ and ε are much larger than lateral interaction parameters, in $(\mu + \varepsilon)/kT$ their "vertical" components are compensated. Hence, remaining lateral part and effective lateral QD interactions play a role of the external field where two-dimensional PT in QD ensemble takes place. The parameter $(\mu + \varepsilon)/kT$ characterizes this external field. Since an external incident flux of the growth components to the surface is considerably lesser than the surface flow that fills in an empty LG cell, changing lateral part of it is considered to be proportional to InAs surface coverage: $(\mu + \varepsilon)/kT \sim \Theta_{\rm InAs}$. A coverage increase by 0.1 results in an increase in $(\mu + \varepsilon)/kT$ by a factor of 20 (see Fig. 7).

The external field changes its sign in the PT point. In our case, it happens with $\Theta_{\rm InAs} \approx 1.6$ Ml (Fig. 7), and the gas branch transforms into the condensed state branch. Subsequent smooth increase in the QD density or, in other words, PT significant asymmetry is a consequence of the lateral interaction parameter relation, i.e., $U \gg V$ $(0.8 > 0.16)$ as it was shown by us in [18, 28, 29].

In Fig. 8, the isotherms calculated for different temperatures are presented. It is known that the critical thickness of the wetting layer increases with the temperature increase [17, 24]. In our theoretical isotherms, there is a PT point shift in a

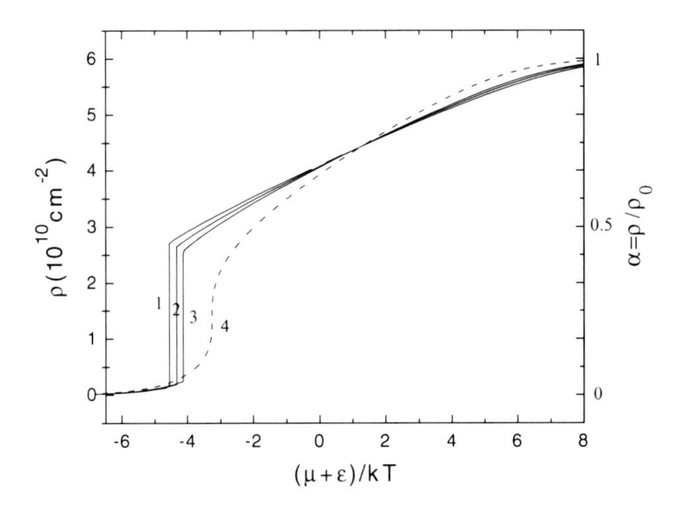

Fig. 8 The theoretical isotherms of phase transition (with QD density jump) at $T = 450, 500, 550$ °C (1, 2, 3 respectively). The isotherm for $T = 840$ °C is also shown (4). For $T > Tc$ the transition is continuous. The parameters are: $E = 0.8$ eV, $V = 0.16$ eV, $E_i = 1.1$ eV

corresponding direction (Fig. 8); however, this shift is negligible. When the temperature changes by 100°C (from 450 to 550°C), an experimental shift is about 0.1 Ml. However, according to the theoretical isotherms, this shift is 0.03 monolayer only. The explanation is that it is impossible to measure experimentally the dependence of the wetting layer critical thickness on the temperature per se due to the fact that the temperature increase changes a GaAs surface also which has a more considerable impact on the formation of QDs than the temperature change itself.

Therefore, according to our hypothesis, the QDs formation occurs in equilibrium conditions. State equations of QD ensemble on GaAs surface, i.e., the QD density as a function of the temperature and InAs surface coverage, describe a two-dimensional first-order PT, where the lateral interactions between QDs play an important role. The value and the nature of the lateral interactions are significantly determined by GaAs initial surface state. Therefore, various processes of QD formation are possible. They can vary in QD density jump value, be continuous, if the temperature is above the critical one, symmetric or asymmetric as in our case, etc. However, the QD density is always an equilibrium quantity determined by usual statistics laws.

3 Droplet Technique of Quantum Dots Fabrication

Droplet epitaxy that has been first implemented by Koguchi et al. in 1991 [30] to produce quantum dots became very popular growth technique in the last years. Due to a wide set of growth parameters beside QDs, it allows fabricating various nanostructures from nanorings and double rings to quantum dot molecules [1–6]. Also droplet epitaxy is a promising line of research to produce low-density QDs for single-photon emitter [31, 32]. Recently, it was demonstrated that low-density QDs can be obtained by using ultra-low InAs growth rate in combination with a proper growth interruption time [33]. Ostwald-type ripening that occurs during the growth interruption time is considered to be the most significant reason for the QD size enlargement and their density decreases. As droplet epitaxy utilizes liquid metal droplets, ripening processes should proceed more efficiently.

Recently a modification of droplet epitaxy was offered [34] to pattern the substrate by etching with metal droplets – so-called local droplet etching (LDE). LDE is demonstrated to be an efficient tool to form templates for QD fabrication, and it allows controlling their properties in the wide range [35–41].

Here, we start with the investigation of metal droplets that are initial stage of droplet epitaxy and then demonstrate the effect of droplet on Stranski–Krastanov growth mode transition.

3.1 *The Investigation of Intermediate Stage of Template Etching with Metal Droplets by Wetting Angle Analysis on (001) GaAs Surface*

In this section, we focus on the investigation of the initial stage of droplet epitaxy and study the distribution and the properties of nucleated metal droplets, which are the centers of subsequent QDs formation after arsenic exposure. As droplet and hole properties are determined by formation, ripening, and etching processes, their investigation is useful for understanding the mechanisms of metal redistribution on the surface.

3.1.1 Experimental Details

Since the phase of metal droplets primarily is a part of QDs growth, we used the parameters with which in our previous experiments QDs with long-wave spectra were reproducibly obtained. Two monolayers of indium were applied to the GaAs surface without As_4 flux and then the sample was immediately quenched. The growth was conducted at temperature of 500°C. To study the possible influence of the deposition rate on droplets properties, two samples with indium deposition rate of 0.04 and 0.16 Ml/s were grown. A sample with gallium droplets on GaAs was produced also to investigate the dependence of materials. The deposition rate of gallium was 0.1 Ml/s.

The morphology of the surface was studied ex situ by the atomic force microscopy (AFM), using a Solver-P47H scanning probe microscope (NT-MDT). Standard silicon cantilevers were used for imaging by AFM. A typical AFM image of indium droplet array is shown in Fig. 9.

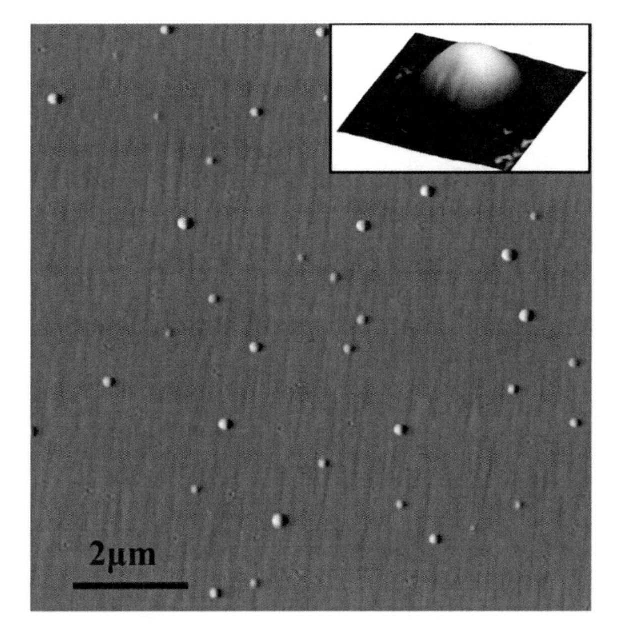

Fig. 9 10 × 10 μm AFM image of GaAs substrate with indium droplets. The single droplet image 300 × 300 nm is shown on the *inset*

3.1.2 Results and Discussion

To study the properties of the droplets, the histograms of the major parameter distributions such as droplet height and radius were plotted. Figure 10 shows the experimental data for the samples with indium droplets.

As it is seen, the distribution of droplet radii is single while the distribution of their heights has two well-pronounced peaks. This interesting result means that there are two different types of droplets on the substrate that may indicate that the droplet nucleation is a double-stage process and during the indium deposition time, the initial surface changes significantly.

The comparison of the height distributions at different deposition rates and consequently deposition times in Fig. 10 evidences that material redistribution on the surface goes actively during this period. With growth rate increased by a factor of four, the height distribution stays bimodal, but the peaks qualitatively change. Therefore, the growth rate does influence the mechanism of the droplet formation.

To investigate the form of the droplets, we used the distributions of the aspect ratio, denoted further as γ, which is the ratio of the droplet height to its radius. Since the lateral droplet size is much larger than the height, the aspect ratio is assumed to be an estimation of the droplet wetting angle. At a high temperature of 500°C, a GaAs substrate can dissolve in the droplet interface region (etching process) and due to the diffusive fluxes there should be material exchange at the interface mostly caused by the diffusion of liquid gallium to the metal droplet (intermixing process). Due to these processes, the composition of the droplets should therefore differ from that of the original material, and consequently the wetting angle of the droplet changes. The investigation of the wetting angle behavior gives us an additional opportunity to analyze and verify the complex processes of the etching.

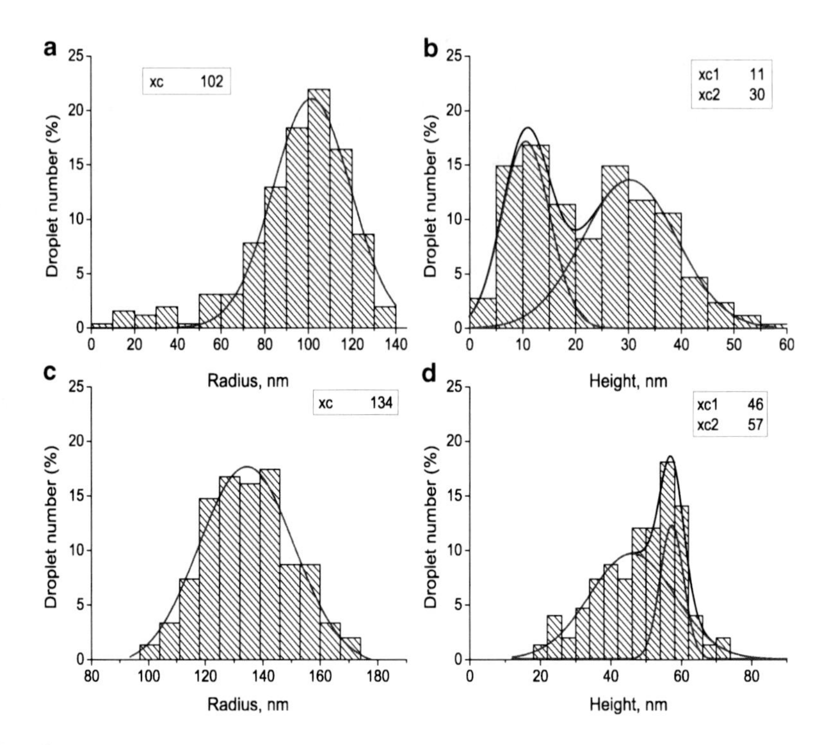

Fig. 10 The histograms of indium droplet geometrical parameter distributions. The histograms (**a**) and (**b**) relate to the sample fabricated with the deposition rate $F_{In} = 0.04$ Ml/s. The histograms (**c**) and (**d**) stand for $F_{In} = 0.16$ Ml/s. The decompositions into Gaussians are shown and the values of the peak centers are presented in the *insets*

The histograms of the aspect ratio for the samples with both indium and gallium droplets are presented in Fig. 11. As it is seen, the distributions related to the indium droplets are bimodal and in comparison with the height and radius distributions, they demonstrate even more pronounced bimodal character. Therefore, we can suppose that the droplet geometry is the same within each group, but quite different in the two obtained groups.

Considering the etching rate to be almost constant, one can assume that the indium content is higher in the sample with the higher deposition rate. Figure 3 evidences that as the growth rate is increased the peak shifts to larger wetting angles and that the peaks of all indium droplets correspond to larger aspect ratios than the only peak of gallium droplets. We therefore conclude that during the deposition time gallium that has dissolved from the substrate mixes with the indium metal droplet and the composition changes from pure indium to In_xGa_{1-x}. Since the aspect ratio measured experimentally correlates with the wetting angle, we can estimate the degree of the intermixing with the indium content x. The wetting angle is determined through the surface tension by the Young equation [42]:

$$\cos\theta = \frac{\sigma_{sl} - \sigma_{gl}}{\sigma_{lg}}, \tag{7}$$

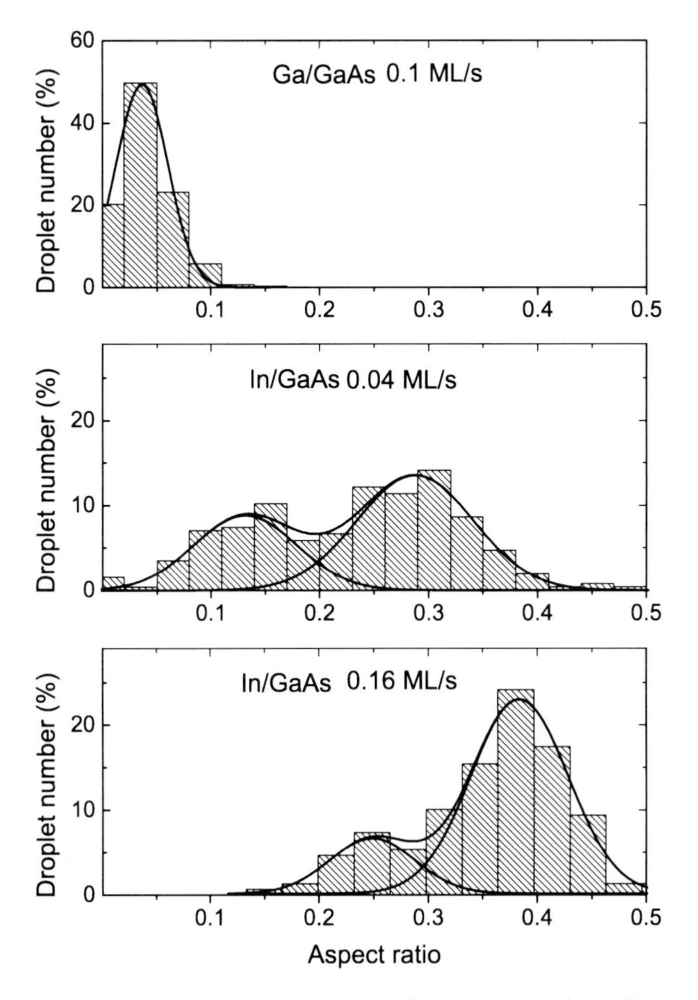

Fig. 11 From *up to bottom*: aspect ratio distributions for the samples with gallium droplets and indium droplets fabricated with indium deposition rate 0.04 Ml/s and 0.16 Ml/s, respectively. The decompositions into Gaussian are shown

where θ is the contact angle, σ_{sg}, σ_{sl}, and σ_{gl} are the surface tensions on the solid–gas, solid–liquid, and gas–liquid boundaries, respectively. As the droplet composition influences σ_{sl} much stronger than σ_{gl}, we neglect the contribution of the gas–liquid boundary tension and consider the wetting angle to be proportional to σ_{sl}. For numerical estimation of the indium content x in In_xGa_{1-x} droplets, we suppose a linear connection between x and γ: $x = a \times \gamma + b$. To determine the constants a and b, we assume $x = 0$ for gallium droplets with $\gamma = 0.04$ and $x \sim 1$ for the largest indium content in our experiments (the deposition rate of 0.16 Ml/s, right peak in Fig. 11), the values of a and b were found to be 2.94 and -0.12, respectively. The results for intermediate values of γ are presented in Table 1.

Table 1 Wetting angle and corresponding indium content

γ	0.38	0.29	0.25	0.13	0.04
x	1	0.74	0.62	0.27	0

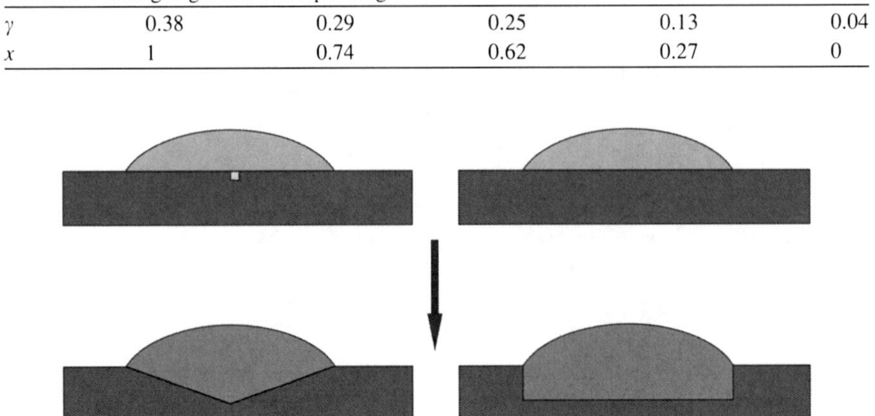

Fig. 12 The sketch of the evolution of metal droplet on the stable substrate (*right*) and in presence of a defect (*left*), resulting in rectangular and triangular profiles, respectively

We therefore can conclude that material intermixing that occurs during the deposition time is very significant. Using these results, we then can estimate the depth of the nanoholes etched by indium in the framework of the following scheme. According to our assumptions, the droplets with $\gamma = 0.38$ consist of pure indium. The droplets with different γ contain some gallium that comes from the dissolved substrate and intermixes with indium. These droplets are supposed to originate from pure indium droplets with the sizes measured by AFM, so the amount of indium was determined. With the experimentally observed wetting angle and estimated indium content, we can calculate the amount of gallium in the droplet and the corresponding volume of dissolved GaAs to provide it. Then, assuming some profile of the substrate etching, it is possible to determine the depth of the formed nanoholes.

We made the estimations for two simple models that are based on the nanohole investigation in [38] where two groups of nanoholes were observed that were called deep and flat due to the depth difference. Simplifying the nanohole geometry, we assume the etching profile to be rectangular or triangular (Fig. 12). The first model might correspond to a low etching rate in the absence of any surface defect, and consequently to a stable substrate providing uniform etching under the droplet. The second model describes the case of defect etching with a high rate, so the etching volume is limited by the crystal plates.

The droplet with a height of 50 nm and a radius of 200 nm (volume of $\sim 1.5 \times 10^6$ nm^3) was considered to be a typical droplet consisting of pure indium as it was mentioned above. For $\gamma = 0.25(x \sim 0.6)$, GaAs volume of 1.8×10^6 nm^3 was found that results in the nanohole depth of 15 nm and 45 nm for cylindrical and conical holes, respectively. The latter value is close to the one obtained for nanoholes in [43] for a high indium content in the etchant. We can suppose that under such droplets, there are conical etched volumes corresponding to the deep nanoholes. Such a simplified consideration can be useful for a further detailed investigation of the connection between the droplets and the nanoholes.

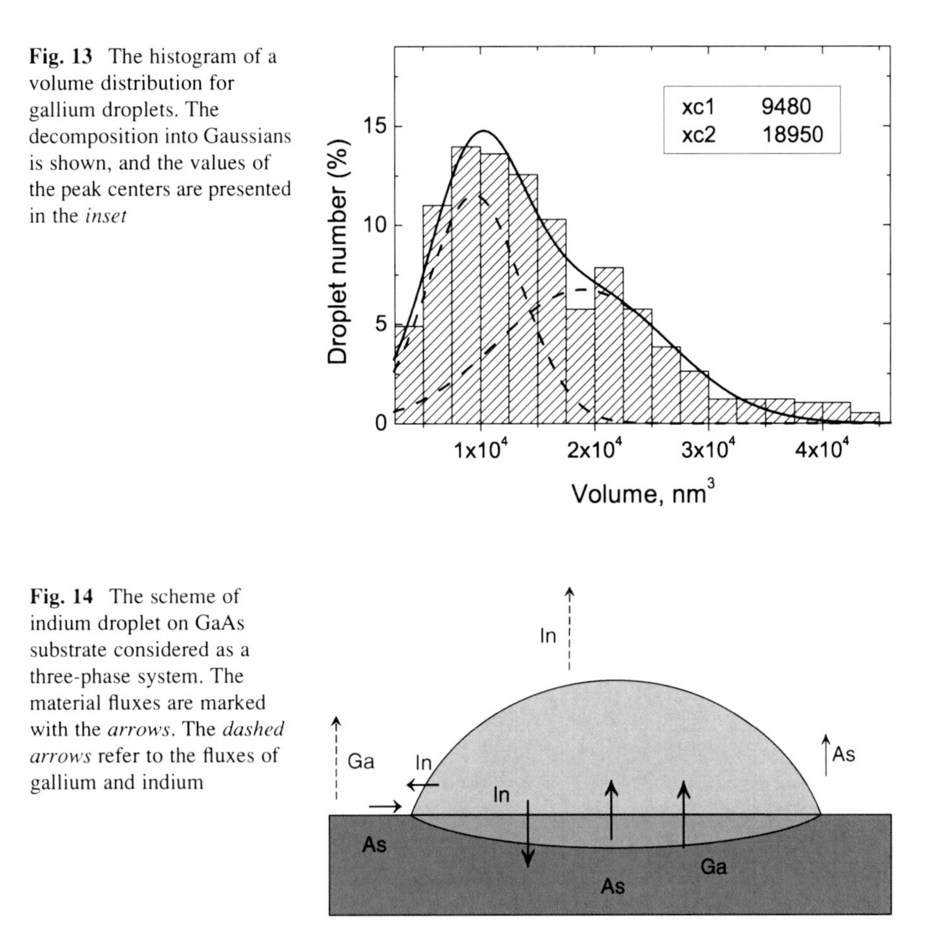

Fig. 13 The histogram of a volume distribution for gallium droplets. The decomposition into Gaussians is shown, and the values of the peak centers are presented in the *inset*

Fig. 14 The scheme of indium droplet on GaAs substrate considered as a three-phase system. The material fluxes are marked with the *arrows*. The *dashed arrows* refer to the fluxes of gallium and indium

While gallium droplets certainly demonstrate a single peak of the aspect ratio, their volume distribution reveals the coexistence of large and small droplets varying by the volume approximately by a factor of two, as it is seen in Fig. 13. The presence of the two pronounced groups of droplets should lead to the subsequent formation of the bimodal nanohole distribution that was obtained in some experiments [35, 38]. Estimating the volume ratio of the deep and flat holes fabricated by the etching with gallium droplets in [38], we found that it is in a good agreement with the droplet volume ratio in our experiment. The bimodality effect was obtained before for different parameters of the nanoholes [35, 38, 43] that are the final result of local droplet etching. Having revealed the bimodality of initial droplet parameters, we suppose that all bimodalities are a consequence of some fundamental mechanisms of droplet formation and evolution in the early stage of LDE. To analyze these processes within thermodynamics approach, we consider the droplet, the substrate, and chamber volume as a three-phase system according to the scheme presented in Fig. 14. This general scheme can be used both for the process of growth where we take into account simultaneous etching and for the post-growth period when an external indium flux should be excluded.

The diffusion flows on the interface between the liquid and solid phases are determined by etching and intermixing processes, and they have a key role in the droplet evolution. The equation of material balance for the droplet has a form:

$$V + V^{\mathrm{WL}} = F^{\mathrm{In}}t + F^{\mathrm{Ga}}t - F^{\mathrm{In}}_{\mathrm{diff}}t - \Delta V^{\mathrm{Ga}}, \tag{8}$$

where F^{In} stands for an external flux, $F^{\mathrm{In}}_{\mathrm{diff}}$, F^{Ga} are the diffusive fluxes, V is the droplet volume, V^{WL} is the volume of the metal wetting layer, and ΔV^{Ga} is the rate of etching related to the specific change of gallium volume that takes place when solid GaAs dissolves to the liquid metal and arsenic under the droplet. The external indium flux is fixed with the growth conditions, while the others can be time dependent due to the changes in the system. $F^{\mathrm{In}}t$ is the total amount of indium supplied to the substrate that was the same for both samples with indium droplets. Our measurements revealed that the droplet volumes per square micrometer are $V_1 \sim 4.3 \times 10^5$ nm^3 for the deposition time of 12 s and $V_2 \sim 1.5 \times 10^5$ nm^3 for 50 s. The total amount of the deposited indium is $V = 12.7 \times 10^5$ nm^3, so it is seen that V_1 and V_2 differ from V significantly and we can expect that a part of material is stored in the wetting layer. The formation of the wetting layer is confirmed by photoluminescence measurements for QDs fabricated by droplet epitaxy in the same growth conditions [44]. A clear peak corresponding to the quantum well and consequently wetting layer was obtained in the spectra measured at 4 K. For quantitative analysis Auger measurements were conducted. Taking into account the AFM data on the QD sizes, we used the model of infinitely thick QDs (the height is much larger than the electron inelastic mean free path) on the surface with a thin wetting layer. The thickness was assumed to be ~1 nm as about the critical value in the Stranski–Krastanov growth mode. The area of the wetting layer was found to be 90% of the total surface, so it is almost uniform and may contain a significant part of deposited material playing a role in (8).

Concerning droplet nucleation, the fluxes on the boundary with vacuum chamber (solid–gas and liquid–gas interfaces) should be considered. There is an uncompensated arsenic flux from the substrate to the chamber, so arsenic evaporation can result in the appearance of additional nucleation centers. It is interesting to notice that the droplet densities were 4.3×10^7 and 3.75×10^7 cm^{-2} for the deposition rates of 0.04 and 0.16 Ml/s, respectively, so the increase in the deposition rate and time from 12 to 50 s does not change the droplet density considerably.

Besides our experimental data, the results of the investigation of the nanoholes fabricated by LDE can be used for the analysis. In [38], the nanohole groups with different geometries were obtained and the possibility of the nanohole formation through different mechanisms was supposed to explain their presence. Taking into account the connection between the initial droplets and the holes formed by the etching, there should be two mechanisms of the droplet formation. The difference obtained for the droplet volumes allows us to assume that the droplets form at different time instants. The appearance of additional nucleation centers leading to the nucleation of the small droplets can be caused by arsenic evaporation (uncompensated arsenic flux in Fig. 14 that is usually neglected in the LDE approach) and a consequent occurrence of surface defects. In our experiment, the deposition time is quite long in comparison with that usual for LDE, so this effect might be more pronounced. The appearance of the defects can be monotonic in time or critical because of a possible surface reconstruction phase transition of the order–disorder type caused by the arsenic evaporation [28].

In the first case, only one droplet group is expected and the droplet density should be proportional to time, which contradicts our data. In the case of a reconstruction transition provided by an arsenic surface content change, a jump of the droplet density is expected, which is not obtained for deposition times of 12 and 50 s. The presence of two droplet groups can be provided by the existence of two stable compounds of the etchant like In_xGa_{1-x}. The bimodality of the aspect ratio is obtained experimentally, but for gallium droplets with a single wetting angle, there are two groups of droplets varying in the volume. The existence of two droplet groups can be explained with the presence of two kinds of surface defects that are etched in different ways, so that the droplets can differ in the composition, volume, and geometry. The origin of these defects should be found out.

Considering our experimental data on the droplet properties and the published results of etched nanohole investigations, we suppose the following scheme of the formation of a metal droplet:

1. The flux of a metal is applied to the substrate and the droplets form on the initial nucleation centers present on the surface.
2. During the deposition time, arsenic evaporates and the substrate surface state changes. A surface reconstruction transition occurs, and additional nucleation centers arise.
3. The droplets evolve; due to different nucleus origin their etching proceeds differently. For indium as an etchant, we expect the presence of metal droplets with two different indium contents. This is corroborated by the bimodality of the aspect ratio obtained in the experiment. For gallium as an etchant, there should be two groups of droplets formed at different moments of time and etched with different rates. Experimentally we observed two droplet groups differing in the volume.

3.1.3 Conclusions

In this section, the array of metal droplets on a semiconductor surface is studied as an initial stage of LDE and droplet epitaxy. The samples with various droplet materials and deposition rates were grown. We have observed a bimodality of the droplet height distribution in the system of indium droplets on a GaAs substrate that is similar to the hole depth bimodality shown in [38]. It indicates that the droplet formation proceeds in two stages. A new method to investigate the intermixing

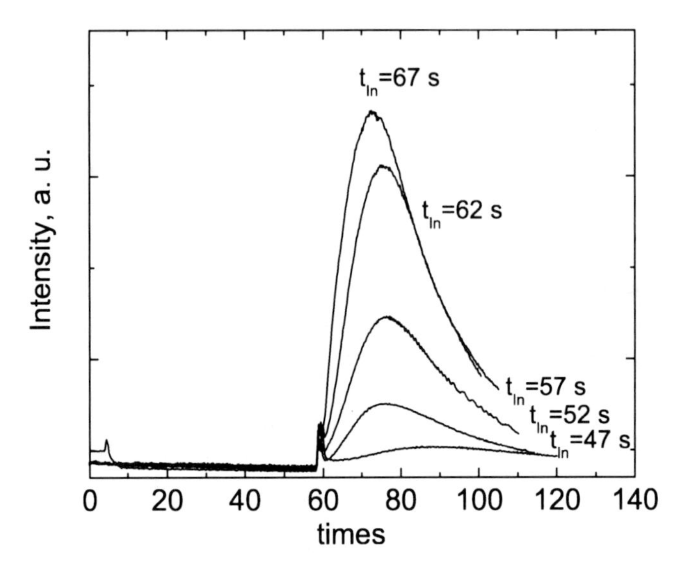

Fig. 15 Kinetic curves for QD formation. Indium deposition time with rate of 0.04 Ml/s is denoted. The moments of growth start are matched

process by a wetting angle analysis is proposed. We used the aspect ratio which is defined as the ratio of the droplet height to its radius as an estimation of the wetting angle depending on the materials. A bimodality of the aspect ratio was found, which indicates the existence of two droplet groups with different compositions. The investigation of the wetting angle and the estimation of indium content revealed a significant material intermixing during the deposition time. On the basis of our experimental results and the nanohole investigations in [38, 43], the existence of two droplet formation mechanisms is suggested. Taking into account the arsenic evaporation and a consequent change in the substrate surface state, we suggested the model of droplet evolution and the formation of two droplet groups.

3.2 Droplet to Stranski–Krastanov Growth Mode Transition in InGaAs Quantum Dot Formation on (001) GaAs Substrate

In this part, we start with the studying of the QD evolution kinetics to find out key parameters and then present the investigation of the dependences of indium dose and growth interruption time.

To regulate QDs density, the kinetics of the formation and destruction of nanostructures on the surface should be considered. We studied the evolution of the diffraction intensity by RHEED. Time-depending measurements of integral RHEED signal from QDs were started before the indium deposition and included the stages of both QD formation and annealing. Dependencies measured for different indium doses are presented in Fig. 15.

The presented curves reveal that the kinetics of the QD evolution changes to the destruction after passing through the intensity maximum. The position and magnitude of this maximum depend on the indium amount deposited on the substrate: the less indium, the lower maximum and slower the height increase. It concerns to the destruction kinetics as well, and in this way can be used to regulate the QD density. Also we can assume that QDs with defects are preferably destroyed by annealing, so that the rest should be high-quality crystalline nanostructures. Therefore, the kinetic curves indicate two main growth parameters which can be used to regulate QDs properties: an amount of deposited indium and growth interruption time.

The dose dependence is of great interest because QDs should enlarge with the increase in the indium amount and red shift with corresponding density decrease. In addition, the role of elastic tension rises and this can influence the QD formation. Thus, indium dependence data are useful for understanding the QD formation mechanisms. Growth interruption is a powerful additional instrument to tune nanostructure properties.

To avoid possible effects of an arsenic excess before the growth, a proper amount of Ga was deposited to the surface which was determined in preliminary tests. Then 2.1 Ml of indium were deposited in the absence of As_4 flux and, after the array of initial droplets formed, the sample was exposed to a beam equivalent pressure of 1.5×10^{-5} torr of As_4. Growth temperature was about 500°C.

To fabricate a sample with the varying indium layer width, a following constructive feature of the growth installation was used. During the growth time, the sample is disposed angularly to the indium source, thus providing a gradient of the indium flux along the axis of the symmetry, while no substrate rotation is used. This results in the indium gradient approximately by a factor of two along the sample. This method gives an important advantage of the continuous dose dependence. By analyzing the evolution of photoluminescence (PL) spectra measured along the sample, we can investigate the changes in quantum-optical properties of our system.

For PL measurements, the layer of QDs was capped between AlGaAs/GaAs layers. Growth interruption dependence was studied by varying the time before QDs capping. The second layer of QDs was grown uncapped to investigate it with surface sensitive methods like AFM imaging. A 533-nm laser with the excitation power of 100 W/cm^2 was used for these measurements.

Figure 16 presents several PL spectra corresponding to different indium doses. Spectra vary with the number and positions of peaks and center of gravity. This evidences that the process of QD formation changes. According to a general nucleation theory, dots should have size distributions close to the normal ones. In postprocessing, the spectra were described as a sum of the Gaussian functions like $A\exp\left(-\left[(x-x_0)^2/2w^2\right]\right)$. Under such decomposition, several parameters are derived; the peak width, its center position, and the amplitude or the area under the peak which is proportional to the density of states. The width is an important parameter of the approximation, because the amplitude and center position are

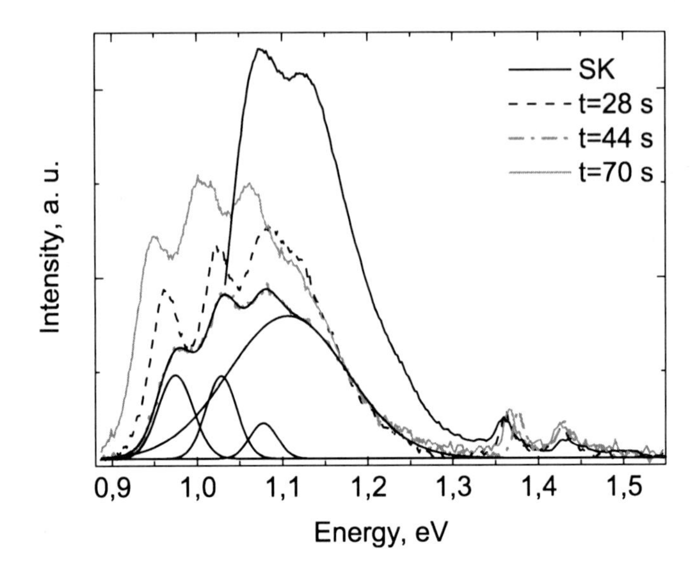

Fig. 16 Room temperature PL spectra of samples differing by growth interruption time (presented on a graph). For spectra with $t_{GI} = 44$ s the decomposition by Gaussians is shown

determined by QD density and sizes, whereas the width is a feature of the distribution and is a characteristic for a group of dots formed by the same mechanism. Its significant potential change is associated with a general modification in the dot formation.

Our analysis revealed that all spectra contain a wide peak with a large area located in the short-wavelength region. Obviously, it corresponds to small dots with a high density. It could be expected that with the increase in growth interruption time, i.e., additional annealing, these small dots evaporate more intensively due to their origin, so a peak center should blue shift. In fact, it does not occur, so we can conclude that the evolution of spectra is associated with kinetics of the dot formation during the growth interruption.

In Fig. 17, PL spectra measured at room temperature are presented from bottom to top in order of increasing indium dose. At the lower spectrum, one can observe the only peak which corresponds to QDs, originating from small drops with a wide size distribution. The next spectrum was measured at a point of the sample with larger amount of indium. As it was expected, dots increase, which can be clearly seen with a peak position shift, and consequently the entire spectrum shifts to the long-wavelength range. The upper spectrum corresponds to a further increased indium dose that is supposedly over some critical value, as considerable qualitative changes take place. Here, some pronounced peaks with rather well-defined vertexes are obtained.

The peak widths are found to differ significantly. This was not obvious before such a mathematical expansion. The largest value is close to the one, which was obtained for lower doses and attributed to small drops. Generally, it could be

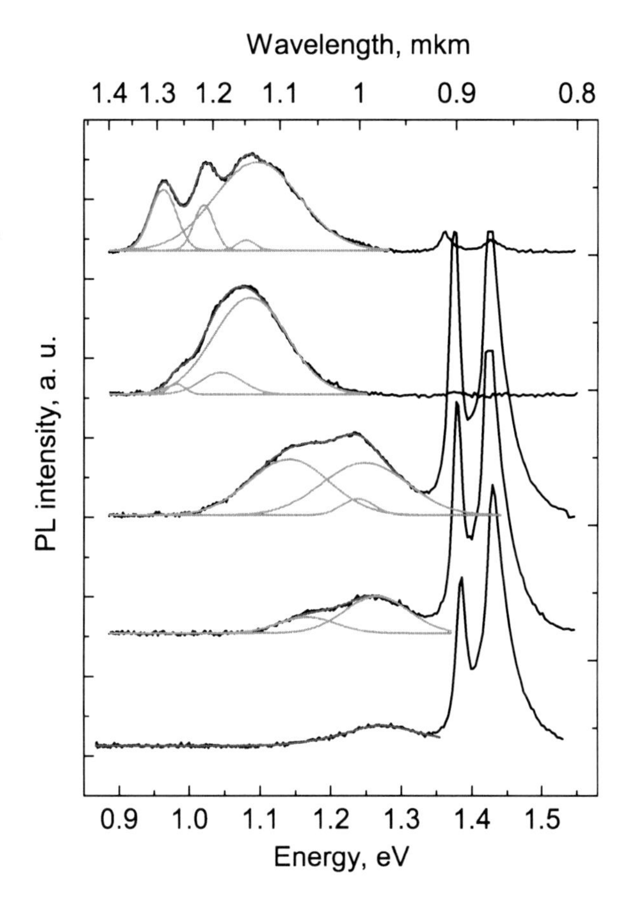

Fig. 17 Room temperature PL spectra corresponding to increase of indium amount (*bottom-up*). The decomposition by Gaussians is presented: *black stands* for the experimental data, *light grey* for a separate peak, and *grey* for an envelope

expected that a size distribution would be rather wide due to the absence of strong interactions between QDs during the formation. At the same time, the upper spectra contain the peaks with a strongly reduced width compared with previous spectra. It is interesting to note that the obtained width is in a good numerical agreement with the data from literature [45]. The presence of several peaks could be interpreted as ground and excited states of quantum dots or the quantization of the dot height [46]. However, the observed peak widths differ in a few times and this implies that peaks relate to different dot groups. As narrow peaks appear with the increase in the indium amount, this group formation should be a critical phenomenon.

Therefore we can conclude that over some critical value of the indium dose, a new dot group arises on the surface. It forms by a mechanism that is different from the common droplet one and then has a different size distribution. We suppose that the group with the narrowest distribution forms under the influence of an additional external force. It significantly decreases the size dispersion and consequently the width of distribution. AFM images display dot complexes with the distinct symmetry of an original drop (high center and some smaller dots around), as it was described in our previous paper [47].

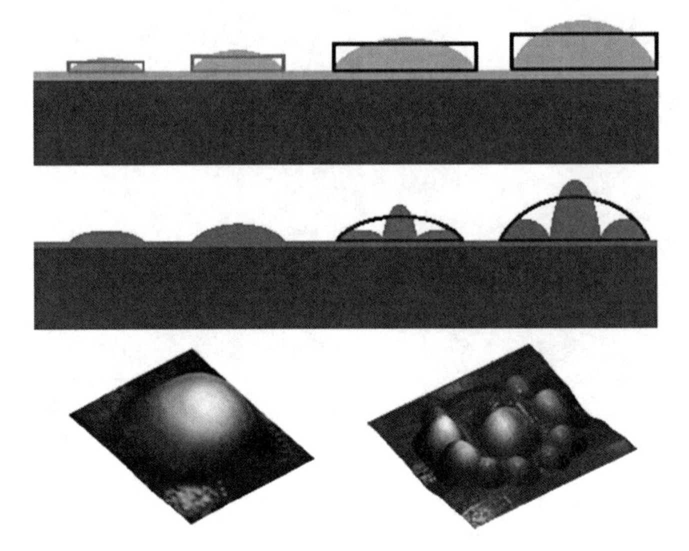

Fig. 18 Scheme of droplet to Stranski–Krastanov growth mode transition. *Upper*: initial metal droplets with different sizes. *Boxes* denote metal film with equivalent volume. *Lower*: droplet transformation to QDs. Subcritical droplets transform to QDs entirely, overcritical ones crack, and form a complex of QDs. Initial droplet shape is denoted. AFM images of a droplet and a QD castle-like structure are shown below

Some peaks have a width that is characteristic for Stranski–Krastanov growth mode. It was measured before on the analogical samples, fabricated in the same conditions by this mode. Also even narrower peaks are obtained in PL spectra. As the width of some peaks is typical for Stranski–Krastanov growth mode, the idealized mechanism of the droplet transformation is not implemented per se. It can be supposed that under some critical size of a metal drop the elastic stress relaxation by Stranski–Krastanov mechanism (as for continuous film) is realized even on the drop area. The proposed mechanism of droplet to Stranski-Krastanov transition for overcritical droplets is schematically presented in Fig. 18. Because of extra tension force, which is absent in case of a continuous film, the size distribution is expected to be significantly narrower. This is in a perfect agreement with the results of presented analysis of the experimental spectra.

4 Optical Properties of Quantum Dots and Wetting Layer

The majority of the studies reported so far has concentrated on the direct gap In(Ga) As/GaAs material system, whereas the technologically similar InAs/AlAs system has received much less attention. The system of InAs quantum dots embedded in an AlAs matrix is very close to the InAs/GaAs QDs system what concerns the Stranski–Krastanov growth mode, since AlAs has practically the same lattice constant as GaAs. Nevertheless, the increase in the QD barrier height leads to a stronger electronic confinement in the InAs/AlAs QDs and to significant changes in their electronic and optical properties compared with the InAs/GaAs QDs system [48–50]. Then an important advantage of such system is shifting the peak of emission into the visible spectral region. In this part, results of investigations of InAs quantum dots in an AlAs matrix grown by Stranski–Krastanov method are presented. To avoid the repetition of experimental details, we summarize common features of growth process here.

InAs QDs samples studied in this section consisted of one or two layers of QDs sandwiched between layers of AlAs about 50 nm thickness grown on top of a 200 nm buffer GaAs layer. The first AlAs layer in all the samples was grown at a substrate temperature 600°C. The QD layers were deposited at a rate $V = 0.04$ Ml/s (as calibrated in the center of the wafer using reference samples) to a nominal thickness of 2.5 monolayers. The typical As_4 beam pressure was taken to be about 8.5×10^{-6} torr. To prevent InAs evaporation, the growth temperature was not increased during the deposition of the first initial monolayers of the second AlAs layer covering the QDs. The rest of the cover layer was grown at 600°C. A GaAs cap layer of 20 nm was grown on top of the sandwiched QD layer to prevent oxidation of AlAs.

The other features of a certain sample like the growth temperature, the growth interruption are mentioned in corresponding subsections.

4.1 Atomic and Energy Structure of InAs/AlAs Quantum Dots

Recently microsecond- [48] and even millisecond-scale [50] nonexponential photoluminescence (PL) decay have been observed in InAs/AlAs QDs at low temperatures. To explain this long decay, several models have been proposed [48–50]. Dawson et al. [48] attributed the long decay times to the recombination involving electrons and holes localized in spatially separated QDs. Later the long PL decay in these QDs was explained in terms of type II indirect transitions of electrons localized in the AlAs matrix with holes localized in the QDs [49]. These models predict a long PL decay at any temperature, since the recombination probability of spatially separated electrons and holes does not depend on temperature. However, we found that an increase in temperature from 50 to 300 K causes a decrease in the PL decay time in such QDs by six orders of magnitude. We demonstrated that this decrease is due to the acceleration of the radiative recombination rate and does not relate to thermal activation of fast nonradiative channels [47]. An alternative model to spatially separate carriers is proposed to explain the thermal dependence of the radiative recombination rate [50]. Excitonic levels in small QDs in a high-barrier AlAs matrix split into an optically active state with a short lifetime (τ_s) and an optically inactive state with a long lifetime (τ_l) [51, 52]. The radiative lifetime at low temperatures when kT is less than this splitting (ΔE) is determined by τ_l and at high temperatures ($kT > \Delta E$) by τ_s. However, recently we observed a prolonged PL decay at low temperatures even in large InAs/AlAs QDs with a small ΔE [53]. Thus, a more adequate model explaining the prolonged PL kinetics of the InAs/AlAs QDs at low temperatures and its temperature dependence is required.

In order to construct such a model, we need to establish the electronic structure of the InAs/AlAs QDs, which has been scarcely studied up to now. Hitherto, investigations devoted to studies of the electronic structure of the InAs/AlAs QDs [49, 54, 55] take only the Γ minimum states of conduction band in InAs into consideration. Therefore, two possible configurations of the QDs electronic structure were proposed (1) band alignment of type I with lowest electron states at the Γ minimum of the InAs conduction band and (2) band alignment of type II, which can be realized if the quantum confinement pushes states at the Γ minimum of the QD conduction band above states at the X minimum conduction band of the matrix [49, 54, 56]. In this study of the electronic structure of InAs/AlAs QDs not only the direct Γ minimum but also the indirect L and X minima of the InAs conduction band are taken into account. The effect of confinement on the InAs QDs structures with an indirect band for the electrons has not been studied up to now. Due to the small electron effective mass of InAs, the electron states in the Γ minimum of the QD conduction band are considerably more shifted by confinement than those of the L and X minima. Therefore, the lowest electron states in the QDs can be at the indirect minimum of the conduction band. Actually, we have recently demonstrated that even ultra thin InAs/AlAs quantum wells (QWs) have type I band alignment with lowest electron states at the X minimum of the conduction band [57]. Thus, it is reasonable to expect that confinement in InAs/AlAs QDs also results in a transformation from direct to indirect lowest state of the conduction band within the type I band alignment.

In this part, we present the results of a study of the atomic structure and the energy spectrum of InAs QDs in an AlAs matrix by transmission electron microscopy (TEM), photoluminescence, and computational work. We demonstrate that (1) these QDs consist of an InAlAs alloy with a fraction of InAs, which is dependent on the growth conditions; (2) the quantum confinement leads to transformation of the lowest electron state from the direct Γ to the indirect X minimum of the InAlAs conduction band, but does not result in a type I to type II transition; and (3) the spectral and temperature dependencies of PL decay time can be explained in terms of a two-level model. This model predicts that the lowest energy level of the QDs with long PL lifetime is located at the indirect X minimum and the excited level with short PL lifetime belonging to the direct Γ minimum of the QDs conduction band.

4.1.1 Experiment

The QDs produced by typical growth procedure (see the description at the beginning of Sect. 4) were formed at temperatures (T_g) varied in the range of 440–540°C with a growth interruption time (t_{GI}) in the range of 10–120 s. The maximum growth interruption time was reduced for samples grown at 530 and 540°C due to evaporation of In ad-atoms from the growth surface at high temperatures [58].

The As_4 beam pressure was 8.5 × 10^{-6} torr except for the sample grown at T_g/t_{GI} = 480°C/120 s, where the beam pressure was 4.5 × 10^{-6} torr. The size and density of the QDs were studied by means of transmission electron microscopy (using a JEM-4000EX operated at 200 keV).

The PL excitation was accomplished above the direct band and under the indirect gaps of the AlAs matrix. The PL was excited by a He–Cd laser ($h\nu$ = 3.81 eV), or a He–Ne laser ($h\nu$ = 1.96 eV) and a red semiconductor laser ($h\nu$ = 1.87 eV) with a power density of 30 W/cm^2. The detection of the macro PL was performed by a double diffraction grating spectrometer equipped with a cooled photomultiplier operated in the photon counting mode. In order to select the PL from a single QD, we used a sample with a small density of QDs (~10^7 cm^{-2}) in such a way that only one QD was excited within the laser spot. The excitation in our micro photoluminescence (μ-PL) system was carried out by a Verdi/MBD266 laser system with a wavelength of 266 nm ($h\nu$ = 4.66 eV). The laser spot was about 1.5 μm in diameter. The excitation power was varied from 5 × 10^{-4} to 6 mW. The emission was collected by a microscope, dispersed by a monochromator, and detected by a CCD.

4.1.2 Intermixing in InAs/AlAs Quantum Dots

It is well known that strong intermixing of the InAs and the barrier materials will take place for In(Ga)As/Ga(Al)As heterostructures due to the strain-driven InAs segregation [59–63]. This intermixing is extensively studied for In(Ga)As/GaAs [54, 59, 61–63] and InAs/AlAs [54, 58, 61] QWs. A quantitative phenomenological model of Muraki et al. [65] was established for describing the indium composition profile across the QWs. There are also many studies demonstrating intermixing in In(Ga)As/GaAs QDs [54, 59, 61–63, 66–72]. While intermixing in InAs/AlAs QDs has been scarcely studied up to now, Ibez et al. have shown that intermixing takes place in InAs/AlGaAs QDs by the Raman scattering technique [73] and Offermans et al. have recently demonstrated a nonuniform InAs/AlAs distribution inside the InAs QDs by cross-sectional scanning tunneling microscopy [54]. However, there is no systematic study of intermixing phenomena in InAs/AlAs QDs as a function of growth conditions up to date. In this work, we demonstrate that the InAs/AlAs intermixing degree in InAs QDs changes with growth temperature and growth interruption times. Figure 19 demonstrates TEM plane view images of QD samples grown under different temperatures and interruption times. All samples except one contain only dislocation-free QDs. The sample grown at T_g/t_{GI} = 480°C/120 s contains both coherent QDs and dislocated clusters of InAs, which appear as a result of an increasing In ad-atoms mobility due to the low As_4 pressure [53, 74]. In order to determine the average diameter (D_{AV}), size dispersion (S_D), and density (D_D) of coherent QDs, the diameters and number of QDs are estimated within an area in the range of 0.3–0.5 μm^2. The results are shown in Fig. 20 as histograms, which are used for determination of S_D and D_{AV} of the QDs and are summarized in Table 2. One can see that D_{AV} increases and D_D decreases with increasing growth

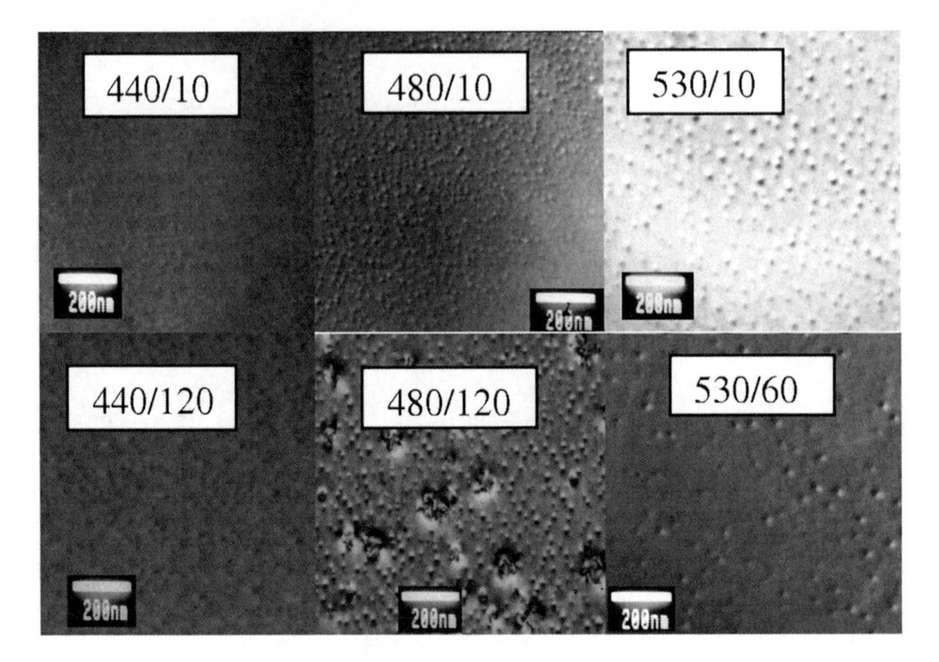

Fig. 19 TEM plane view images of QD samples grown under different temperatures (temperature/interruption time in each figure)

temperature and growth interruption time. These changes in D_{AV} and D_D become especially apparent at higher growth temperatures. It should be noted that the observed values of D_D are larger and the corresponding values of D_{AV} are smaller than for InAs/GaAs QDs formed at similar conditions. These facts have earlier been observed for InAs/AlAs QDs by many groups [75–78]. However, a specific feature noted here is a higher value of the dot size dispersion for InAs/AlAs QDs.[1] As seen in Table 2, the width of Gaussian distribution of the dot size reaches up to almost 40% of the mean QDs diameter, which is considerably larger than typical S_D values (5–20%) for InAs/GaAs QDs [83–85].

The formation of the QDs and the InAs segregation in both InAs/GaAs and InAs/AlAs systems are determined by kinetic effects on the growth surface [59, 75, 86]. Therefore, a strong variation in D_{AV}, D_D, and S_D between the InAs/GaAs and InAs/AlAs QDs systems denotes a different intermixing rate in these QDs systems. The discussion about mechanisms of intermixing and their difference in InAs/GaAs and InAs/AlAs QDs is out of the framework of our study. We rather estimate the effect of temperature and growth interruption time on the intermixing degree in InAs/AlAs QDs. For this purpose, we first make a rough estimation by defining an

[1] The large size dispersion appears also as a large linewidth (FWHM achieves 180 meV) of the QDs PL band. In our case, the FWHM of PL bands and, therefore, dispersion are in several times larger than for InAs/GaAs QDs (see, e.g.: [79–82].)

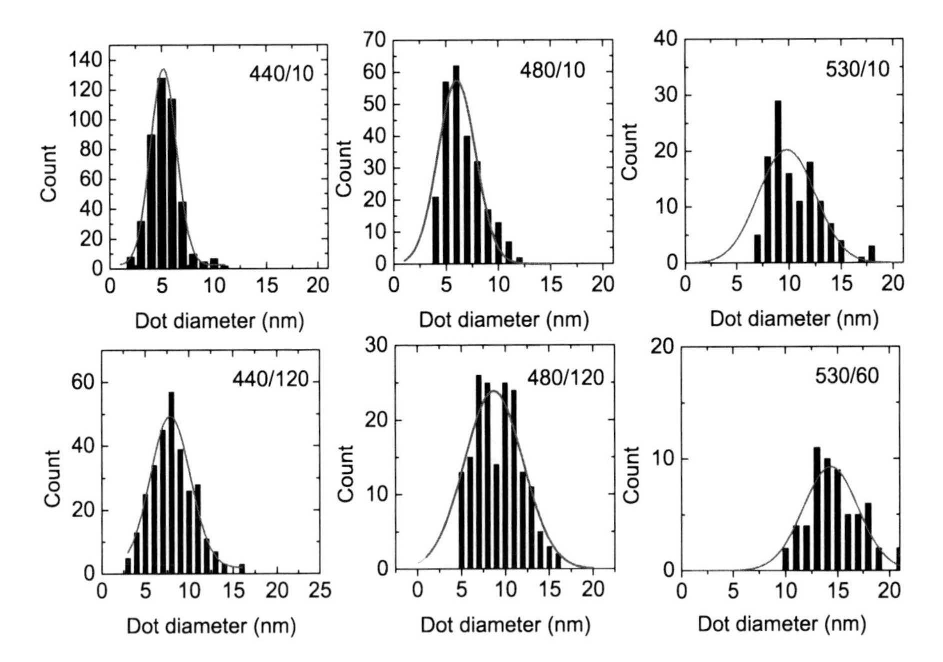

Fig. 20 Histograms of the QD size distribution for InAs/AlAs QDs samples grown under different temperatures (temperature/interruption time in each figure). The distribution of the size dispersion is fitted by Gaussian curves. The average dots diameters and the widths of the dots diameters distribution (S_D) obtained from these curves are given in Table 2

Table 2 Average diameter, size dispersion, density, composition, and spectral parameters of the PL emission for InAs/AlAs QDs

T/t_{GI}	D_{AV} (nm)	S_D (nm)	S_D in % of D_{AV}	$D_D\times$ (10^{10} cm^{-2})	Volume of QDs on cm^{-2} $\times$ (10^{11} nm^3)	Maximum of PL band (eV)	Fraction of In As (x)
440/10	5.2 ± 0.03	1.7 ± 0.08	33	11.1	20	1.805	0.99
440/120	8.0 ± 0.15	3.0 ± 0.15	37	6.3	40	1.755	0.78
480/10	6.3 ± 0.15	1.8 ± 0.2	30	10	31	1.750	0.93
480/120	8.7 ± 0.05	3.3 ± 0.14	38	6.1	52	1.751	0.73
510/30	10.4 ± 0.04	3.9 ± 0.13	36	3.7	58	1.680	0.75
510/120	12.7 ± 0.01	2.5 ± 0.06	21	3.5	90	1.685	0.68
530/10	9.8 ± 0.06	2.7 ± 0.25	28	2.8	33	1.695	0.73
530/60	14.3 ± 0.03	2.7 ± 0.15	19	0.9	33	1.875	0.42
540/30	18.3			0.01		1.856	0.41

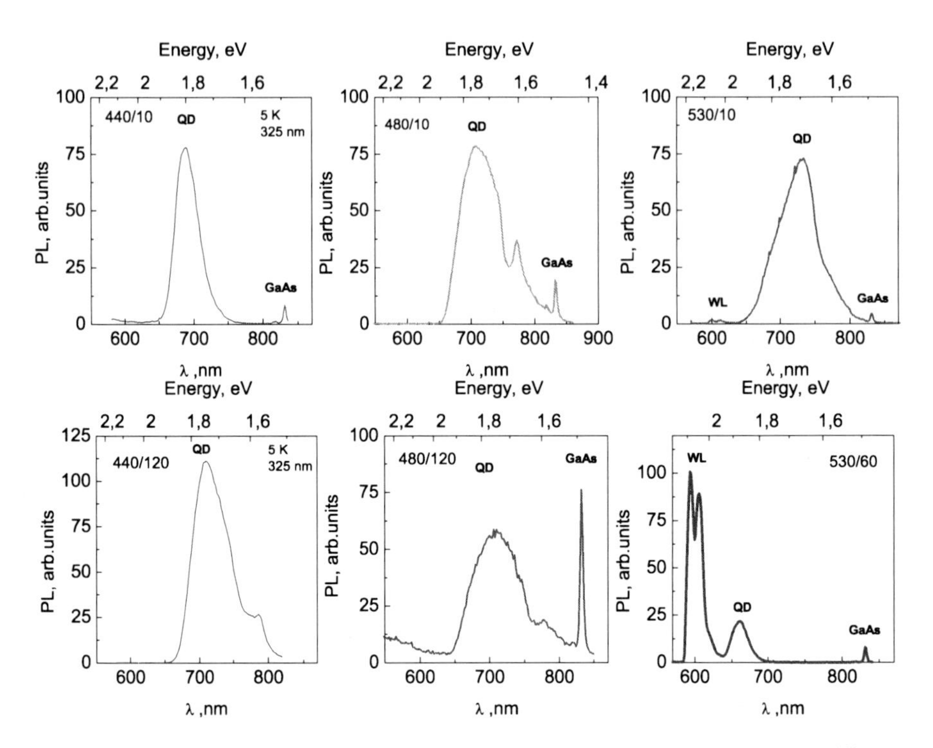

Fig. 21 Low-temperature (5 K) PL spectra of InAs/AlAs QDs samples grown under different temperatures (temperature/interruption time in each figure)

effective volume occupied by all QDs located within an area of 1 cm^2 as a function of the growth conditions. This effective volume is calculated as the product of D_D and the volume of the QD (with the diameter D_{AV}). To calculate the QD volume, lens-shaped QDs with the aspect ratio of 4:1 were selected. In the resulting data (presented in Table 2), one can see that in spite of the equal amount of deposited InAs, the effective volume increases by about a factor of 3.5 times with increasing temperature from 440 to 510°C and a raised growth interruption time, from 10 to 120 s. We assume that this increase in the effective volume is a result of an increasing intermixing with increasing growth temperature and interruption time. (However, a further increase in T_g results in a decreasing effective volume due to evaporation of In ad-atoms from the surface of the sample [58].)

Low-temperature PL spectra of the InAs/AlAs QDs grown at different conditions (Fig. 21) confirm this assumption. Three bands marked WL, QD, and GaAs presented in the spectra refer to recombination in the wetting layer [57], the QDs, and the GaAs buffer layer, respectively. The energy positions at the maximum ($h\nu_{QD}$) of the QD PL band as a function of growth condition are collected in Table 2. For QDs grown with a fixed growth interruption time, an increasing T_g (from 440 to 530°C) results in red shift of $h\nu_{QD}$. For a fixed T_g, $h\nu_{QD}$ (1) is red shifted for lower growth temperatures ($\leq T_g = 440$°C), (2) remains almost unshifted for T_g in the range of 480–510°C, and (3) is blue shifted at higher T_g with increasing growth interruption time. The energy of the PL emission from QDs is primarily determined by the confinement of charge carriers. An invariance or even a blue shift of the PL band with increasing QD size implies that InAs/AlAs intermixing results in an increasing AlAs fraction in the alloy composition of QDs.

Below, we estimate the alloy composition of the studied QDs by comparing the energy of the PL transition in the QDs with the calculated energy for the optical transition in the QD using the AlAs fraction of the QD as the variable parameter of the calculations.

4.1.3 Electronic and Atomic Structure of InAs/AlAs Quantum Dots

Calculations

The electronic structure of the studied QDs has been calculated by using the nanodevice simulation tool NEXTNANO3 [87]. The energy levels of the holes and the electrons in the Γ, X, and L minima of the conduction bands have been calculated by means of a simple band effective mass approach. The strain, deformation potentials, and the nonparabolic form of the electron dispersion [88] have been taken into account in the calculations. For simplicity, the exciton correction for the energy levels was neglected.

The electronic structure of an InAs/AlAs QD is determined not only by the size, shape, and chemical composition of the QD, but also by the InAs and AlAs parameters. The diameters of the QDs grown were determined from plane view TEM micrographs, whereas the shape of the QDs was determined from cross-section TEM micrographs [89]. The QDs were found to be lens like with an aspect ratio varying from 3:1 to 5:1. For our calculations, the aspect ratio of 4:1 was selected.

The chemical composition of the QDs is not known. A nonhomogeneous distribution of the InGaAs alloy across the QDs has been demonstrated in numerous investigations of the composition of In(Ga)As/GaAs QDs [54, 59, 62, 63, 68, 71, 90]. Up to date, there are several experimental descriptions of the InGaAs alloy composition across the QDs. Litvinov et al. [59] demonstrated that QDs look like nuclei with a high fraction of InAs in the center surrounded by a shell in which the InAs fraction decreases from the center to the edge of the QDs. On the other hand, Offermans et al. [54] demonstrated a monotonic increase in the InAs fraction from the base to the top of the InGaAs/GaAs QD, to reach an InAs fraction of 100% at the dot top. The difficulty in describing the In diffusion and segregation during the formation of the capped QDs hampers the development of a theoretical model of the alloy composition distribution in the QDs even in the well-studied InGaAs/GaAs QDs system. Consequently, the alloy composition distribution is less known in the scarcely studied InAlAs/AlAs QDs. Here, we accept that the $In_xAl_{1-x}As$ alloy composition changes inside QD according to a two-dimensional Gaussian function. An InAs concentration maximum is located at the center of the QD, two Mls above its bottom. The InAs fraction (x) decreases along the growth axis through the center of the QD down to a value of 15% below the maximum x, while in the plane two

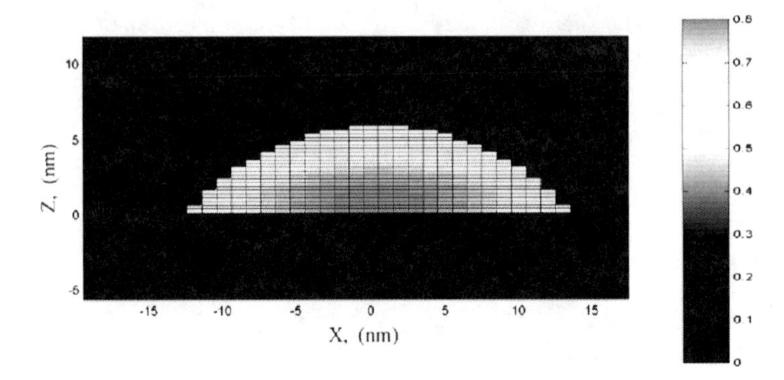

Fig. 22 An example of the distribution of the alloy composition across a QD used in the calculations. The *scale bar* shows the fraction of InAs in the $In_xAl_{1-x}As$ alloy

Mls above the QDs bottom perpendicularly to the growth axis, the x value decreases down to a value of 10% below the maximum x. An example of such an alloy composition distribution is demonstrated in Fig. 22 for a QD with a maximum InAs fraction of 0.8 and a diameter of 26 nm.

Unfortunately, together with accurately experimentally determined or calculated parameters for InAs and AlAs such as lattice constants, elastic and piezoelectric constants, spin-orbit splitting, and deformation potentials [91–96], there are parameters, which are poorly known. These parameters are, e.g., the low-temperature energy gaps of the conduction band minima at X (E_X) and at L (E_L) in InAs, and the valence band offset (VBO) for an InAs/AlAs heterojunction. For instance, Vurgaftman et al. [94] proposed $E_X = 1.433$ eV and $E_L = 1.133$ eV, whereas Ridley et al. [97] suggested $E_X = 1.39$ eV and $E_L = 0.98$ eV, and Boykin [98] and Landolt–Bornstein tables [99] give $E_X = 2.27$ eV and $E_L = 1.152$ eV. Also the VBO value for an InAs/AlAs heterojunction is also not well determined. There are many calculations for strained InAs layers on unstrained AlAs substrates, which give various values for the VBO in the range from 0.29 to 0.83 eV [57].

In order to choose relevant parameters, we calculated the energy spectrum for a 1.4 Ml thick InAs/AlAs QW with a chemical composition profile described by the Muraki model [65] and select available parameters on the E_X, E_L, and VBO (from the literature), which ensure type I band alignment for the QW. Other parameters used in the calculations were taken from [57]. However, the energy of the optical transition in the QW calculated using the selected parameters differs from the experimentally determined energy [57]. We found that the energy for the optical transition in the QW is more sensitive to E_X than to the VBO value. The best agreement between the calculated and the experimental results on the energy for the optical transition in the QW was reached for $E_X = 1.580$ eV, which is used in the following calculations. The parameters used in calculations are given in Table 3.

Table 3 Selected band structure parameters for InAs and AlAs used in the calculations

Parameters	InAs	AlAs
E_{Γ}	0.417 [94]	3.099 [94]
E_X	1.580	2.24 [94]
E_L	1.133 [94]	2.46 [94]
Valence band edge energy	0.44 [95]	−0.093 [95]

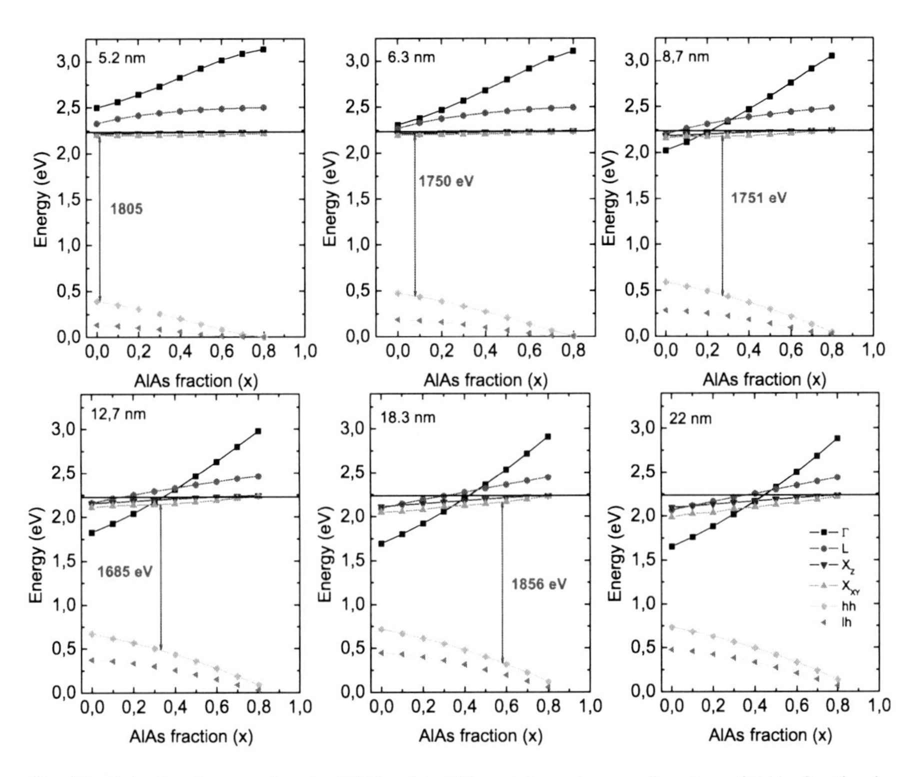

Fig. 23 Calculated energy levels of QDs with different diameters as a function of AlAs fraction in the QD. QD diameters: (**a**) 5.2 nm, (**b**) 6.3 nm, (**c**) 8.7 nm, (**d**) 12.7 nm, (**e**) 18.3 nm, (**f**) 22 nm. The reference level is at the top of the AlAs valence band. The labels of the different levels are presented in the figure (**f**). The *horizontal solid line* is the bottom edge of the AlAs matrix conduction band. The *vertical lines* refer to the alloy composition at which the PL emission energy corresponds to the energy of the optical transition in the QD with a corresponding diameter

Calculations of Electronic Structure of the Quantum Dots

Calculated energy spectra of the studied QDs with different diameters as a function of composition are presented in Fig. 23. All QDs have a type I band alignment. One can see that related positions of the electronic levels belonging to different minima of the QD conduction band depend on size and composition of the QD. In small QDs (diameter < 7.5 nm), the lowest electronic level is at the X_XY minimum of the conduction band independently on the composition. In large QDs with a low AlAs

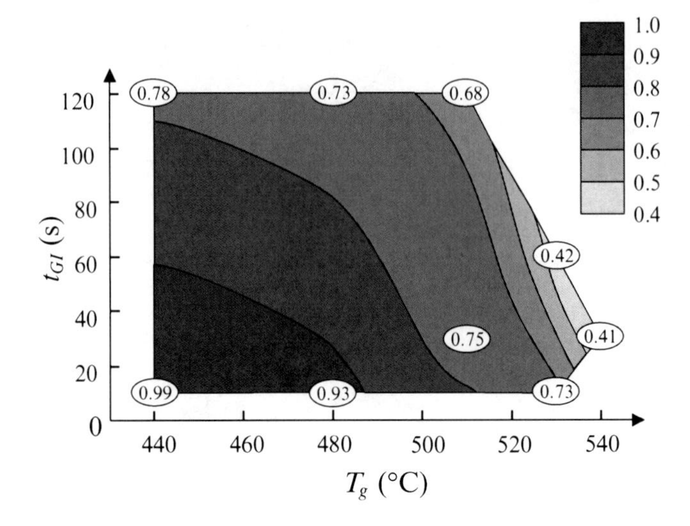

Fig. 24 A nomogram of InAs fraction in QDs as a function of growth temperature and interruption. The *scale bar* shows fraction of InAs (x) in the $In_xAl_{1-x}As$ alloy. The *numbers* in the plane of the picture mark the InAs fraction in the studied QDs

fraction, the electronic level is at the Γ minimum of the conduction band. With increasing AlAs fraction, the electronic states at the Γ minimum shift considerably more than of the states at the X minimum. Accordingly, the states at the Γ and X minima intersect for some composition and the X_XY minimum becomes the lowest electronic state in the QDs. The composition corresponding to this intersection depends on the diameter of the QD. Thus, InAs/AlAs QDs can have direct or indirect band structures within the type I band alignment.

Atomic Structure of the Quantum Dot

A comparison of the QD PL emission energy with predicted transition energies from the calculations allows us to estimate the composition of the QD. Assuming that photons with the peak energy in the PL spectra (Fig. 21) are emitted from QDs with diameters corresponding to the maximum of the size distribution, we can estimate the InAs fraction in the QDs as a function of the growth conditions (see Table 2). We interpolate these data and build a nomogram, as shown in Fig. 24.[2] One can see that QDs grown at low temperature (440°C) and small interruption time (10 s) consist essentially of pure InAs. An increase in T_g with a maintained

[2] Composition of QDs form at fixed temperature with a t_{GI} determines via linear interpolation between the minimal and the maximal values of t_{GI}. The interruption times correspond to QDs with compositions, x equal to $0.1n$, where n can be taken from 4 to 10, were selected as reference points for each T_g. Then the points with equal composition x are connected by B-spline lines.

growth interruption time causes a decrease in the InAs fraction in the QD down to 0.73 at $T_g = 530°C$. In addition, an increase in the growth interruption time for a fixed T_g leads to a decreasing InAs fraction. Accordingly, QDs grown at higher temperatures always contain a higher concentration of AlAs.

4.1.4 Relationship Between Calculated Spectra and Experimental Data on the Carriers Recombination in InAs/AlAs Quantum Dots

In this section, we demonstrate that the available experimental data on the carriers recombination for InAs/AlAs QDs are consistent with a type I band alignment based on the calculations in the previous section.

Micro-PL (μ-PL) of InAs/AlAs Quantum Dot

Recently μ-PL of a single InAs/AlAs QD has been reported by Sarkar et al. [56].[3] They observed sharp peaks below 1.8 eV as well as a continuous spectrum above this energy and concluded that this continuous spectrum is due to type II band alignment in small QDs. However, this conclusion is in contradiction to our calculations resulting in a type I band alignment in any InAs/AlAs QDs. In this section, we experimentally demonstrate that InAs/AlAs QDs have type I band alignment even for energies above 1.8 eV. We will also discuss the possible origin of the continuous spectrum.

A method for determination of the band lineup of heterostructures based on the PL technique has been proposed by Ledentsov et al. [100]. They theoretically calculated and experimentally demonstrated that the energy position of the PL band should blue shift proportionally with the cube root of the excitation power density (P) for any structures with type II alignment. The blue shift is characteristic to all type II heterostructures (both QWs and QDs) and reflects the dipole layer formation caused by a spatial separation of nonequilibrium holes confined in the QW or QD and electrons confined in the nearby matrix region [100, 101]. This PL technique was successfully used for many type II systems such as GaSb/GaAs QWs and QDs [100, 101], GaAs/AlAs QWs [57], InAs/GaAs QDs at high hydrostatic pressure [102] and ZnMnTe/ZnSe QDs [103].

μ-PL spectra of the InAs/AlAs QDs specimen grown at $T_g/t_{GI} = 540/30$ as a function of excitation power density are depicted in Fig. 25. A PL band (labeled as QD_0) with a maximum at 1.815 eV and with a full width at half maximum (FWHM) of 14 meV, related to the recombination of electrons and holes in the QD, is observed in the μ-PL spectrum at an excitation power of $P = 1$ kW/cm^2. We believe that the QD_0 band is due to the emission of a single QD since (1) TEM

[3] In order to select PL of a single QD in array the samples studied in [56] were covered by an aluminum mask containing small square apertures fabricated by electron beam lithography.

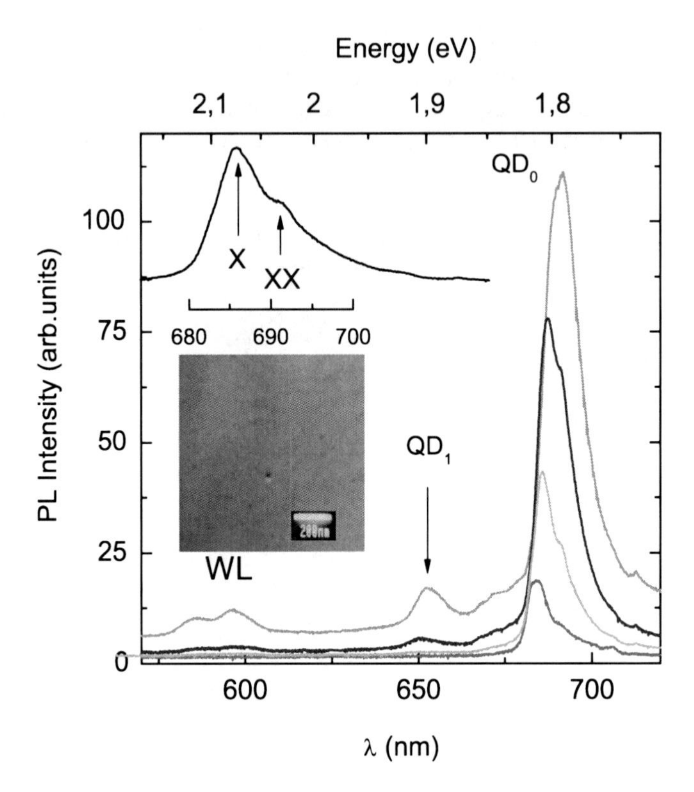

Fig. 25 μ-PL spectra of a single InAs/AlAs QD as a function of power density *P*. *top-down* kW/cm^2: 80, 19, 9, 1. *Insets* demonstrate: (*upper*) QD_0 and XX lines at $P = 9$ kW/cm^2, (*lower*) TEM plane view images of the sample. A temperature of 60 K was established

data exhibit a dots density of about 10^7 cm^{-2} (inset in Fig. 25) and (2) a small shift (~1 μm) of the excited laser spot results in a jump in the energy position of the QD_0 band by 70 meV due to the fact that another QD is monitored. A large PL linewidth from a single QD was previously observed in different QD systems such as InP/InGaP [104], CdSe/ZnS [105], and Si [106]. It has been demonstrated that the PL line width associated with the indirect band gap QD is larger than for the corresponding PL associated with a QD with a direct band gap [106].

Upon an increasing excitation power, additional bands (labeled WL, XX, and QD in Fig. 25) will appear in the μ-PL spectra. The intensities of these bands will increase more with increasing excitation power than the QD_0 band. The WL energy position is consistent with the recombination of the wetting layer [57]. The XX band exhibits a square dependence on the excitation power and becomes dominating the PL spectra for $P > 40$ kW/cm^2. This fact together with the 9-meV red shift of the XX band with respect to the QD_0 strongly indicates that the XX band is due to the recombination of the bi-exciton [56]. The QD_1 emission is proposed to be due to the recombination of excitons originating from the excited levels of the studied QD.

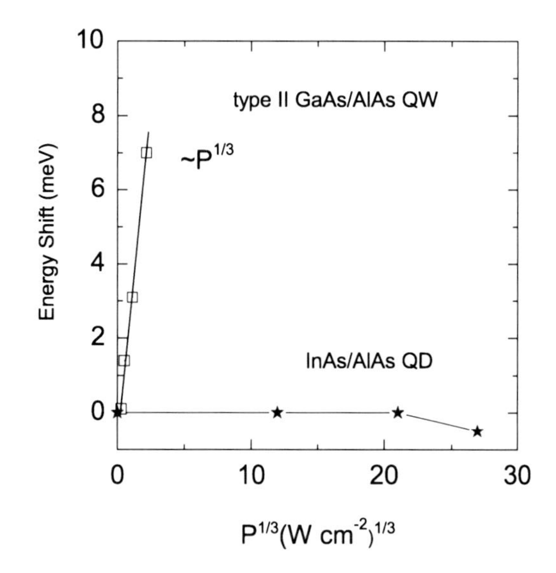

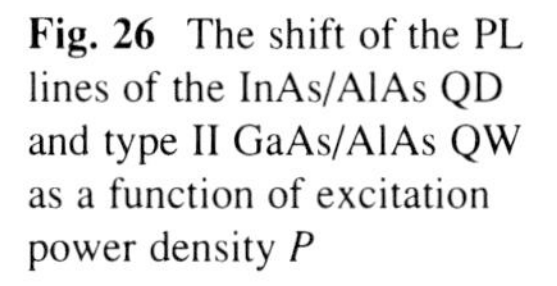

Fig. 26 The shift of the PL lines of the InAs/AlAs QD and type II GaAs/AlAs QW as a function of excitation power density *P*

The dependencies of the PL band shift on the excitation power for the InAs/AlAs QD and type II GaAs/AlAs QW taken from our recent study [57] are shown in Fig. 26. The energy position of the QD_0 band is independent of the excitation power for $P < 9\,kW/cm^2$. A small red shift of the PL band for further increased excitation power is due to an increase in the relative intensity of multiparticle exciton PL bands.[4] On the other hand, the PL of the type II GaAs/AlAs QW exhibits a blue shift proportional to the cubic root of the excitation power, which is in good agreement with predictions [100, 101].

Accordingly, the observed dependence of the PL band shift on the excitation density unambiguously implies a type I alignment for the band structure of the studied InAs/AlAs QD, which is in agreement with our calculations, but in contrast with the conclusion by Sarkar et al. [56]. We believe that the reason for this discrepancy is due to the fact that the indirect minima of the QD conduction band were not taken into account by Sarkar et al. [56]. The continuous spectrum observed in [56] is not due to the type II alignment of the QD, but rather associates with overlapping broad bands of a few QDs with indirect band gap.

It is interesting that the PL spectrum of the InAs/AlAs QD with an indirect band gap exhibits only one (zero-phonon) band, whereas the corresponding PL spectrum of an InAs/AlAs QW with a similar electronic structure contains several phonon replicas.

[4] It was recently demonstrated that type I In(Ga)As/GaAs QDs exhibit a red shift of the PL band with increasing excitation density. This effect was observed for a QDs array [107]. Besides, for a single QD increasing excitation density leads to appearance of additional PL bands red shifted relatively the X_0 excitonic transition [108]. The relative intensity of these bands strongly increases with increasing excitation power density [109]. These bands were explained in terms of the recombination of multiparticle neutral and charged excitons (see, e.g., [110, 111]). In our case, a red shift is due to the appearance of additional low energy PL bands with high FWHM.

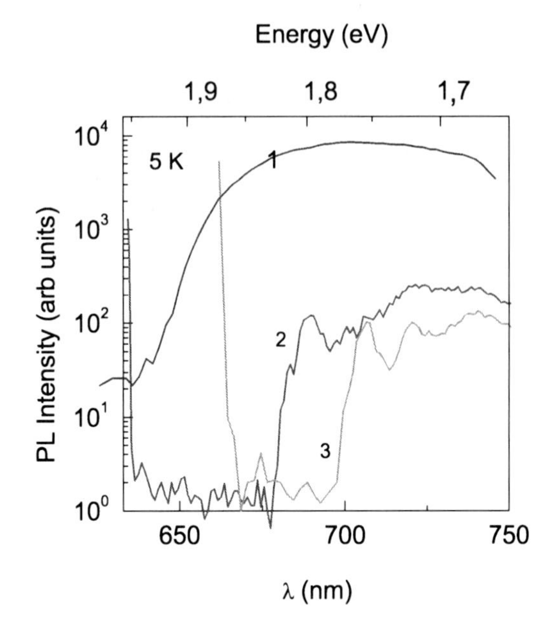

Fig. 27 Low temperature (5 K) PL spectra of InAs/AlAs QDs measured at different energy of excitation light: *1* – 3.81 eV, *2* – 1.96 eV, *3* – 1.87 eV

A single emission band without phonon replicas was recently observed in PL spectra for an indirect band gap single Si QD [106]. That is typical for indirect band gap QDs due to a strong 3D confinement, which breaks down the *k*-conservation rule and results in an increased intensity of the zero-phonon PL band [106, 112] and could explain the predominance of the zero-phonon band in the PL spectra of small InAs QDs.

Steady State PL as a Function of Excitation Energy

The PL of InAs/AlAs QDs as a function of excitation energy (E_{EX}) has been investigated. For excitation with energies above the band gap of the AlAs matrix, the PL band involving the QDs will appear in the spectra of all studied samples. However, decreasing the excitation energy to $E_{EX} < 1.96$ eV divides the samples into two groups: (1) the samples showing high-energy PL bands (1.805, 1.856, and 1.875 eV) for above band gap excitation but do not show any PL for below band gap excitation. (2) The other samples demonstrate PL spectra similar to those depicted in Fig. 27 (for the sample grown at $T_g/t_{GI} = 480°C/10$ s). For decreasing excitation energy, E_{EX}, from 3.81 to 1.96 eV and below for constant excitation power, the PL intensity will decrease by several orders of magnitude. The shape of the PL spectra also changes with the excitation energy. A broad unstructured PL band observed upon excitation above the AlAs band gap changes to a series of relatively sharp lines superimposed on a background band upon excitation below the band gap.

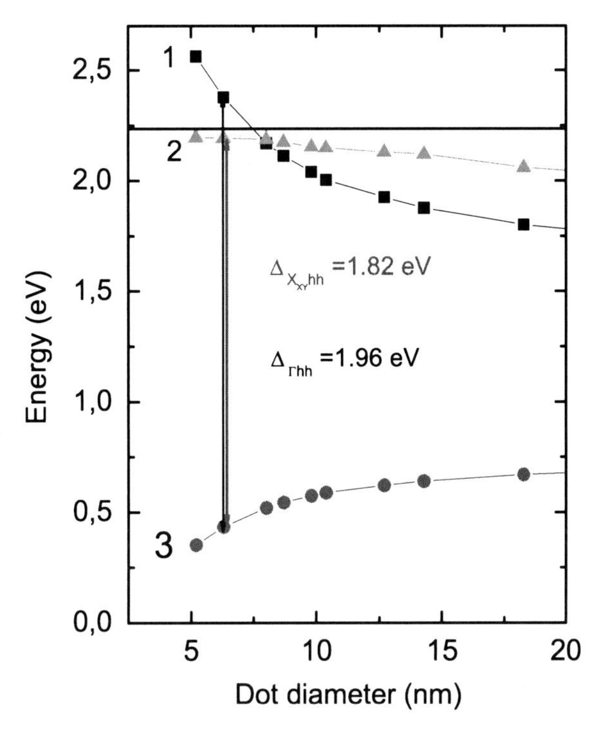

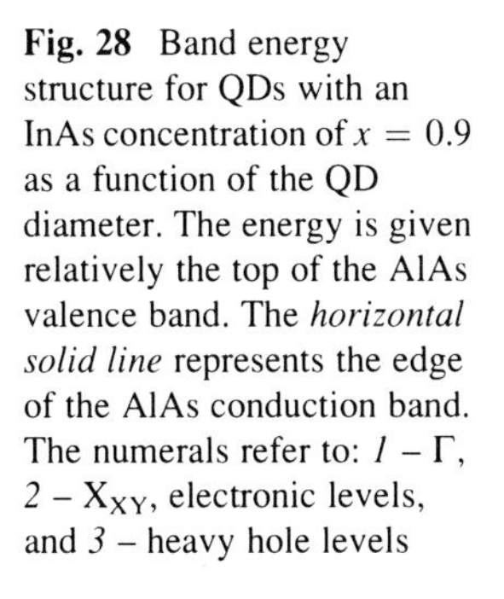
Fig. 28 Band energy structure for QDs with an InAs concentration of $x = 0.9$ as a function of the QD diameter. The energy is given relatively the top of the AlAs valence band. The *horizontal solid line* represents the edge of the AlAs conduction band. The numerals refer to: *1* – Γ, *2* – X_{XY}, electronic levels, and *3* – heavy hole levels

The lines demonstrate a Stokes shift, which decreases with decreasing excitation energy E_{EX}. These facts were observed earlier for InAs/AlAs QDs by Dawson et al. [49]. They demonstrated that these sharp lines appearing in the spectra are due to the resonant absorption in excited levels of the QDs.

In order to explain these experimental results, we construct an energy diagram of the InAlAs/AlAs QD as a function of the dot's diameter. An energy diagram for QDs with different sizes with an InAs fraction $x = 0.9$, i.e., close to the composition of the sample grown at $T_g/t_{GI} = 480°C/10$ s is shown in Fig. 28. We will next consider the generation and recombination of charge carriers in the framework of this diagram. When the excitation energy E_{EX} exceeds the band gap of AlAs, electrons and holes are generated in the matrix. Subsequently, the carriers will become captured and recombine in QDs of all sizes, which result in a broad unstructured band in the PL spectra. On the other hand, for E_{EX} below the band gap of AlAs, carriers can only be excited from QDs levels. Light absorption in thin QDs layer is much smaller than for a thick AlAs matrix, which subsequently results in a strong decrease in the PL intensity for the low-energy excitation. According to selection rules, a significant contribution of the absorption of photons with an energy below the barrier band gap gives rise to transitions between the heavy hole level (E_{hh}^{QD}) and the electronic level at the Γ minimum of the QDs conduction band (E_{Γ}^{QD}). The energy difference between these levels is $\Delta_{\Gamma hh} = E_{\Gamma}^{QD} E_{hh}^{QD}$. Meanwhile for small size QDs, the recombination occurs between the lowest electronic

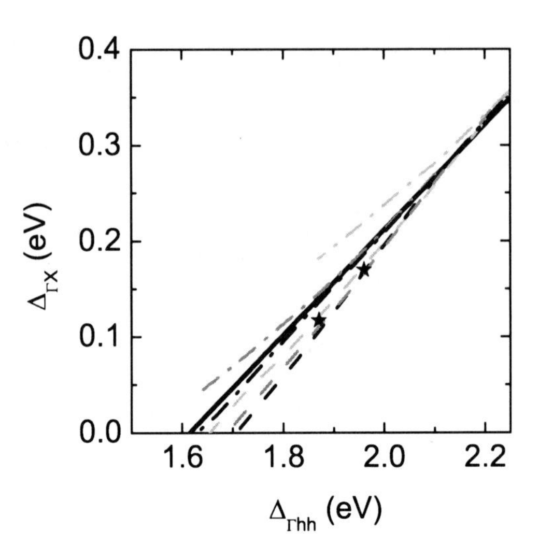

Fig. 29 $\Delta_{\Gamma X}$ versus $\Delta_{\Gamma hh}$ in an array of $In_xAl_{1-x}As/AlAs$ QDs with different sizes. The different InAs fractions x are illustrated by the *dashed lines*: *black* – 1, *grey* – 0.9, *light grey* – 0.8 and the *dash dotted lines: black* – 0.7, *grey* – 0.6, *light grey* – 0.5. The *black solid line* corresponds to experimental data from [46]. The *stars* refer to our experimental data for $In_{0.93}Al_{0.07}As/AlAs$ QDs

level at the X_{XY} minimum of the QDs conduction band (E_{XY}^{QD}) and the heavy hole level. The energy difference between these levels is $\Delta_{Xhh} = E_{XY}^{QD}E_{hh}^{QD}$. Therefore, QDs with $\Delta_{\Gamma hh} > E_{EX}$ do not emit even if E_{EX} exceeds Δ_{Xhh}. For QDs with $\Delta_{Xhh} = E_{EX}$, a resonant excitation can occur, which results in the appearance of a sharp PL line. The Stokes shift of the line equals the difference between E_{Γ}^{QD} and $E_{XY}^{QD}(\Delta_{\Gamma X})$. For large-size QDs, the resonant excitation in the excited states at the Γ minimum of the QDs conduction band gives rise to several sharp lines in the PL spectrum of the QDs array [49].

Figure 29 shows $\Delta_{\Gamma X}$ versus $\Delta_{\Gamma hh}$ for QDs with different diameters and composition, based on our calculations. In a QDs array, $\Delta_{\Gamma X}$ decreases with an increasing QDs diameter due to decreasing $\Delta_{\Gamma hh}$ down to zero at the intersection points of the Γ and X_{XY} minima of the QDs conduction band. For comparison, the curve obtained by Dawson et al. [49] with fitting of their experimental data as well as our experimental data are presented in the same figure. As can be seen, a nice agreement between the experimental results and the predictions based on our calculations is achieved. A best fit to our experimental data is achieved by the curve related to the QDs with an InAs fraction of $x = 0.9$, while the experimental data of Dawson et al. [49] represents an InAs fraction of $x = 0.70$, which reflects the difference in growth conditions for the arrays of QDs.

Spectral and Temperature Dependencies of PL Kinetics

The indirect–direct transition in the band energy structure with increasing QDs size, appears also in the PL kinetics of an InAs/AlAs QDs array as an abrupt acceleration of PL decay at low-energy region of QD PL band [49, 113]. It is interesting that for a fixed temperature, the PL decay demonstrated a monotonic acceleration with an associated red shift of the spectrum (i.e., an increasing QDs size) before slump into the direct band gap QDs [49]. A similar monotonic acceleration of the PL decay observed at the high-energy region of the QD PL band for increasing temperatures [50]. Calculated energy spectra of InAs/AlAs QDs allow us to qualitatively explain the observed monotonic acceleration of the PL kinetics. In accordance with the data on QDs sizes and composition presented in Table 2, most QDs in the studied structures have the lowest electronic level at the X_{XY} minimum and the excited electronic level at the Γ minimum of the conduction band. For small QDs, the radiative lifetime at low temperatures ($kT = \Delta_{\Gamma X}$) is determined by a long lifetime of an electron located in the indirect QD conduction band. For an increasing kT relatively $\Delta_{\Gamma X}$ with increasing QDs diameter or increasing temperature, a redistribution of electrons between the lower and excited states in the QD will occur. Due to an increasing fraction of carriers that will recombine from an electronic level at the Γ minimum of the conduction band, the PL decay will be shortened to the nanosecond region.

Conclusion

The atomic and electronic structure of InAs QDs in an AlAs matrix has been investigated. We have shown that only QDs grown at low temperatures and with short growth interruption times consist of essentially pure InAs, while an increase in any or both of these parameters results in a strong InAs–AlAs intermixing. It has been concluded that the QDs have a type I band alignment, while the lowest electron state of the QDs can be either at the direct Γ or at the indirect X minima of the QD conduction band, depending on the size and composition of the QD. The spectral and temperature dependencies of the PL kinetics of the InAs/AlAs QDs are explained in terms of an electron redistribution between long-lived indirect and short-lived direct states of the QDs conduction band.

4.2 *Coexistence of Direct and Indirect Band Structures in Arrays of InAs/AlAs Quantum Dots*

In spite of important advantages of InAs/AlAs QDs, their energy structure is still studied only scarcely. On the basis of recombination dynamics study, Dawson et al. [49] proposed that these QDs have type II band alignment with electrons localized in the AlAs matrix and holes localized inside the QDs. On the other hand, recently we demonstrated experimentally that a single InAs/AlAs QD and even an InAs/AlAs QW with a nominal thickness of 1.4 monolayers have type I band alignment [57]. Meanwhile, the exact knowledge of band structure is a key moment in device design.

Here, we study the energy structure of InAs/AlAs QDs by stationary and transient photoluminescence and model calculations. We reveal that the QDs with

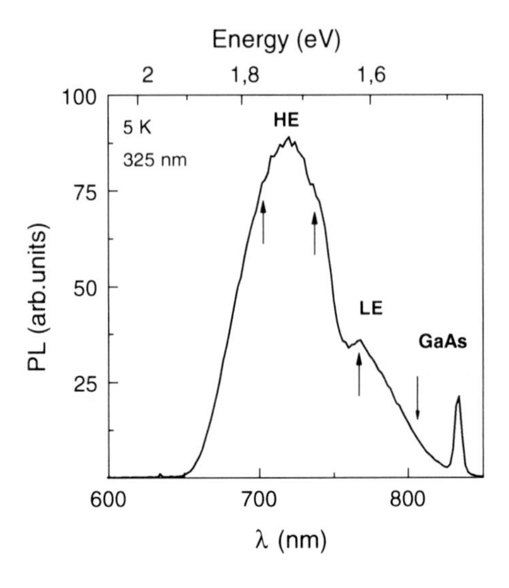

Fig. 30 Low-temperature PL spectra of InAs/AlAs QDs. *Arrows* mark spectral points where PL kinetics presented in Fig. 31 were measured

direct and indirect band gap structures coexist in QD arrays, both having the type I band alignment.

In these samples, the thickness of AlAs buffer layers was 25 nm. The QDs were formed at a temperature of 495°C. The stationary PL was excited by a He–Cd laser ($h\nu = 3.81$ eV) with a power density of 50 W/cm^2. The excitation of transient PL was accomplished by 2 ms pulses of semiconductor laser ($h\nu = 1.87$ eV). The detection of stationary and transient PL was performed by a double diffraction grating spectrometer equipped with a cooled photomultiplier operated in the photon counting mode.

Figure 30 shows a stationary PL spectrum of the InAs/AlAs QDs measured at 5 K. The spectrum contains three bands peaked at 1.51 eV, 1.61 eV, and 1.73 eV marked in the Fig. 1 as GaAs, LE, and HE, respectively. The GaAs band is related to the recombination in the GaAs buffer layer, whereas the LE and HE bands are connected with the recombination in QDs. The LE and HE PL bands were observed by many groups [50, 114–116]; however, the origin of these bands has not been finally established. The interpretation given by Ma et al. [115] connects the high-energy PL band with excited electronic states, which, however, contradicts to the experimental observation of this band at a low excitation density [50] and to the appearance of the low-energy band at a high excitation density [116].

We found that the PL kinetics of these LE and HE bands is considerably different. Actually, one can see in Fig. 31 that the HE band (curves 1 and 2) demonstrates a prolonged decay similar to that observed earlier in [49] and [50], whereas the PL intensity of the LE band first drops quickly (curve 3), the decay time being shorter than the time resolution of our registration system (less than 20 ns), and then demonstrates a prolonged decay. This fast decay appears as a step on the PL kinetics curve, and the amplitude of this step increases with decreasing PL energy.

The calculation of the energy structure of InAs/AlAs QDs was made using a nanodevice simulation tool NEXTNANO3 [87]. To further develop the

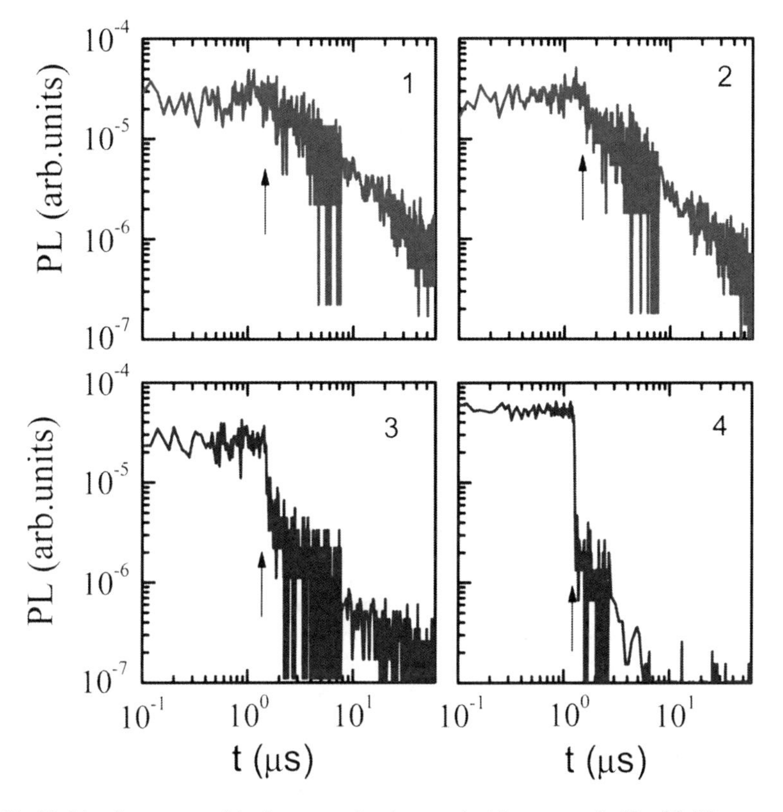

Fig. 31 PL kinetics measured in the spectral points marked by *arrows* in Fig. 30. The *arrows* in this figure designate the end of the laser pulse

understanding of the energy structure of real InAs/AlAs QDs, in this work we take into account: (1) intermixing of InAs and AlAs due to strain-driven segregation of InAs [71, 73], (2) inhomogeneous composition of InAlAs alloy across the QDs [54], and (3) the energy states of the indirect minima in the QD conduction band. The calculation technique described in detail elsewhere [117]. Cross-section TEM micrographs of the InAs/AlAs QDs grown in our group presented previously [89] show that the QDs have lens-shaped structure with an aspect ratio varying from 3:1 to 5:1. The aspect ratio of 4:1 was selected for calculation.

The energy structure for QDs with an alloy composition of $In_{0.7}Al_{0.3}As$ typical for the QDs grown at conditions used in this study [117] is presented in Fig. 32. One can see that, in a good agreement with the results of our recent experimental investigation on a single QD [57], the QDs have the type I band alignment. The lowest electronic level in large QDs belongs to the Γ minimum of the conduction band. A decrease in the QD size results in a direct–indirect transition in band energy structure, with a lowest electronic level switching to the X_{XY} minima of the conduction band. The calculated optical transition energy near the point of the direct–indirect transition in the band energy structure is close to the energy of

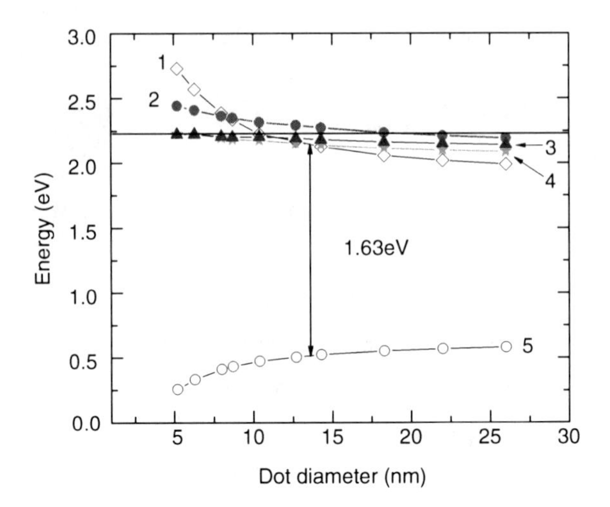

Fig. 32 Band structure of $In_{0.7}Al_{0.3}As/AlAs$ QDs as a function of the QD size. The curves are related to: *1* – Γ, *2* – L, *3* – X_Z, and *4* – X_{XY} electronic levels of conduction band, and *5* – the heavy hole level of valence band. Zero energy corresponds to the top of the AlAs valence band. The *horizontal line* marks the bottom of the AlAs conduction band. The *vertical line* shows the energy of optical transition near the point of the direct-indirect transition in the band energy structure

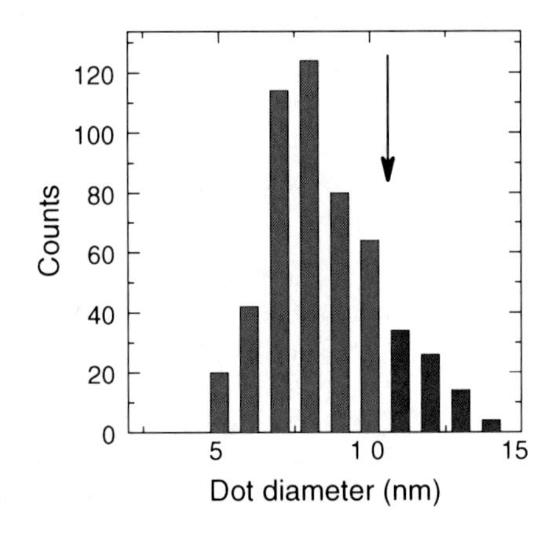

Fig. 33 Dispersion in sizes of InAs/AlAs QDs. The *arrow marks* the QD diameter corresponding to the direct–indirect transition in the band energy structure

the crossover between the LE and HE bands in the PL spectrum. Figure 33 demonstrates the dispersion in QD sizes observed in transmission electron microscopy (TEM) data. We find that the relative amounts of QDs with direct and indirect band gap structures are close to the relative intensities of the HE and LE bands in the PL spectrum. Thus, we conclude that the prolonged and fast PL decays result from recombination in indirect and direct band gap QDs, respectively.

In arrays of QDs, we have dispersion not only of QD sizes but also of shapes and material compositions [117–120]. The calculation demonstrates that the energy of the optical transition near the direct–indirect transition in the energy structure is a

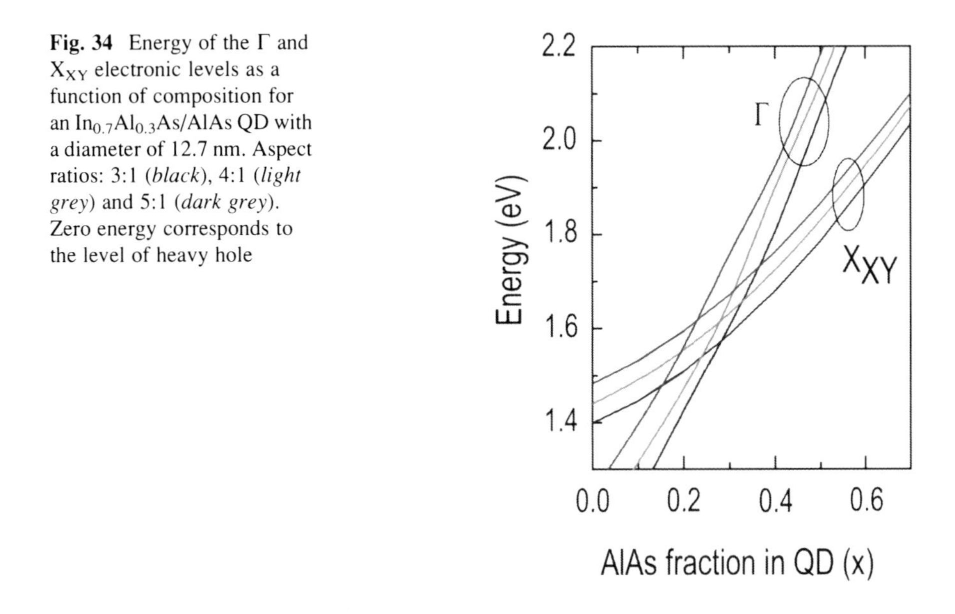

Fig. 34 Energy of the Γ and X_{XY} electronic levels as a function of composition for an $In_{0.7}Al_{0.3}As$/AlAs QD with a diameter of 12.7 nm. Aspect ratios: 3:1 (*black*), 4:1 (*light grey*) and 5:1 (*dark grey*). Zero energy corresponds to the level of heavy hole

function not only of the QD size but also of the QD composition and aspect ratio. Therefore, we have a distribution of QDs with different band energy structures but similar recombination (PL) energies. Figure 34 shows the calculated energy of the Γ and X_{XY} electronic levels for a QD with the diameter of 12.7 nm as a function of composition and QD aspect ratio. One can see that the energy region of the direct–indirect transition is in a good agreement with our experimental data. Consequently, we conclude that the increase of the amplitude of the initial step in the PL decay curves with decreasing PL energy is a result of an increase in the fraction of direct band gap QDs with the QD size increasing.

In conclusion, we demonstrated that the InAs/AlAs QD system has the type I band alignment. In arrays of such QDs fractions with direct as well as indirect band gap structures coexist. Since the QDs radiating around 780 nm are mainly of direct band gap structure, the system of InAs/AlAs QDs with a strong localization of charge carriers offers an improvement in performance for optical reading devices.

4.3 Nonradiative Energy Transfer Between Vertically Coupled Indirect and Direct Band Gap InAs Quantum Dots

The energy transfer (ET) between quantum dots has been intensively studied during the recent years due to their potential significance for fabrication of novel sensors [121, 122] energy of exciton localized in one QD (donor) can nonradiatively transfer to another QD (acceptor) via resonant dipole–dipole Coulomb interaction called Förster process. The efficiency of the ET is determined by the lifetime of the exciton localized in the donor (τ_D) and the donor to acceptor transfer time (τ_{DA}): $E_{ET} = 1 - (\tau_{DA}/\tau_D)$ [123]. For typical direct band gap QDs, the value of τ_D is

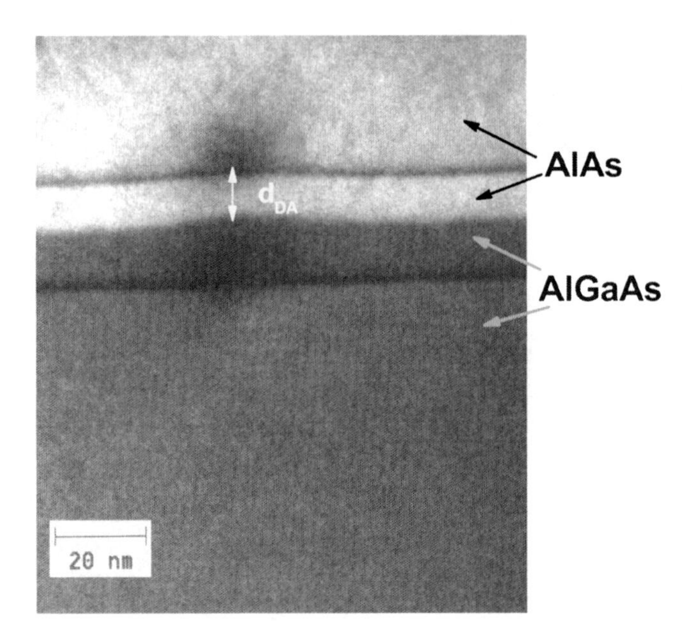

Fig. 35 Cross section TEM image of the structure with coupled QDs

restricted by the radiative recombination of exciton and lies in the nanosecond region. Therefore, efficiency of the ET between the direct band gap QDs is determined by τ_{DA} value, which decreases strongly ($\sim d_{DA}^{-6}$) with an increase in the distance between the donor and acceptor QDs (d_{DA}) [124]. The typical distance for efficient ET between coupled direct band gap QDs, such as colloidal CdS or CdTe QDs, does not exceed 5 nm as it has been demonstrated in [123, 125]. Recently, we demonstrated that strong electronic confinement in small InAs/AlAs QDs leads to the transition of the lowest state of the conduction band from the Γ to X valley within the type I band alignment that results in a strong increase in exciton lifetime in such QDs up to five orders of magnitude above that in direct band gap InAs quantum dots [50, 113]. Below we show that an increase in the exciton lifetime in the donor QDs allows a strong increase in the distance of effective ET between coupled QDs.

The structures with coupled self-assembled InAs QDs studied in this subsection consisted of two vertically coupled layers of InAs QDs sandwiched between the layers of AlAs (indirect band gap donor QDs) and $Al_{0.36}Ga_{0.64}As$ (direct band gap acceptor QDs).Two structures with $d_{DA} = 9$ and 13 nm were grown. The QDs were formed at a temperature 510°C. Typical cross-section image for the grown structure, obtained by means of transmission electron microscopy with using a JEM-4000EX operated at 350 keV is presented in Fig. 35. One can see vertical coupling of the QDs in the lower and upper layer of the structure. Test structures containing the only one layer of donor-like or acceptor-like QDs were also grown in the same manner.

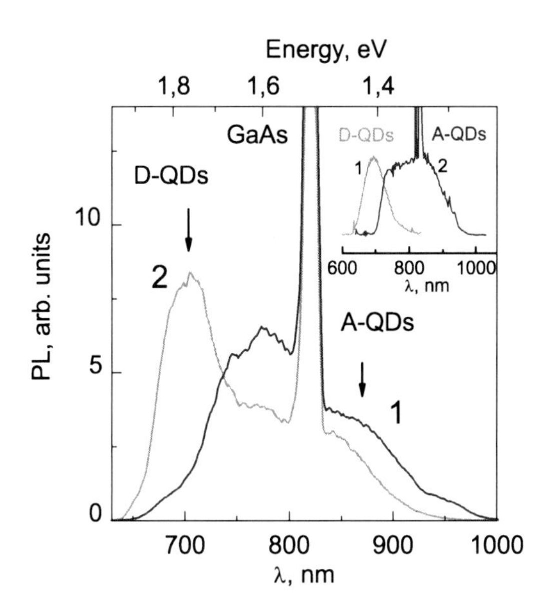

Fig. 36 Low-temperature PL spectra of the structures with coupled QDs, d_{DA}: *1* – 9 nm, *2* – 13 nm. *Arrows* mark spectral points where PL kinetics presented in Fig. 37 were measured. The *inset* contains PL spectra of the structures with single layer of donor-like (1) and acceptor-like (2) QDs

Steady-state PL was excited with an ultraviolet He–Cd laser ($h\nu = 3.81$ eV) with the power density (P) of 10 W/cm^2. The excitation of transient PL was accomplished by a pulsed red semiconductor ($h\nu = 1.96$ eV) laser. The detection of steady state and transient PL was performed using a double diffraction grating spectrometer equipped with a cooled photomultiplier operated in the photon counting mode. All measurements of the PL were made at 77 K.

Figure 36 demonstrates the PL spectra of the structures with coupled QDs measured under ultraviolet excitation, which is mainly absorbed in AlAs layer. The spectra contain three PL bands marked in the figure as D-QDs, A-QDs, and GaAs. A comparison of the spectra of the structures with coupled QDs with the spectra of the structures with single layer of donor-like or acceptor-like QDs presented in the inset to Fig. 36 PL allows us to identify these bands as related to recombination in donor and acceptor layers of the QDs, and GaAs buffer layer, respectively. One can see that the D-QDs band dominates in the spectrum of the structure with larger ($d_{DA} = 13$ nm) distance between QDs. In the structure with smaller d_{DA} value the PL intensity of the D-QDs band strongly decreases, while the intensity of the AQDs band increases. The results suggest nonradiative ET from indirect band gap donor QDs to direct band gap acceptor QDs in the structure with smaller d_{DA} value.

Changes in the steady state PL spectra of the structures with coupled QDs with increased d_{DA} value are accompanied by changes in transient PL of these structures. The PL kinetics of the A-QDs and D-QDs bands for both structures measured under excitation with the red laser are shown in Fig. 37, and in the inset to Fig. 37, respectively. The data are presented in double logarithmic coordinates, which are convenient to present a nonexponential law of the QD band decay [50]. Under such condition of excitation, the only intra-dot excitation in both the donor and acceptor

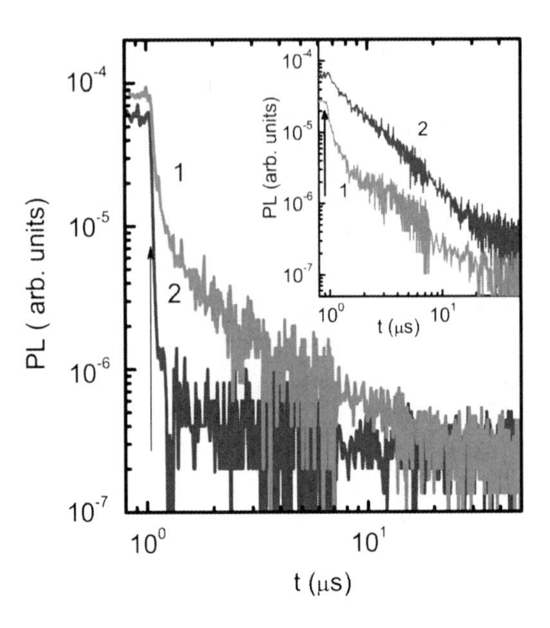

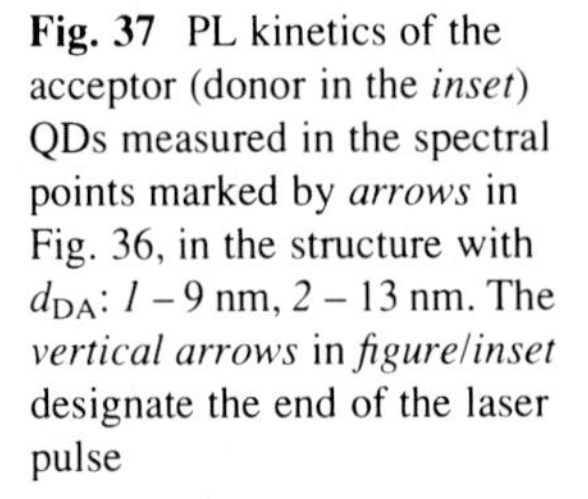
Fig. 37 PL kinetics of the acceptor (donor in the *inset*) QDs measured in the spectral points marked by *arrows* in Fig. 36, in the structure with d_{DA}: *1* – 9 nm, *2* – 13 nm. The *vertical arrows* in *figure*/*inset* designate the end of the laser pulse

QDs occurs. One can see that PL of the D-QDs band in both structures demonstrates long (up to 50 μs) decay duration, which is typical for indirect band gap InAs/AlAs QDs [117]. However, the several microseconds initial period of the PL decay in the structure with smaller d_{DA} is appreciably faster. The PL kinetics of the A-QDs band also strongly depends on the value of d_{DA}. In the structure with the larger d_{DA}, the PL of the A-QDs band decays faster than 20 ns (time resolution of our setup) which is typical for direct band gap InAs QDs [113], while in the structure with the smaller d_{DA} the PL intensity of the direct band gap acceptor QDs first drops quickly and then demonstrates a prolonged decay similar to that in the indirect band gap donor QDs. The fast initial PL decay of InAs/AlGaAs QDs in the structure with the smaller d_{DA} is explained by direct excitation of these QDs with red light, while the prolonged decay reflects nonradiative ET from long-lived excitons of InAs/AlAs QDs. The structure with larger d_{DA} value InAs/AlGaAs QDs demonstrates only the fast PL decay; therefore, the excitation does not transfer from the InAs/AlAs QDs to the InAs/AlGaAs QDs. Thus, transient PL confirms our assumption about excitation transfer from indirect band gap D-QDs to direct band gap A-QDs in the structure with smaller d_{DA} value.

Let us discuss the possible mechanisms for nonradiative ET between coupled QDs. It can be the tunneling of charge carriers or the long-range Förster transfer of the exciton energy. It is known that the carrier tunneling time strongly depends on the height and width of the barrier and the effective mass of the carrier. Using the Wentzel–Kramers–Brillouin approximation [126] and taking into account the effective masses of the electron and heavy hole in the QDs, we find that 9 nm is too long distance in order that electron (hole) tunnel between the QDs within even during hundred microseconds, which is an overestimation of the transfer time from the PL decay data. Thus we conclude that ET can occur via Förster process. It is known that Förster energy transfer takes place under the resonant conditions when the energy levels of exciton in donor and acceptor QDs coincide [124]. Difference in electronic confinement for InAs/AlAs and InAs/AlGaAs QDs in our structures results in significant difference in the excitonic levels in the donor and acceptor QDs. However, as it was calculated in [127] efficient ET occurs for resonant conditions between the ground state of exciton in donor QDs and the excited state of exciton in acceptor QDs. The data presented in Fig. 36 provide evidence that the coincidence between the ground states in donor QDs and the excited states in acceptor QDs occurs in our structure for wide range of the QDs sizes.

In conclusion, the energy transfer between indirect band gap InAs/AlAs donor QDs and direct band gap acceptor InAs/AlGaAs QDs was studied by means of photoluminescence. It was demonstrated that the long excitonic lifetime in the donor QDs allows increased distance for efficient nonradiative energy transfer between QDs up to 9 nm.

4.4 Carrier Dynamics in InAs/AlAs Quantum Dots: Lack in Carrier Transfer from Wetting Layer to Quantum Dots

In this part structures with self-assembled InAs quantum dots (QDs) embedded in an AlAs matrix have been studied by steady state and transient photoluminescence, and calculations. It has been shown that in contrast to InAs/GaAs QDs system carriers are mainly captured by quantum dots directly from the AlAs matrix, while transfer of carriers captured by the wetting layer far away from QDs to the QDs is suppressed. At low temperatures, the carriers captured by the wetting layer are localized by potential fluctuations at the wetting layer interface, while at high temperatures the carriers are delocalized but captured by non-radiative centers located in the wetting layer.

4.4.1 Introduction

To design optoelectronic devices study of relaxation and recombination of non-equilibrium carriers in structures with such QDs is required [128]. Recently special attention was attracted to the investigation into the process of carriers collection in QDs [129].

In an array of QDs grown in the Stranski–Krastanov mode, the states of the wetting layer (WL) lying under the QDs play an important part in carrier capture. The wetting layer as a thin quantum well (QW) acts as a strong attractive potential for electrons and holes [57, 130, 131]. For typical sizes and density, QDs cover an area smaller than WL [75, 117]; as a result, most of the carriers are captured from the matrix in the WL. Therefore, investigation of the transfer of carriers captured in the wetting layer to the QDs attracts particular attention. Carriers captured in the WL located in the vicinity of QDs then are quickly captured by the QDs. Several mechanisms are proposed to explain the effective capture of such carriers in QDs: multiphonon emission [131, 132], Auger carrier–carrier scattering [131–134], and relaxation of carriers via continuum tail of the energy states extended down to the ground states of QDs [135–138]. Carriers captured in the wetting layer far away from QDs can recombine here or move along its plane before capture in the QDs [139].

To date, many investigations demonstrated for the structures with In(Ga)As QDs in a GaAs matrix that the carriers are efficiently captured in the QDs by both ways: (1) directly from the matrix and (2) via the WL [133, 140–146]. However, carriers capture in InAs QDs embedded in an AlAs matrix, which are very similar to the InAs/GaAs QDs from the point of view of the Stranski–Krastanov growth mode due to very close lattice constant values of AlAs and GaAs, has not been studied yet. Meanwhile, the energy structure of the InAs/AlAs QDs is quite different from that of the In(Ga)As/GaAs QDs. Actually, stronger electronic confinement in the InAs/AlAs QDs leads to the transition of the lowest state of the conduction band from Γ to X valley within type I band alignment [113, 117], and to significant change in their electronic and optical properties as compared with the In(Ga)As/GaAs QDs system [48, 50]. Unfortunately, peculiarities of the energy spectrum of structures with the InAs/AlAs QDs do not allow us to study the transfer of carriers from the WL to QDs via direct carrier generation in the WL at resonance excitation just as it was made for In(Ga)As/GaAs QDs [131, 144]. It is because effective light absorption in the WL takes place at an energy related to transitions between states of heavy holes and electrons at the direct Γ minimum of the conduction band, which is larger than the band gap of the AlAs matrix [130].

In this part, we present the results of investigation into carrier transfer from the WL to QDs in type I InAs/AlAs QDs structures with the lowest electronic level at the X minimum of the conduction band by steady state and transient photoluminescence under excitation above the band gap of the AlAs matrix and calculation of the potential relief around the QDs. It was demonstrated that the major part of carriers captured by the wetting layer do not get transferred into QDs.

4.4.2 Experimental Details

The QDs were formed by standard growth procedure (see refsec:4) at temperatures varied within the range 520–530°C. The density (9×10^9 and 9×10^{10} cm^{-2}) and average diameter (14.3 and 13.5 nm) of the QDs were determined using transmission electron microscopy [117]. All the InAs/AlAs structures have similar energy spectra with lowest electronic level of QDs and WL belonging to the X minimum of the conduction band [57, 113, 117]. For comparison, the structures of InAs/GaAs QDs consisting of one InAs QDs layer deepened to 50 nm from the surface with the QD density of 3×10^9 and 2×10^{10} cm^{-2} were also grown.

In the case of InAs/AlAs QDs steady state PL was excited with a He–Cd laser ($h\nu = 3.81$ eV) with the power density (P) of 0.1–25 W/cm^2 adjusted with neutral filters. The excitation of transient PL in the microsecond range was accomplished by a pulsed N_2 laser ($h\nu = 3.68$ eV) with the pulse duration of 10 ns and power density of 10^4 W/cm^2, or 0.1 mJ/cm^2 density of the pulse energy. The detection of steady state and transient PL in the microsecond range was performed using a double diffraction grating spectrometer equipped with a cooled photomultiplier operated in the photon counting mode. In the case of InAs/GaAs QDs PL was excited by a He–Ne laser ($h\nu = 1.96$ eV) with the power density of 15 W/cm^2, analyzed by the double diffraction grating spectrometer and detected with a Ge detector (Edinburgh Instrument) cooled with liquid nitrogen.

4.4.3 Experimental Results

Non-equilibrium carriers generated during light absorption in an AlAs matrix can then recombine in the matrix, to be captured in the wetting layer and QDs, recombine in the wetting layer or move along the wetting layer and get transferred from the wetting layer to the QDs, and finally they recombine in the QDs. In order to estimate which fraction of the carriers captured by the WL then get transferred from the wetting layer to QDs, we analyze steady-state and transient PL in the structures with InAs/AlAs QDs and compare these results with similar experimental data for InAs/GaAs QDs structures, which as it is well known from literature demonstrate a high rate of carriers transfer from the wetting layer to QDs [147].

Low-Temperature Steady State Photoluminescence of InAs/AlAs and InAs/GaAs QDs

Low-temperature steady state PL spectra of the structures with different density of InAs/GaAs and InAs/AlAs QDs are presented in Fig. 38a, b, respectively. All the spectra contain PL bands related to recombination in the wetting layer and quantum dots marked as WL and QD, respectively. Both the structures with InAs/GaAs QDs demonstrate PL spectra with similar integrated intensity of the QD band, which is much larger than that of the WL band. The QD band in the spectrum of the structures InAs/GaAs with the lesser density of the QDs has a shoulder related to recombination from excited state of the QDs. In the structures with InAs/AlAs QDs, the WL band is a sum of strongly overlapped bands, which relate to no-phonon transition and its TOA_{ALAS}, LO_{InAS} and LO_{ALAS} phonon replicas [57]. The integrated intensity of the WL and QD bands in these structures strongly depends on the QD density. In the structure with a low QDs density (9×10^9 cm^{-2}), the WL band dominates, while the intensity of the QD band is several times smaller. In the structure with a high QDs density (9×10^{10} cm^{-2}), the QD band dominates. Nevertheless, the relative intensity of the WL band in this structure is much higher than that for the InAs/GaAs QDs structure having two orders of magnitude lower QD density. The integrated intensity of the QD band in the structure with the high QDs density is 8.2 times larger than that in the structure with the low QDs density.

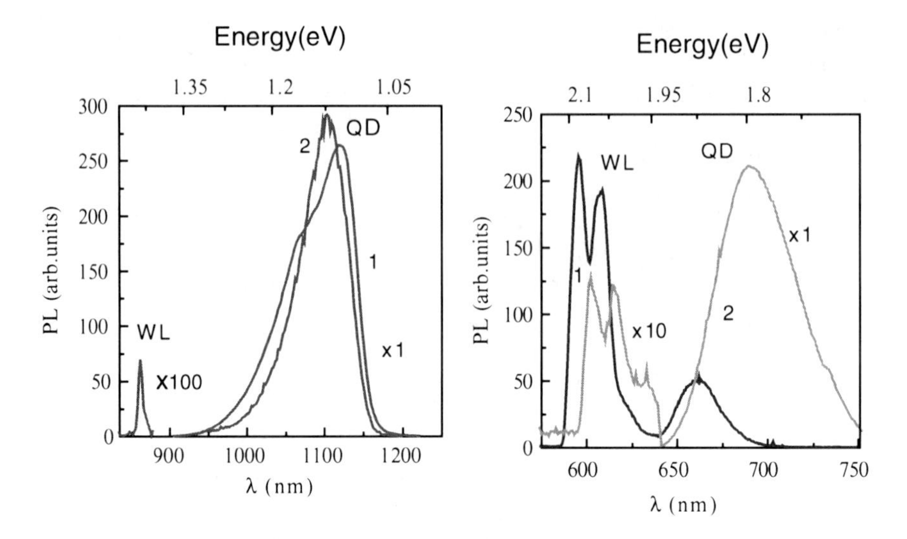

Fig. 38 Low-temperature (5 K) PL spectra of structures with (**a**) InAs/GaAs QDs with density of (*1*) 3×10^9 cm^{-2} and (*2*) 2×10^{10} cm^{-2} measured at the excitation power density of 15 W/cm^2; (**b**) InAs/AlAs QDs with densities of (*1*) 9×10^9 cm^{-2} and (*2*) 9×10^{10} cm^{-2} measured at the excitation power density of 25 W/cm^2

The number of carriers captured by QDs (or wetting layer) per unit time is proportional to the product of the area occupied by the QDs (wetting layer) and the probability of carrier capture per unit area; the latter is assumed similar for both WL and QDs.[5] Using transmission electron microscopy plane view images [117], we estimate that share of the area occupied by QDs is less than 2% for both the InAs/GaAs QDs structures, and equals to ~15% and ~14% for the structures with the low and the high InAs/AlAs QDs density, respectively. Therefore, in all the QDs in any studied structure, the number of carriers captured per unit time directly from the matrix is much smaller than that in the wetting layer. Absolute domination of the QD band in the PL spectra of both the InAs/GaAs QDs structures indicates that carriers captured in the wetting layer leave this layer via any channels rather than recombine with photon emission here. This is a well-known result of the transfer of carries captured in the wetting layer to the QDs [147]. Invariability of the integral intensity of the QD band with increase in the QDs density in order of magnitude evidences that majority of the carriers captured in the wetting layer get transferred to the QDs and nonradiative recombination in the wetting layer of the structures with InAs/GaAs QDs is negligible. The recombination from excited state of QDs in the structures InAs/GaAs with the lesser density of QDs evidences larger filling of

[5] Rise time of PL is similar in structure with a separate thin QW and in structure with QDs (In(Ga)As/GaAs, InAs/InP, and GaAs/AlGaAs) equals to several tens of picosecond (see [148–151]). Therefore, one can assume similar probability of carriers capture from matrix to QW and QDs.

the QDs by carriers in this structure. For both the InAs/AlAs QDs structures, the ratio of the WL to the QD band intensity is also much smaller than the ratio of the areas occupied by the WL and the QDs. Therefore, the majority of carriers captured in the wetting layer leaves this layer rather than recombines via photon emission. At the same time, we showed recently that the carriers captured in the InAs/AlAs QDs recombine solely via radiative recombination.[6] Contrariwise to the case of InAs/GaAs QDs increase in the area occupied by the InAs/AlAs QDs (about 9 times) results in a proportional increase (8.2 times) in the integrated intensity of the QD band. This fact allows us to suggest that a portion of carriers captured in the wetting layer do not get transferred from the WL to the QDs, but leaving the WL via another channel such as nonradiative recombination in the WL or escape to the matrix. Recently we determined that binding energies of electron and hole in the InAs/AlAs wetting layer lie within the range of several tens meV [117], therefore at the low temperature used here we can exclude escape of the carriers from the WL in the matrix.

Carrier transfer from WL to QDs manifests in the dependence of the intensity of WL and QD PL bands on the excitation power density. It is well known that in the structures with InAs/GaAs QDs the ratio of the WL to the QD band integrated intensity is linear at low excitation powers and one starts increasing with an increase in excitation power as a result of filling the QDs with carriers, which prevents further carriers transfer from the wetting layer to QDs. The effect of the filling is especially appreciable for structures with a small QDs density. It was demonstrated that in the InAs/GaAs QDs structure with a low dot density 5×10^9 cm^{-2}, an increase in the relative intensity of the WL band starts at a power density above 100 W/cm^2 [152].

Lifetimes of indirect (in the momentum space) excitons in InAs/AlAs QDs are four to five orders of magnitude longer than that of direct excitons in InAs/GaAs QDs [50, 117]. Therefore, filling of the QDs by carriers is expected in the structures with InAs/AlAs QDs at appreciably smaller power densities than that in InAs/GaAs structures with similar QDs density. Hence, we can expect an increase in the relative intensity of the WL band in the structure with the low InAs/AlAs QDs density at very low excitation densities. However, dependences of the WL and QD band intensities on excitation power density presented for the low density InAs/AlAs structure in Fig. 39 demonstrate an opposite behavior. We fitted the experimental data by power functions and found the best agreement for power indexes equal to 0.77 ± 0.07 and 1.05 ± 0.07 for the WL and QD bands, respectively. In other words, the relative intensity of the WL band decreases with an increase in excitation density. In addition, since the intensity of the QD band is only several percents of the WL band intensity, the total PL intensity of the structure increases sublinearly with an increase in the excitation power. This means that a portion of the

[6] Longtime PL decay in the InAs/AlAs QDs observed in this investigation demonstrates that carriers captured in QDs recombine via photon emission only. Actually, we demonstrated recently in the [53] that even a small part (~5%) of QDs with nonradiative centers decreases the duration of QDs PL decay down to about 1 μs due to the long-range transfer of the exciton energy to the nonradiative centers.

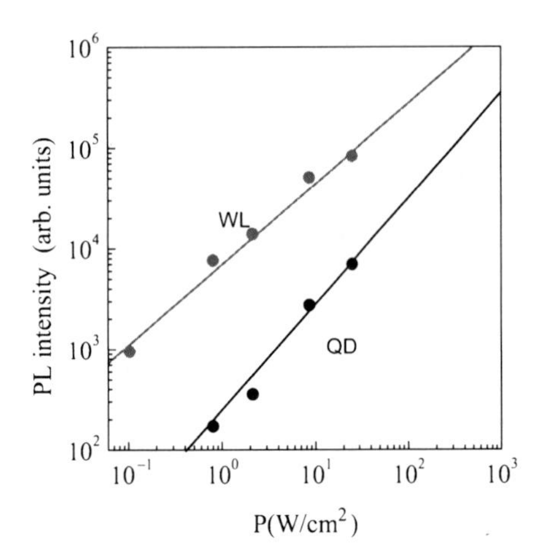

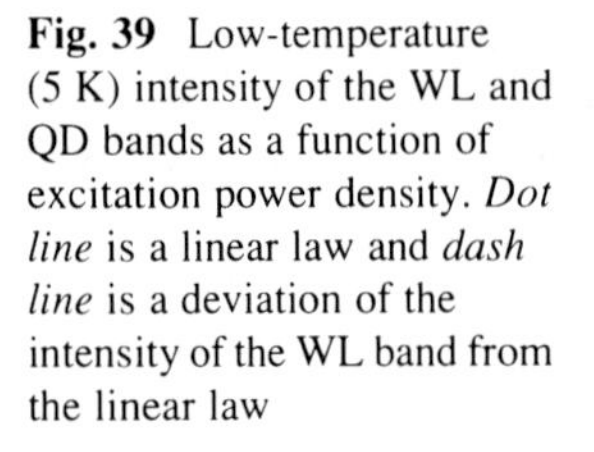

Fig. 39 Low-temperature (5 K) intensity of the WL and QD bands as a function of excitation power density. *Dot line* is a linear law and *dash line* is a deviation of the intensity of the WL band from the linear law

carriers recombined in the wetting layer as well as in the whole structure via photon emission strongly decreases with an increase in the excitation power. As in the InAs/AlAs QDs, carriers recombine via radiative recombination only (see footnote 6), a close-to-linear dependence of the QD band intensity on the excitation power indicates that the portion of the carriers captured in the QDs does not change with variation of the excitation power. A simple estimation demonstrates that the number of carriers leaving the wetting layer nonradiatively (which is a deviation of the intensity of the WL band from linear law presented in Fig. 39) exceeds a total amount of the carriers recombined in the QDs in order of magnitude. Even if we suppose that rather small superlinearity (within an error of the measurement) in the dependence of the QD band intensity on the excitation power is a result of carrier transfer from WL to QDs, the same estimation restricts the maximal portion of carriers coming from the WL to QDs as less than 2.5% of the total amount of carriers leaving the wetting layer nonradiatively. It follows from these data that most of the carriers leaving the wetting layer are not transferred into the QDs.

Thus, a comparison of the steady state PL in the structures with InAs/GaAs and InAs/AlAs QDs reveals that (1) in the structures of both types, only a small portion of carriers captured in the WL recombine here via photon emission, (2) moreover, in the InAs/AlAs QD structures this portion in much larger than that in the InAs/GaAs QD structures with similar QDs density; (3) in the structures with the InAs/AlAs QDs, in contrast to that with InAs/GaAs QDs, the major part of the carriers leaving the wetting layer are not transferred in the QDs but vanish via nonradiative recombination.

We propose the following model to explain the peculiarities of the carrier behavior in InAs/AlAs QDs structures. Heavy electrons in the X minimum of the conduction band are captured by potential minima arising due to fluctuations of the WL width [153]. These localized carriers cannot leave the wetting layer and recombine here via photon emission, which leads to relatively high PL intensity of the WL band at low excitation powers. With an increase in the excitation power, the mobile states of the wetting layer begin to get populated. The mobile carriers diffuse to the defects that act as nonradiative centers. We assume that the nature of these nonradiative centers in the wetting layer is the same as in GaAs/AlAs superlattices [154, 155] and is connected with the high chemical activity of Al atoms [156]. The rate of carrier capture in nonradiative centers is much higher than that in QDs, so that the carriers are rather captured by these centers than transferred in QDs. Due to filling of localized states, the fraction of the mobile carries increases with an increase in the excitation power, which results in the sublinear dependence of the WL band intensity.

Hence, the structures with InAs/AlAs QDs differ from those of InAs/GaAs QDs in the following way: (1) although the atomic structures of InAs/GaAs and InAs/AlAs WL are similar [61], holes and electrons in the Γ minimum of conduction band with small effective mass are not localized in InAs/GaAs WL; (2) the InAs/GaAs WL does not contain nonradiative centers connected with the high chemical activity of Al atoms.

In order to test the model, in the next sections we study the carriers recombination in InAs/AlAs QDs by transient and steady state PL at different temperatures and briefly discuss possible reasons promoting capture of mobile carriers by the nonradiative centers.

Transient and Steady State Photoluminescence of InAs/AlAs QDs at Different Temperatures

PL decay of the WL and QD bands for the InAs/AlAs structure with the low QDs density is shown in Fig. 40. The data presents in double logarithmic coordinates, which are convenient for presentation of a nonexponential law of the QD band decay [50]. One can see that the intensities of both the PL bands demonstrate prolonged decay resulting from indirect structure of the conduction band of QDs and WL. The intensity of the QD band decreases by more than three orders of magnitude for 70 μs (see inset to Fig. 40), while the intensity of the WL band decreases less than threefold during the same time interval.[7] This longer duration of the WL band decay accords with our statement about localization of some fraction of the carries within this layer, which suppresses carrier transport (1) to nonradiative centers and (2) from the wetting layer to the QDs.

In order to delocalize the carries in mobile states, we increase temperature. Temperature dependences of the transient and steady state PL in the structure

[7] The large difference in the prolonged recombination decay of localized carriers in the wetting layer and QDs is explained by k-conservation rule. Both the QDs and wetting layer have similar energy structure with the lowest electronic level at the X_{XY} minimum of the conduction band (see [57, 117]). The k-conservation rule is applied for weakly localized carriers in the wetting layer, where the carriers recombine mainly via scattering on phonons (see [57]); however, this rule is broken in QDs, where, as we have shown in [117], faster no-phonon recombination dominates.

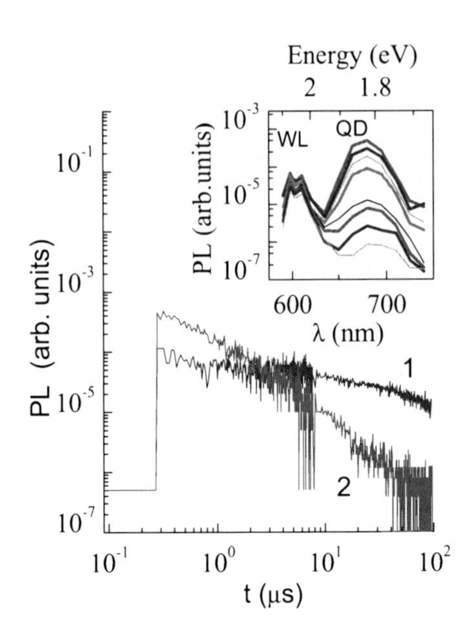

Fig. 40 Microsecond-range PL dynamics of the (*1*) WL and (*2*) QD bands in the structure with the low density of InAs/AlAs QDs. *Inset* shows transient PL spectra of the structure. From *top to bottom* PL integrated in time windows after the end of excitation pulse (μs): 0.4 + 0.2; 0.6 + 0.3; 1 + 0.5; 2 + 0.5; 3 + 1; 5 + 2; 8 + 3; 10 + 5; 25 + 5; 40 + 10; 70 + 10

with the low density of InAs/AlAs QDs are depicted in Figs. 41 and 42, respectively. Comparing the kinetics obtained at different temperatures, one can see that an increase in temperature results in a steep acceleration of the WL band kinetics, while the decay of the QD band (shown in inset to Fig. 41) remains constant within this temperature range. In the steady state spectra, the intensity of the WL band also strongly decreases with temperature rise, with the activation energy of $E_a = 8 \pm 1$ meV. The activation energy of the WL band quenching is much smaller than the electron or hole localization energies in the InAs/AlAs wetting layer [57]. While the integrated intensity of the WL band decreases from a value of order of magnitude larger to a value of order of magnitude smaller than the integrated intensity of the QD band, the latter is practically constant up to 110 K as it is shown in the inset to Fig. 42. Further temperature rise leads to a decrease in the integrated intensity of the QD band with the activation energy of 120 ± 20 meV due to escape of carriers from the QDs in the matrix.[8]

Acceleration of the WL band kinetics, together with a strong decrease in the intensity of this PL band, as well as constant kinetics and very small change in the intensity of the QD band at temperature rise confirm that, in good agreement with our model: (1) with temperature rise, carriers in the wetting layer actually delocalize into the mobile states, and (2) the mobile carriers leaving the wetting layer mainly via nonradiative recombination do not get transferred from the wetting layer to the QDs.

[8] According to the calculation of the InAs/AlAs QDs energy structure made in our recent study [117], the energy of electron localization in an array of QDs varies from 70 to 150 meV. The latter value is close to the activation energy of the QD band quenching.

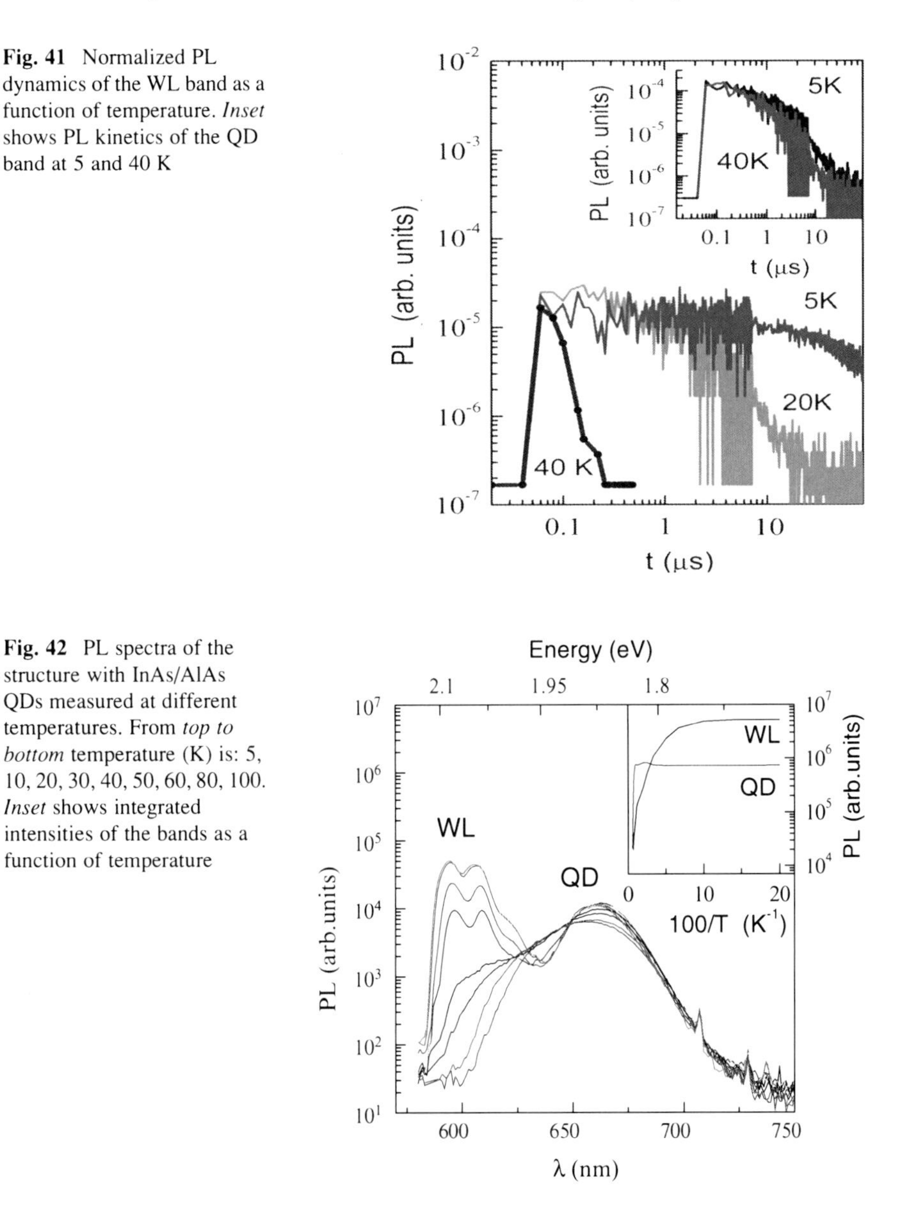

Fig. 41 Normalized PL dynamics of the WL band as a function of temperature. *Inset* shows PL kinetics of the QD band at 5 and 40 K

Fig. 42 PL spectra of the structure with InAs/AlAs QDs measured at different temperatures. From *top to bottom* temperature (K) is: 5, 10, 20, 30, 40, 50, 60, 80, 100. *Inset* shows integrated intensities of the bands as a function of temperature

Indeed, intensity of a PL band is proportional to number of carriers recombined with photon emission. At the same time, disappearance of a large fraction of the carries (in order of magnitude larger than the total amount of carriers recombine in the QDs) from the WL with temperature increase does not influence on intensity of the QDs band. The invariable decay law and the intensity of the QD band up to the temperature of 110 K also confirm the absence of nonradiative recombination in the QDs. In addition, we can conclude that the QDs mainly capture carriers directly from the matrix (and from WL in the nearest vicinity of QDs) and the capture rate does not depend on temperature at least up to 110 K.

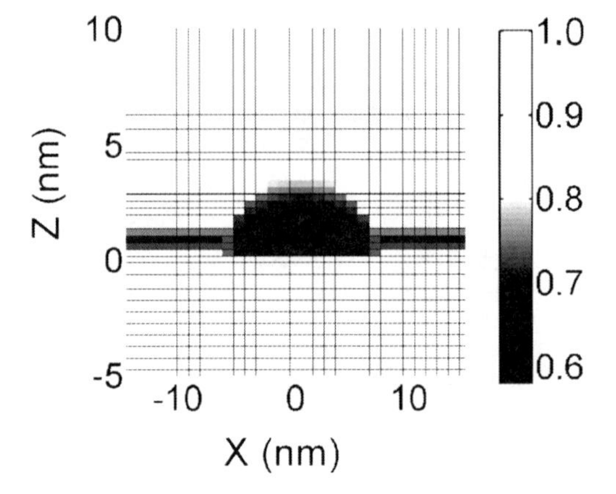

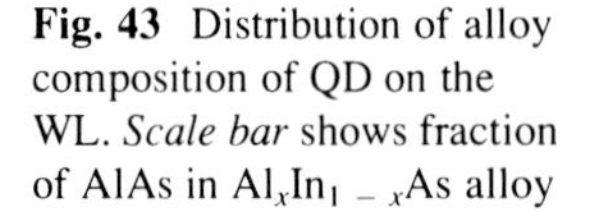
Fig. 43 Distribution of alloy composition of QD on the WL. *Scale bar* shows fraction of AlAs in $Al_xIn_{1-x}As$ alloy

4.4.4 Calculation of Potential Relief Around InAs/AlAs QD

The higher rate of carrier capture in nonradiative centers in comparison with that in QDs can be a result of either high density of the centers, which can be much higher than the QDs density, or peculiarity of their structure, which gives rise to a large value of the center capture cross section. Here, we cannot distinguish between these cases. It is possible, however, that a small rate of carrier capture in QDs is due to a barrier at the interface of the wetting layer and the QDs. Such a barrier would decrease the probability for the mobile carriers to be captured in QDs and increase the residence time of the carriers in the WL where they recombine mainly via nonradiative centers. In this section, in order to examine if the distribution of strain around the QD results in the appearance of a barrier between the wetting layer and QDs we calculate the potential relief around an InAs/AlAs QD.

The potential relief around a QD has been calculated using the nanodevice simulation tool NEXTNANO3 [87]. The relief is determined by size, shape, and chemical composition of the QD and the wetting layer, as well as by the parameters of InAs and AlAs. The atomic structure of InAs/AlAs QDs was determined in our recent paper [117]. An average QD in the structure has a lens shape with the diameter of 14.3 nm and the aspect ratio of 4:1. Matrix and QD materials intermix strongly, so $Al_xIn_{1-x}As$ alloy composition changes inside QD according to a two-dimensional Gauss function with the maximum AlAs concentration of 0.58 determined early for the QDs in the structure with the low QDs density [117]. Such QD was put in a wetting layer selected as an InAs/AlAs QW with chemical composition profile described by Muraki model [57, 65]. Parameters of the InAs and AlAs used in the calculation were taken from [57] and [117]. The resulting distribution of alloy composition for the QD located on the wetting layer is demonstrated in Fig. 43. The potential relief around the InAs/AlAs QD was calculated in the two minima of the conduction band: (1) Γ minimum, where photo-excited carriers are generated, and (2) X_{XY} minimum, which is the lowest electronic state for the matrix, wetting layer, and QD. The results of calculations are shown in Figs. 44 and 45 for Γ and X_{XY}

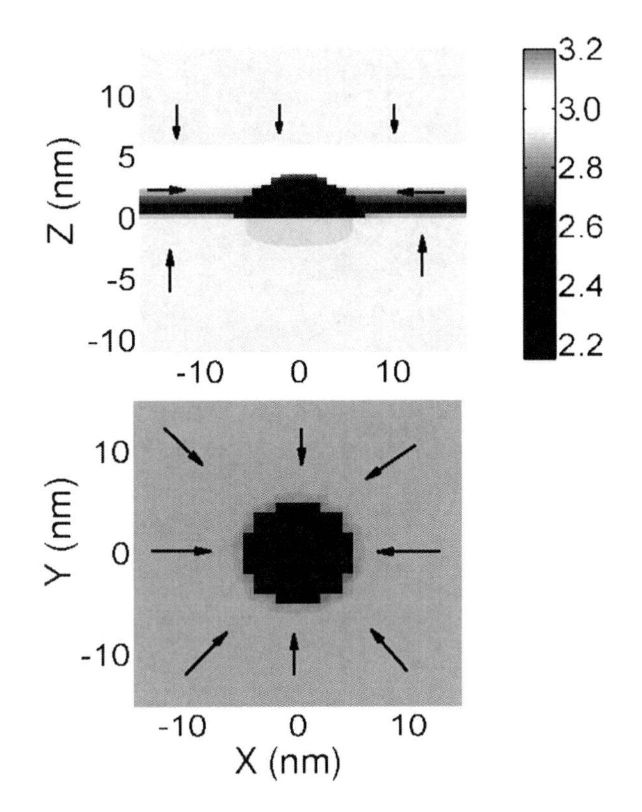

Fig. 44 Calculated potential relief around InAs/AlAs QD in Γ minimum of the conduction band. XY section corresponds to energy of electronic level in WL. *Scale bar* shows energy of the conduction bang bottom in eV. *Arrows* mark easy ways for electron to be captured in the QD

minima, respectively. The scale bar shows energy of the conduction band bottom in eV. Due to confinement, the energy of electron in WL and QDs is larger than the energy of the conduction band bottom. XY section in the figures corresponds to the energy of electrons at the bottom of the WL sub-band.

One can see that for Γ minimum there is a small potential barrier under the QD; however, carriers have an opportunity to be easily captured in the QD from the matrix. The strain around the QD influences the Γ and X_{XY} conduction band minima in different ways. The difference results in the appearance of the barriers above and under the QD for electrons moving in the X_{XY} minima of the matrix. Hence, carriers flying in the X_{XY} minima of the matrix hit in the QD only from some azimuth sector. Therefore, an effective area from which electrons can penetrate in the QDs from the matrix is smaller than the geometrical area of the QDs. On the other hand, there is no obstacle for capturing the electrons of Γ and valleys of the conduction band from the matrix to the WL.

Calculation also demonstrates that there are no barriers at the WL/QD interface for Γ electrons, as well as for X_{XY} electrons traveling in X_X direction from the wetting

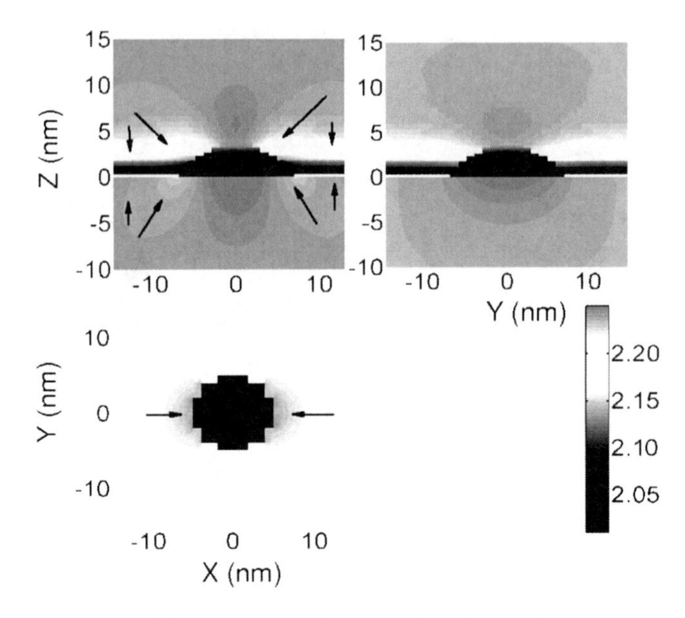

Fig. 45 Calculated potential relief around InAs/AlAs QD in X_{XY} minima of the conduction band. XY section corresponds to energy of electronic level in WL. *Scale bar* shows energy of the conduction bang bottom in eV. *Arrows* mark easy ways for electron to be captured in the QD

layer into the QDs. A weak barrier at the WL/QD interface exists for X_{XY} electrons traveling in X_Y direction. One can see that this barrier comprises a sector of about 150° around the QD. Therefore, only about 40% of electrons aspired to a QD from all directions have to be reflected by the barrier, which prolongs the duration of electron stay in the WL. Nevertheless, these electrons can hit into the QD after a few scattering events. Hence, the majority of electrons can reach QDs without any hindrances.

Thus, our calculation demonstrates that there are no significant barriers at the WL/QD interface, which would suppress the transfer of the majority of mobile carriers from the WL to the QDs. Therefore, the high rate of carrier capture in nonradiative centers in the WL, which hampers the transfer of mobile carriers from the WL to the QDs in the InAs/AlAs structures, is a result of either high density of the centers or peculiarity of their structure, which gives rise to the large value of capture cross section. Investigations of the reasons for the high carrier capture rate in nonradiative centers at WL of InAs/AlAs structures will be the subject of our future study.

4.4.5 Conclusions

The steady state and transient PL of structures with InAs/AlAs QDs grown in the Stranski–Krastanov mode have been studied. We demonstrate suppression carrier transport along wetting layer at low temperatures and low excitation power densities due to localization on roughness of InAs/AlAs heterointerface. Carriers populated mobile states of the wetting layer with increase in temperature or excitation power density do mainly not transfer in QDs but are rather captured by nonradiative centers located in the wetting layer.

5 Comparison of Quantum Dots Fabricated by Stranski–Krastanov Mechanism and Droplet Epitaxy

We have shortly summarized here some drawbacks of QDs grown by SK, and we show in this part that they can be overcome with a help of the droplet epitaxy. The wetting layer in structures formed by droplet molecular beam epitaxy has a high smoothness of the heterointerface and a low concentration of defects – nonradiative centers. As a result high efficiency of the carrier transfer from WL to the QDs is obtained in the structures.

5.1 *High-Quality Structures with InAs/$Al_{0.9}$ $Ga_{0.1}$As Quantum Dots Produced by Droplet Epitaxy*

As it is known (shown in previous paragraph), there is problem to produce effective light-emitting devices based on InAs/AlAs QDs due to insufficient capture of charge carriers from wetting layer to QDs [157]. This problem is caused by a heterointerface roughness that complicates a motion of charge carriers in the wetting layer and high concentration of nonradiative recombination centers localized in the wetting layer that capture charge carriers impeding their penetration into QDs. In this section, we demonstrate that perfect InAs/$Al_{0.9}Ga_{0.1}$As QDs structures with significantly flatter heterointerface and low concentration of nonradiative recombination centers in the wetting layer can be grown by droplet molecular beam epitaxy.

After removal of native oxide, a thick $Al_{0.9}Ga_{0.1}$As buffer layer was grown. Then the surface of $Al_{0.9}Ga_{0.1}$As was optimized for QD growth in the way described below. To avoid the effects caused by arsenic excess on the growth surface, extra aluminum was applied to the $Al_{0.9}Ga_{0.1}$As before QD growth. Simultaneously, the quality of the surface was controlled by means of reflection high-energy electron diffraction (RHEED). The diffraction patterns observed just before QD growth are presented in Fig. 46a, b. One can clearly see (5×2) surface reconstruction (with the periodicity of 5 and 2 for [110] and $[\bar{1}10]$ azimuth in *a* and *b*, respectively) that corresponds to metal-enriched surface structure [158]. Such reconstruction is maximum flat, contains less nucleation centers, and therefore is best suited for droplet epitaxy. Interesting to notice that this metal-enhanced reconstruction cannot be used for standard Stranski–Krastanov growth mode, because an arsenic valve is open all the time due to growth conditions.

Sharp diffraction spots of RHEED pattern in Fig. 46a indicate the presence of large coherent terraces. The estimations based on an obtained longitudinal width of

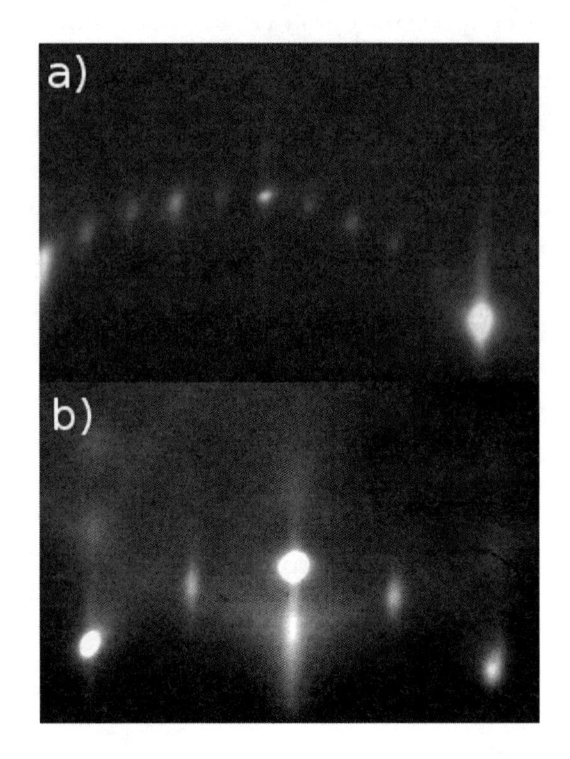

Fig. 46 RHEED of the surface before QD growth in two azimuths: [110] (**a**) and [$\bar{1}$10] (**b**) azimuths are shown. (5 × 2) AlAs surface reconstruction evidences surface high quality

the spots reveal an average terrace size to be about 100 nm. In Fig. 46b, Kikuchi lines can be clearly observed. As they are formed by diffusely scattered electrons, e.g., result from thermal atom vibrations, their presence is an additional proof of a perfect crystalline structure we obtained. Therefore RHEED pattern analysis confirms a high quality of the $Al_{0.9}$ $Ga_{0.1}As$ initial surface that was used for subsequent QDs formation in our experiments. In addition a surface prepared for QD growth was investigated by AFM. For AFM measurements $Al_{0.9}$ $Ga_{0.1}As$ surface was capped with 5 nm of GaAs to prevent the oxidization and then the smoothness of a surface was verified by AFM measurements in air. The AFM image of the surface is presented in Fig. 47. One can see the presence of atomic steps with an average terrace size about 100 nm that coincides with the estimations made by RHEED.

Droplet epitaxy was used to fabricate QDs [159]. Two monolayers of indium were deposited to the $Al_{0.9}$ $Ga_{0.1}As$ surface at 510°C in the absence of arsenic flux and the array of initial In droplets formed. Then the sample was exposed to a beam equivalent pressure of 1.5×10^{-5} torr of As_4 and InAs quantum dots crystallized. The growth was interrupted for 60 s after In valve had been closed and then QDs were capped with 25 nm of $Al_{0.9}$ $Ga_{0.1}As$. The structure with such QDs will be referred to as DMBE in the following. The density (~10^8 cm^{-2}) of the DMBE $InAl_{0.9}$ $Ga_{0.1}As$ QDs was determined using transmission electron microscopy. A test structure with InAs/AlAs QDs was grown by molecular beam epitaxy in Stranski–Krastanov mode for comparison. A detailed description of a growth

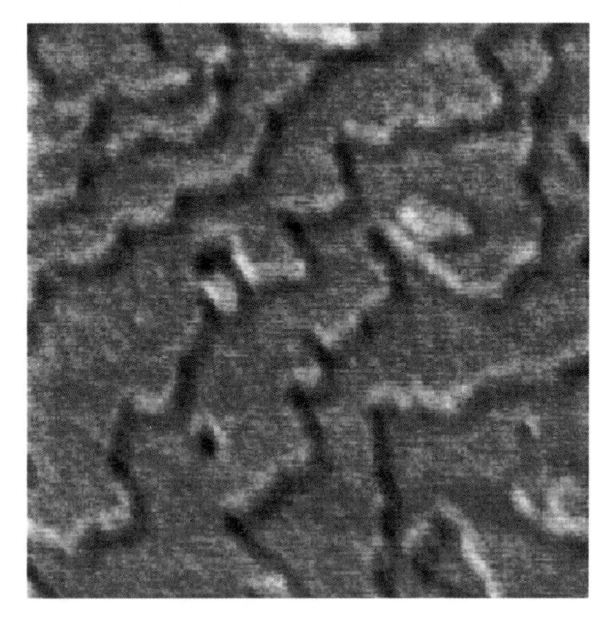

Fig. 47 AFM image with clearly observed atomic steps. Image size is 1×1 μm

procedure can be found in previous section and our work [157]. We will refer structures with such QDs as MBE. The density (9×10^9 cm^{-2}) of the MBE InAs/AlAs QDs was also determined using transmission electron microscopy [117].

The PL was excited by a He–Cd laser ($h\nu = 3.81$ eV) with the power density of 10 W/cm^2. The detection of the PL was performed by a double diffraction grating spectrometer equipped with a cooled photomultiplier operated in the photon counting mode. PL spectra of the MBE and DMBE structures are shown in Fig. 48 a, b, respectively. All the spectra contain PL bands related to the recombination in the wetting layer (WL) and quantum dots marked as WL and QD, respectively [117]. In the spectrum of MBE structure with a higher QDs density (9×10^9 cm^{-2}), the intensity of WL band exceeds that of QD band in order of magnitude smaller, while for DMBE structure with the two order of magnitude lower QDs density (~10^8 cm^{-2}) intensity of the QD band opposite is order of magnitude higher than that of WL band. Usually, it is considered that WL does not form in droplet molecular beam epitaxy growth mode; however, well-defined WL band is observed in all our spectra. Additionally, Auger measurements confirms that the uniform WL is presented in our DMBE structures [44]. Recently, we demonstrate that WL band is a sum of strongly overlapped bands related to no-phonon transition and its TOA_{AlAs}, LO_{InAs} and LO_{AlAs} phonon replicas [57]. In the PL spectrum of MBE structure, the WL bands of separate phonon replicas are broadened and overlapped that the spectrum degenerates in two wide peaks, while in the PL spectrum of DMBE structure all phonon replica lines are clearly observed (Fig. 49). Since the broadening of the WL lines is caused by a rough interface of a wetting layer [160], one can conclude that the smoothness of the wetting layer interface is considerably better in the case of DMBE structure.

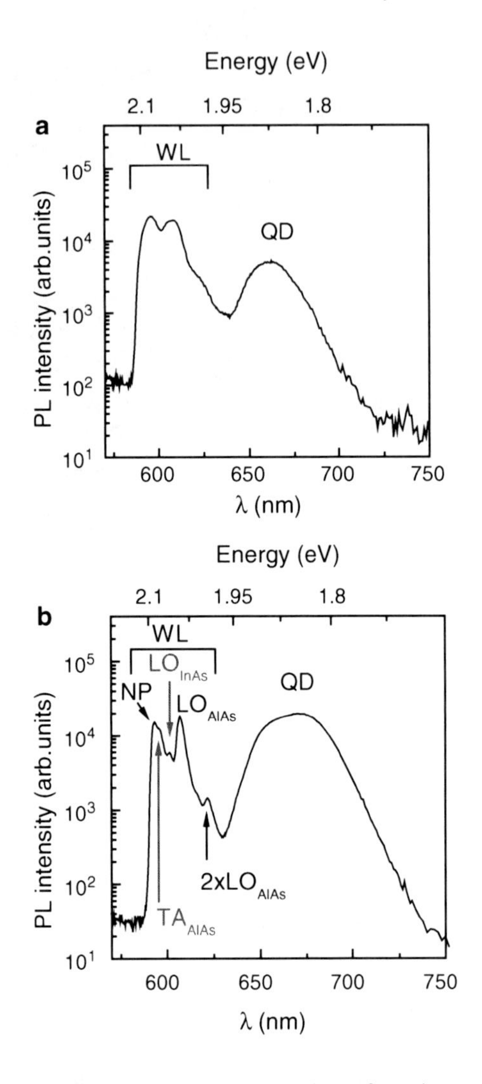

Fig. 48 Low-temperature (5 K) PL spectra of the structures with InAs/Al(Ga) As QDs grown by (**a**) MBE and (**b**) DMBE

Due to low density of the QDs in both studied structure, the number of carriers captured per unit time directly from the matrix is much smaller than that in the wetting layer [157]. Domination of the WL band in spectrum of the MBE structure reflects the fact shown in our previous study that the major part of the carriers captured by the wetting layer are not transferred in the QDs but vanish via nonradiative recombination [157]. On the other hand, absolute domination of the QD band in the PL spectrum of the DMBE structure indicates that carriers captured from matrix in the wetting layer than efficiently collected to QDs. To evaluate the effect of nonradiative recombination centers in the WL of the DMBE structure on the capture of carries from the WL to QDs, the temperature dependence of integral intensity of WL and QD lines was measured. The dependence is presented in Fig. 50. One can see that with increase in temperature the intensity of the WL band falls with activation energy 5 $\pm$ 1 meV, while intensity of the QD band

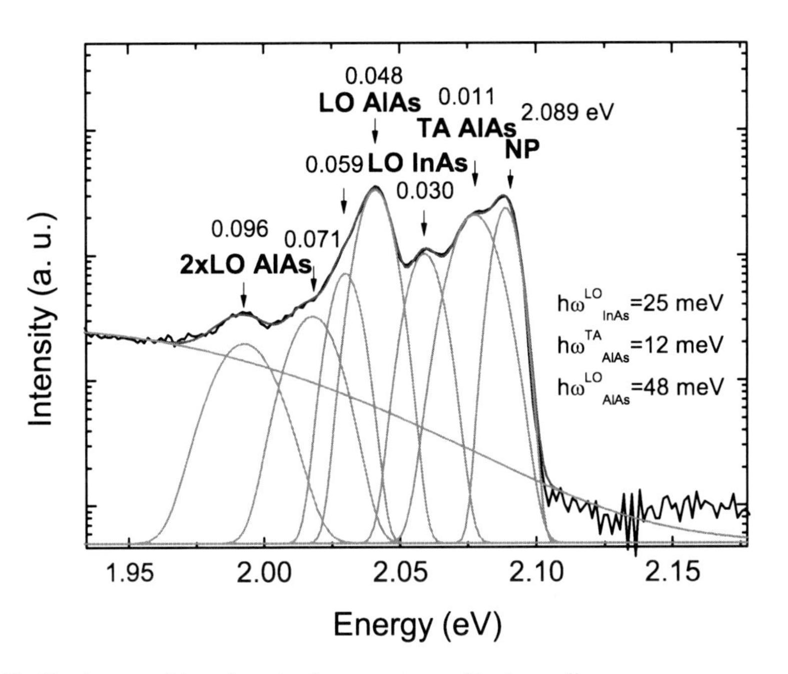

Fig. 49 The decomposition of wetting layer spectrum with phonon lines

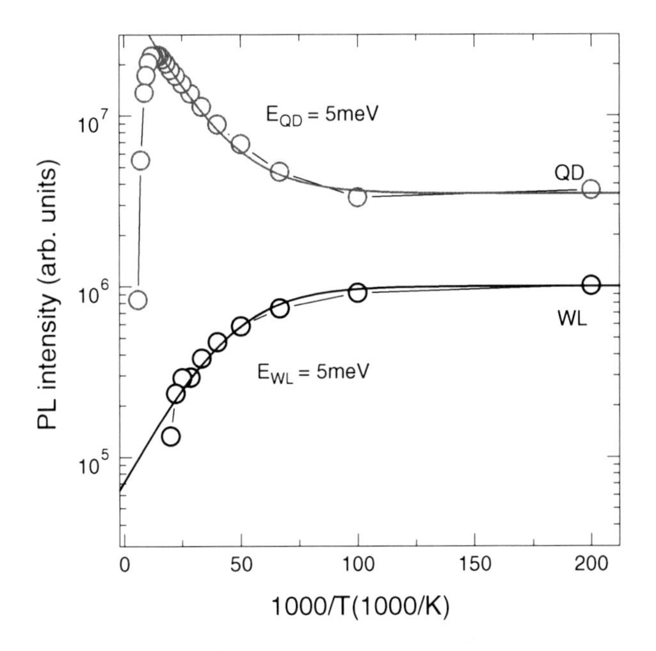

Fig. 50 Temperature dependence of integrated intensity of the WL and QD bands in the structure grown by DMBE. Activation energy obtained from fitting the curves by Arrhenius function

increases (with the same activation energy 5 ± 1 meV) in antiphase to decrease intensity of the WL band from 5 to 100 K. Further temperature increase results in the decrease in intensity QD band due to escape carriers from QDs to the matrix [157]. This coincident increase intensity of the QD band and decrease intensity of WL band unambiguously evidences that nonequilibrium carriers, de-localized in WL with a temperature increase, are captured in QDs mainly and do not recombine in nonradiative way contrary to that in MBE structures [157].

Advantages of the DMBE structure when compared with the MBE can be explained with growth specificities. An important feature of the droplet epitaxy is the ability to use the flat initial (5×2) reconstruction of the growths surface. As it was mentioned above, such surface is not used in standard Stranski–Krastanov growth mode. Another great growth advantage of the droplet epitaxy is softness of QD formation. It is well known that significant mass transfer takes place in Stranski–Krastanov mechanism [161, 162]. This process is accompanied by elastic relaxation that can create defects (worked as nonradiative centers) in a WL/lower layer interface. Otherwise, in the droplet epitaxy a conversion of an In droplet to InAs QD is arsenic driven. Assuming that crystallization spreads from origin InAs nucleus similar to autocatalysis, it involves upper layers mostly. Thus, in droplet epitaxy mode WL/lower layer interface does not change and it can be saved in this case defect-free. Therefore, we assume that the decrease in density of nonradiative recombination centers in WL of the DMBE structure is a feature of the droplet molecular beam epitaxy.

Therefore, we demonstrate that droplet molecular beam epitaxy allows to grow high-quality InAs/$Al_{0.9}$ $Ga_{0.1}As$ QDs structures with a perfect heterointerface and a low concentration of defects nonradiative recombination centers located in WL. The structures demonstrate high efficiency of the carrier capture from WL to QDs.

6 Conclusions

The system of InAs quantum dots in AlAs matrix is of great interest for the investigation of spin relaxation mechanism of charge carriers and excitons in quantum dots. For indirect band gap, the PL decay is prolonged that makes possible to study spin relaxation. The atomic and energy structure of InAs/AlAs QDs was investigated in different growth conditions and the coexistence of direct and indirect band structures within the one sample is revealed for Stranski–Krastanov method. However, all the samples demonstrate a lack of carrier transfer due to a low quality of a heterointerface and a high concentration of nonradiative recombination centers. To overcome these problems, we consistently investigated a technique of droplet epitaxy.

For droplet epitaxy, quantum dot density is a parameter of great importance and it is defined by general nucleation mechanisms. We started to study the nucleation with an initial stage of homoepitaxy in the model system of GaAs.

It was found that under low surface coverage (below 0.03 Ml), the metastable phase α-(2×4)-n forms which exists in an ordered state at low temperatures. Under coverage of 0.03–0.09, the transition centers with the β1-(2×4) local structure form. Finally, under coverage above 0.09 Ml, stable crystalline nuclei of a new phase form. For statistical analysis, an ensemble of such centers with the same size can be considered as an intermediate lattice gas.

This thermodynamic model turned out to be fruitful also for QD formation in Stranski–Krastanov transition. It was proved that the QD density, being a function of the temperature and InAs surface coverage, can be described as a two-dimensional first-order PT, where the lateral interactions between QDs play an important role. Three-parameter isotherms are in a perfect agreement with experimental data which demonstrates that QD formation can be treated in the framework of a general statistical approach.

Turning to droplet epitaxy itself, we analyzed the array of metal droplets on a semiconductor surface being an initial stage of LDE and droplet epitaxy and grew the samples with various droplet materials and deposition rates. A new method to investigate the intermixing process by a wetting angle analysis is proposed. The aspect ratio which is defined as the ratio of the droplet height to its radius was used as an estimation of the wetting angle depending on the materials. Experimentally observed bimodality of the aspect ratio indicates the existence of two droplet groups with different compositions. The estimation of indium content by wetting angle revealed a significant material intermixing during the deposition time. On the base of our experimental results, we suggested the model of droplet evolution and the formation of two droplet groups.

Then the formation and destruction of QDs formed from the droplet array was studied by means of RHEED. Kinetic curves of QD evolution during the growth and annealing, obtained by RHEED, enable to reveal and optimize growth parameters. The influence of the growth interruption time and indium dose was shown and the dependence of QD properties on indium dose was investigated in details.

The analysis of PL data revealed that over a critical indium amount droplet to Stranski–Krastanov growth mode transition takes place. Initial droplet cracks and forms a castle-like complex of QDs, thus combining both mechanisms under certain conditions. It is confirmed by PL spectra transformation and AFM images with clearly observed quantum dot molecules with the round symmetry.

Finally, we demonstrate that droplet epitaxy allows growing high-quality InAs/ $Al_{0.9}Ga_{0.1}As$ QDs structures with a perfect heterointerface and a low concentration of defects – nonradiative recombination centers located in WL. Fabricated structures reveal a high efficiency of the carrier capture from WL to QDs.

Acknowledgment This work has being partially supported by RFBR via Grants 10-02-00513 and 10-08-00851.

References

1. Alonso-Gonzalez, P., Alen, B., Fuster, D., Gonzalez, Y., Gonzalez, L.: Appl. Phys. Lett. **91**, 163104 (2007)
2. Lee, J.H., Wang, Zh.M., Strom, N.W., Mazur, Yu.I., Salamo, G.J.: Appl. Phys. Lett. **89**, 202101 (2006)
3. Lee, C.D., Park, C., Lee, H.J., Lee, K.S., Park, S.J., Park, C.G., Noh, S.K., Koguchi, N.: Jpn. J. Appl. Phys. **37**, 7158 (1998)
4. Mano, T., Kuroda, T., Sanguinetti, S., Ochiai, T., Tateno, T., Kim, J., Noda, T., Kawabe, M., Sakoda, K., Kido, G., Koguchi, N.: Nano Lett. **5**, 425 (2005)
5. Wang, Zh.M., Holmes, K., Shultz, J.L., Salamo, G.J.: Phys. Stat. Sol. **202**, R85 (2005)
6. Watanabe, K., Koguchi, N., Gotoh, Y.: Jpn. J. Appl. Phys. **39**, L79 (2000)
7. Bimberg, D. (ed.): Nanoscience and Technology: Semiconductor Nanostructures. Springer-Verlag, Berlin (2008)
8. Itoh, M.: Prog. Surf. Sci. **66**, 53 (2001)
9. Avery, A.R., Dobbs, H.T., Holmes, D.M.: Phys. Rev. Lett. **70**, 3938 (1997)
10. Ishii, A., Kawamura, T.: Surf. Sci. **436**, 38 (1999)
11. Itoh, M., Bell, G.R., Avery, A.R.: Phys. Rev. Lett. **81**, 633 (1998)
12. Kangawa, Y., Ito, T., Taguchi, A.: Appl. Surf. Sci. **190**, 517 (2002)
13. Kratzer, P., Morgan, C.G., Scheffler, M.: Phys. Rev. B **59**, 15246 (1999)
14. Kratzer, P., Penev, E., Scheffler, M.: Appl. Phys. A **75**, 79 (2002)
15. Kratzer, P., Scheffler, M.: Phys. Rev. Lett. **88**, 036102 (2002)
16. Penev, E., Kratzer, P., Scheffler, M.: Appl. Surf. Sci. **216**, 436 (2003)
17. Joyce, B., Vvedensky, D.: Mater. Sci. Eng. **R46**, 127 (2004)
18. Galitsyn, Yu, Dmitriev, D., Mansurov, V., Moshchenko, S., Toropov, A.: JETP Lett. **86**, 482 (2007)
19. Xue, Q.K., Hashizume, T., Ohno, A.T., Hasegawa, Y., Sakurai, T.: Sci. Rep. RITU A **44**, 113 (1997)
20. Koduvely, H.M., Zangwill, A.: Phys. Rev. B **60**, R2204 (1999)
21. Kobayashi, N.P., Ramachandrah, T.R., Chen, P., Madhukar, A.: Appl. Phys. Lett. **68**, 3299 (1996)
22. Dobbs, H.T., Zangwill, A., Vvedensky, D.D.: Surface diffusion
23. Leonard, J.D., Pond, K., Petroff, P.M.: Phys. Rev. B **50**, 11687 (1994)
24. Placidi, E., Arciprete, F., Fanfoni, M., Patella, F., Orsini, E., Balzarotti, A.: J. Phys. Cond. Matt. **19**, 225006 (2007)
25. Sun, J., Jin, P., Wang, Z.: Nanotechnology **15**, 1763 (2004)
26. Marchenko, V.I.: JETP Lett. **33**, 397 (1981)
27. Zhdanov, V., Kasemo, B.: Surf. Sci. Rep. **20**, 111 (1994)
28. Galitsyn, Yu.G., Dmitriev, D.V., Mansurov, V.G., Moshchenko, S.P., Toropov, A.I.: JETP Lett. **81**, 629 (2005)
29. Galitsyn, Yu, Dmitriev, D., Mansurov, V., Moshchenko, S., Toropov, A.: JETP Lett. **84**, 505 (2006)
30. Koguchi, N., Takahashi, S., Chikyow, T.: J. Crystal Growth **111**, 688 (1991)
31. Sholz, M., Buttner, S., Benson, O., Toropov, A.I., Bakarov, A.K., Lochmann, A., Stock, E., Schulz, O., Hopfer, F., Haisler, V.A., Bimberg, D.: Opt. Express **15**, 9107 (2007)
32. Stock, E., Warming, T., Ostapenko, I., Rodt, S., Schliwa, A., Töfflinger, J.A., Lochmann, A., Toropov, A.I., Moshchenko, S.P., Dmitriev, D.V., Haisler, V.A., Bimberg, D.: Appl. Phys. Lett. **96**, 093112 (2010)
33. Li, L.H., Chauvin, N., Patriarche, G., Alloing, B., Fiore, A.: J. Appl. Phys. **104**, 083508 (2008)
34. Wang, Zh.M., Liang, B.L., Sablon, K.A., Salamo, G.J.: Appl. Phys. Lett. **90**, 113120 (2007)
35. Alonso-Gonzalez, P., Fuster, D., Gonzalez, L., Martin-Sanchez, J., Gonzalez, Y.: Appl. Phys. Lett. **93**, 183106 (2008)

36. Gong, Z., Niu, Z.C., Huang, S.S., Fang, Z.D., Sun, B.Q., Xia, J.B.: Appl. Phys. Lett. **87**, 093116 (2005)
37. Heyn, Ch, Stemmann, A., Koppen, T., Strelow, Ch, Kipp, T., Grave, M., Mendach, S., Hansen, W.: Appl. Phys. Lett. **94**, 183113 (2009)
38. Heyn, Ch, Stemmann, A., Hansen, W.: J. Crystal Growth **311**, 1839 (2009)
39. Heyn, Ch, Stemmann, A., Eiselt, R., Hansen, W.: J. Appl. Phys. **105**, 054316 (2009)
40. Liang, B.L., Wang, Zh.M., Lee, J.H., Sablon, K., Mazur, Yu.I., Salamo, G.J.: Appl. Phys. Lett. **89**, 043113 (2006)
41. Stemmann, A., Heyn, Ch, Koppen, T., Kipp, T., Hansen, W.: Appl. Phys. Lett. **93**, 123108 (2008)
42. Landau, L.D., Lifshits, E.M.: Statistical Physics. Nauka, Moscow (1964)
43. Stemmann, A., Koppen, T., Grave, M., Wildfang, S., Mendach, S., Hansen, W., Heyn, Ch: J. Appl. Phys. **106**, 064315 (2009)
44. Lyamkina, A.A., Dmitriev, D.V., Galitsyn, Yu.G., Kesler, V.G., Moshchenko, S.P., Toropov, A.I.: Nanoscale Res. Lett. **6**, 42 (2011)
45. Heitz, R., Guffarth, F., Potschke, K., Schliwa, A., Bimberg, D.: Phys. Rev. B **71**, 045325 (2005)
46. Pohl, U.W., Pötschke, K., Schliwa, A., Guffarth, F., Bimberg, D.: Phys. Rev. B **72**, 245332 (2005)
47. Lyamkina, A.A., Moshchenko, S.P., Haisler, V.A., Galitsyn, Yu.G., Toropov, A.I.: Proceedings of Collaborative Conference on Interacting Nanostructures, San Diego, 9–13 November 2009, p. 23.
48. Dawson, P., Ma, Z., Pierz, K., Göbel, E.O.: Appl. Phys. Lett. **81**, 2349 (2002)
49. Dawson, P., Göbel, E.O., Pierz, K.: J. Appl. Phys. **98**, 013541 (2005)
50. Shamirzaev, T.S., Gilinsky, A.M., Toropov, A.I., Bakarov, A.K., Tenne, D.A., Zhuravlev, K. S., von Borczyskowski, C., Zahn, D.R.T.: JETP Lett. **77**, 389 (2003)
51. Fu, H., Wang, L.-W., Zunger, A.: Phys. Rev. B **59**, 5568 (1999)
52. Goupalov, S.V., Ivchenko, E.L.: Phys. Sol. State **42**, 2030 (2000)
53. Shamirzaev, T.S., Gilinsky, A.M., Kalagin, A.K., Toropov, A.I., Gutakovskii, A.K., Zhuravlev, K.S.: Semicond. Sci. Technol. **21**, 527 (2006)
54. Offermans, P., Koenraad, P.M., Wolter, J.H., Pierz, K., Roy, M., Maksym, P.A.: Phys. Rev. B **72**, 165332 (2005)
55. Williamson, A.J., Franceschetti, A., Fu, H., Wang, L.W., Zunger, A.: J. Electron. Mater. **28**, 414 (1999)
56. Sarkar, D., van der Meulen, H.P., Calleja, J.M., Becker, J.M., Haug, R.J., Pierz, K.: Phys. Rev. B **71**, 081302R (2005)
57. Shamirzaev, T.S., Gilinsky, A.M., Kalagin, A.K., Nenashev, A.V., Zhuravlev, K.S.: Phys. Rev. B **76**, 155309 (2007)
58. Heyn, Ch, Hansen, W.: J. Crystal Growth **251**, 218 (2003)
59. Litvinov, D., Gerthsen, D., Rosenauer, A., Schowalter, M., Passow, T., Feinaugle, P., Hetterich, M.: Phys. Rev. B **74**, 165306 (2006)
60. Martini, S., Quivy, A.A., Lamas, T.E., da Silva, E.C.F.: Phys. Rev. B **72**, 153304 (2005)
61. Offermans, P., Koenraad, P.M., Notzel, R., Wolter, J.H., Pierz, K.: Appl. Phys. Lett. **87**, 111903 (2005)
62. Rosenauer, A., Oberst, W., Litvinov, D., Gerthsen, D., Förster, A., Schmidt, R.: Phys. Rev. B **61**, 8276 (2000)
63. Rosenauer, A., Gerthsen, D., Van Dyck, D., Arzberger, M., Böhm, G., Abstreiter, G., Schmidt, R.: Phys. Rev. B **64**, 245334 (2001)
64. Schowalter, M., Rosenauer, A., Gerthsen, D., Arzberger, M., Bichler, M., Abstreiter, G.: Appl. Phys. Lett. **79**, 4426 (2001)
65. Muraki, K., Fukatsu, S., Shiraki, Y., Ito, R.: Appl. Phys. Lett. **61**, 557 (1992)
66. Lemaitre, A., Patriarche, G., Glas, F.: Appl. Phys. Lett. **85**, 3717 (2004)
67. Liao, X.Z., Zou, J., Cockayne, D.J.H., Leon, R., Lobo, C.: Phys. Rev. Lett. **82**, 5148 (1999)

68. Liu, N., Tersoff, J., Baklenov, O., Holmes Jr., A.L., Shih, C.K.: Phys. Rev. Lett. **80**, 334 (1997)
69. Passow, T., Li, S., Feinäugle, P., Vallaitis, T., Leuthold, J., Litvinov, D., Gerthsen, D., Hetterich, M.: J. Appl. Phys. **102**, 073511 (2007)
70. Quinn, P.D., Wilson, N.R., Hatfield, S.A., McConville, C.F., Bell, G.R., Noakes, T.C.Q., Bailey, P., Al-Harthi, S., Gard, F.: Appl. Phys. Lett. **87**, 153110 (2005)
71. Walther, T., Cullis, A.G., Norris, D.J., Hopkinson, M.: Phys. Rev. Lett. **86**, 2381 (2001)
72. Wang, P., Bleloch, A.L., Falke, M., Goodhew, P.J., Ng, J., Missous, M.: Appl. Phys. Lett. **89**, 072111 (2006)
73. Ibánez, J., Cuscó, R., Artús, L., Henini, M., Patané, A., Eaves, L.: Appl. Phys. Lett. **88**, 141905 (2006)
74. Cherkashin, N.A., Maksimov, M.V., Makarov, A.G., Shchukin, V.A., Ustinov, V.M., Lukovskaya, N.V., Musikhin, Yu.G., Cirlin, G.E., Bert, N.A., Alferov, Zh.I.: Semiconductors **37**, 861 (2003)
75. Ballet, P., Smathers, J.B., Yang, H., Workman, C.L., Salamo, G.J.: J. Appl. Phys. **90**, 481 (2001)
76. Ferdosa, F., Wanga, S., Weia, Y., Sadeghib, M., Zhaoc, Q., Larsson, A.: J. Crystal Growth **251**, 145 (2003)
77. Park, S.K., Tatebayashi, J., Arakawa, Y.: Appl. Phys. Lett. **84**, 1877 (2004)
78. Pierz, K., Ma, Z., Hapke-Wurst, I., Keyser, U.F., Zeitler, U., Haug, R.J.: Physica E **13**, 761 (2002)
79. Heidemeyer, H., Kiravittaya, S., Müller, C., Jin-Phillipp, N.Y., Schmidt, O.G.: Appl. Phys. Lett. **80**, 1544 (2002)
80. Le Ru, E.C., Howe, P., Jones, T.S., Murray, R.: Phys. Rev. B **67**, 165303 (2003)
81. Song, H.Z., Usuki, T., Nakata, Y., Yokoyama, N., Sasakura, H., Muto, S.: Phys. Rev. B **73**, 115327 (2006)
82. Yang, T., Tatebayashi, J., Tsukamoto, S., Nishioka, M., Arakawa, Y.: Appl. Phys. Lett. **84**, 2817 (2004)
83. Dubrovskii, V.G., Cirlin, G.E., Ustinov, V.M.: Phys. Rev. B **68**, 075409 (2003)
84. Joyce, P.B., Krzyzewski, T.J., Bell, G.R., Jones, T.S., Malik, S., Childs, D., Murray, R.: Phys. Rev. B **62**, 10891 (2000)
85. Songmuang, R., Kiravittaya, S., Sawadsaringkarn, M., Panyakeow, S., Schmidt, O.G.: J. Crystal Growth **251**, 166 (2003)
86. Pierz, K., Ma, Z., Keyser, U.F., Haug, R.J.: J. Crystal Growth **249**, 477 (2003)
87. The NEXTNANO3 software package can be downloaded from http://www.wsi.tum.de/nextnano3; http://www.nextnano.de
88. Kane, O.E.: J. Phys. Chem. Solids **1**, 249 (1957)
89. Milekhin, A.G., Toropov, A.I., Bakarov, A.K., Tenne, D.A., Zanelatto, G., Galzerani, J.C., Schulze, S., Zahn, D.R.T.: Phys. Rev. B **70**, 085314 (2004)
90. Biasiol, G., Heun, S., Golinelli, G.B., Locatelli, A., Mentes, T.O., Guo, F.Z., Hofer, C., Teichert, C., Sorba, L.: Appl. Phys. Lett. **87**, 223106 (2005)
91. Gironcoli, S., Baroni, S., Resta, R.: Phys. Rev. Lett. **62**, 2853 (1989)
92. Munoz, M.C., Armelles, G.: Phys. Rev. B **48**, 2839 (1993)
93. Van der Walle, C.: Phys. Rev. B **39**, 1871 (1989)
94. Vurgaftman, I., Meyer, J.R., Ram-Mohan, L.R.: J. Appl. Phys. **89**, 5815 (2001)
95. Wei, S.-H., Zunger, A.: Appl. Phys. Lett. **72**, 2011 (1998)
96. Wei, S.-H., Zunger, A.: Phys. Rev. B **60**, 5404 (1999)
97. Ridley, B.K.: J. Appl. Phys. **48**, 754 (1977)
98. Boykin, T.B.: Phys. Rev. B **56**, 9613 (1997)
99. Madelung, O., Weiss, H., Schulz, M. (eds.): Numeral Data and Functional Relationships in Science and Technology, Landolt-Bornstein, New Series, Group III (Crystal and Solid State Physics), vol. 17. Springer, Heidelberg (1982)

100. Ledentsov, N.N., Böhrer, J., Beer, M., Heinrichsdorff, F., Grundmann, M., Bimberg, D., Ivanov, S.V., Meltser, B.Y., Shaposhnikov, S.V., Yassievich, I.N., Faleev, N.N., Kopev, P.S., Alferov, Zh.I.: Phys. Rev. B **52**, 14058 (1995)
101. Hatami, F., Ledentsov, N.N., Grundmann, M., Heinrichsdorff, F., Bimberg, D., Ruvimov, S. S., Werner, P., Gosele, O., Heydenreich, J., Richter, U., Ivanov, S.V., Meltser, B.Ya., Kopev, P.S., Alferov, Zh.I.: Appl. Phys. Lett. **67**, 656 (1995)
102. Itskevich, I.E., Lyapin, S.G., Troyan, I.A., Klipsteinet, P.C., Eaves, L., Main, P.C., Henini, M.: Phys. Rev. B **58**, R4250 (1998)
103. Kuo, M.C., Hsu, J.S., Shen, J.L., Chiu, K.C., Fan, W.C., Lin, Y.C., Chia, C.H., Chou, W.C., Yasar, M., Mallory, R., Petrou, A., Luo, H.: Appl. Phys. Lett. **89**, 263111 (2006)
104. Blome, P.G., Wenderoth, M., Hbner, M., Ulbrichet, R.G., Porsche, J., Scholz, F.: Phys. Rev. B **61**, 8382 (2000)
105. Empedocles, S.A., Norris, D.J., Bawendi, M.G.: Phys. Rev. Lett. **77**, 3873 (1996)
106. Sychugov, I., Juhasz, R., Valenta, J., Linnros, J.: Phys. Rev. Lett. **94**, 087405 (2005)
107. Raymond, S., Guo, X., Merz, J.L., Fafard, S.: Phys. Rev. B **59**, 7624 (1999)
108. Dekel, E., Regelman, D.V., Gershoni, D., Ehrenfreund, E., Schoenfeld, W.V., Petroff, P.M.: Phys. Rev. B **62**, 11038 (2000)
109. Regelman, D.V., Mizrahi, U., Gershoni, D., Ehrenfreund, E., Schoenfeld, W.V., Petroff, P.M.: Phys. Rev. Lett. **87**, 257401 (2001)
110. Dekel, E., Gershoni, D., Ehrenfreund, E., Spektor, D., Garcia, J.M., Petroff, P.M.: Phys. Rev. Lett. **80**, 4991 (1998)
111. Lomascolo, M., Vergine, A., Johal, T.K., Rinaldi, R., Passaseo, A., Cingolani, R., Patan, S., Labardi, M., Allegrini, M., Troiani, F., Molinari, E.: Phys. Rev. B **66**, 041302(R) (2002)
112. Kovalev, D., Heckler, H., Ben-Chorin, M., Polisski, G., Schwartzkopff, M., Koch, F.: Phys. Rev. Lett. **81**, 2803 (1998)
113. Shamirzaev, T.S., Nenashev, A.V., Zhuravlev, K.S.: Appl. Phys. Lett. **92**, 213101 (2008)
114. Jung, S.I., Yoon, J.J., Park, H.J., Park, Y.M., Jeon, M.H., Leem, J.Y., Lee, C.M., Cho, E.T., Lee, J.I., Kim, J.S.: Physica E **26**, 100–104 (2005)
115. Ma, Z., Pierz, K., Hinze, P.: Appl. Phys. Lett. **79**, 2564 (2001)
116. Ma, Z., Pierz, K., Keyser, U.F., Haug, R.J.: Physica E **17**, 117 (2003)
117. Shamirzaev, T.S., Nenashev, A.V., Gutakovskii, A.K., Kalagin, A.K., Zhuravlev, K.S., Larsson, M., Holtz, P.O.: Phys. Rev. B **78**, 085323 (2008)
118. Convertino, A., Cerri, L., Leo, G., Viticoli, S.: J. Crystal Growth **261**, 458 (2004)
119. Fu, Y., Ferdos, F., Sadeghi, M., Wang, S.M., Larsson, A.: J. Appl. Phys. **92**, 3089 (2002)
120. Stangl, J., Holy, V., Bauer, G.: Rev. Mod. Phys. **76**, 725 (2004)
121. Lin, C.-A.J., Liedl, T., Sperling, R.A., Fernández-Argüelles, M.T., Costa-Fernández, J.M., Pereiro, R., Sanz-Medel, A., Chang, W.H., Parak, W.J.: J. Mater. Chem. **17**, 1343 (2007)
122. Medintz, I.L., Clapp, A.R., Mattoussi, H., Goldman, E.R., Fisher, B., Mauro, J.M.: Nat. Mater. **2**, 630 (2003)
123. Lunz, M., Bradley, A.L., Chen, W.-Y., Gunko, Yu.K.: Superlatt. Microstruct. **47**, 98 (2010)
124. Förster, T.: Ann. Phys. **2**, 55 (1948)
125. Kim, D., Okahara, S., Nakayama, M., Shim, Y.: Phys. Rev. B **78**, 153301 (2008)
126. Tackeuchi, A., Kuroda, T., Mase, K., Nakata, Y., Yokoyama, N.: Phys. Rev. B **62**, 1568 (2000)
127. Govorov, A.O.: Phys. Rev. B **68**, 075315 (2003)
128. Arakawa, Y., Sakaki, H.: Appl. Phys. Lett. **40**, 939 (1982)
129. Sun, K.W., Chen, J.W., Lee, B.C., Lee, C.P., Kechiantz, A.M.: Nanotechnology **16**, 1530 (2005)
130. Brandt, O., Tapfer, L., Cingolani, R., Ploog, K., Hohenstein, M., Phillipp, F.: Phys. Rev. B **41**, 12599 (1990)
131. Siegert, J., Marcinkeviius, M., Zhao, Q.X.: Phys. Rev. B **72**, 085316 (2005)
132. Ohnesorge, B., Albrecht, M., Oshinowo, J., Forchel, A., Arakawa, Y.: Phys. Rev. B **54**, 11532 (1996)

133. Narvaez, G., Bester, G., Zunger, A.: Phys. Rev. B **74**, 075403 (2006)
134. Piwonski, T., O'Driscoll, I., Houlihan, J., Huyet, G., Manning, R.J., Uskov, A.V.: Appl. Phys. Lett. **90**, 122108 (2007)
135. Bogaart, E.W., Haverkort, J.E.M., Mano, T., van Lippen, T., Notzel, R., Wolter, J.H.: Phys. Rev. B **72**, 195301 (2005)
136. Mazur, Yu.I., Wang, Zh.M., Kissel, H., Zhuchenko, Z.Ya., Lisitsa, M.P., Tarasov, G.G., Salamo, G.J.: Semicond. Sci. Technol. **22**, 86 (2007)
137. Moskalenko, E.S., Donchev, V., Karlsson, K.F., Holtz, P.O., Monemar, B., Schoenfeld, W. V., Garcia, J.M., Petroff, P.M.: Phys. Rev. B **68**, 155317 (2003)
138. Toda, Y., Moriwaki, O., Nishioka, M., Arakawa, Y.: Phys. Rev. Lett. **82**, 4114 (1999)
139. Uskov, A.V., McInerney, J., Adler, F., Schweizer, H., Pilkuhn, M.H.: Appl. Phys. Lett. **72**, 58 (1998)
140. Deppe, D.G., Huffaker, D.L.: Appl. Phys. Lett. **77**, 3325 (2000)
141. Ding, F., Chen, Y.H., Tang, C.G., Xu, B., Wang, Z.G.: Phys. Rev. B **76**, 125404 (2007)
142. Fafard, S., Leonard, D., Merz, J.L., Petroff, P.M.: Appl. Phys. Lett. **65**, 1388 (1994)
143. Le Ru, E.C., Fack, J., Murray, R.: Phys. Rev. B **67**, 245318 (2003)
144. Leon, R., Fafard, S., Piva, P.G., Ruvimov, S., Liliental-Weber, Z.: Phys. Rev. B **58**, R4262 (1998)
145. Markussen, T., Kristensen, P., Tromborg, B., Berg, T.W., Mrk, J.: Phys. Rev. B **74**, 195342 (2006)
146. Matthews, D.R., Summers, H.D., Smowton, P.M., Blood, P., Rees, P., Hopkinson, M.: IEEE J. Quantum Electron. **41**, 344 (2005)
147. Müller, T., Schrey, F.F., Strasser, G., Unterrainer, K.: Appl. Phys. Lett. **83**, 3572 (2003)
148. Brübach, J., Silov, A.Yu., Haverkort, J.E.M., van der Vleuten, W., Wolter, J.H.: Phys. Rev. B **61**, 833 (2000)
149. Hinooda, S., Loualiche, S., Lambert, B., Bertru, N., Paillard, M., Marie, X., Amand, T.: Appl. Phys. Lett. **78**, 3052 (2001)
150. Lobo, C., Perret, N., Morris, D., Zou, J., Cockayne, D.J.H., Johnston, M.B., Gal, M., Leon, R.: Phys. Rev. B **62**, 2737 (2000)
151. Marcinkeviius, S., Leon, R.: Phys. Rev. B **59**, 4630 (1999)
152. Raymond, S., Hinzer, K., Fafard, S., Merz, J.L.: Phys. Rev. B **R16**, 331 (2000)
153. Polimeni, A., Patané, A., Henini, M., Eaves, L., Main, P.C.: Phys. Rev. B **59**, 5064 (1999)
154. Krivorotov, I.N., Chang, T., Gilliland, G.D., Fu, L.P., Bajaj, K.K., Wolford, D.J.: Phys. Rev. B **58**, 10687 (1998)
155. Zhuravlev, K.S., Petrakov, D.A., Gilinsky, A.M., Shamirzaev, T.S., Preobrazhenskii, V.V., Semyagin, B.R., Putyato, M.A.: Superlatt. Microstruct. **28**, 105 (2000)
156. Naritsuka, S., Kobayashi, O., Mitsuda, K., Nishinaga, T.: J. Crystal Growth **254**, 310 (2003)
157. Shamirzaev, T.S., Abramkin, D.S., Nenashev, A.V., Zhuravlev, K.S., Trojánek, F., Dzurnák, B., Malý, P.: Nanotechnology **21**, 155703 (2010)
158. Dabiran, A.M., Cohen, P.I.: J. Crystal Growth **150**, 23–27 (1995)
159. Koguchi, N., Ishige, K.: Jpn. J. Appl. Phys. **32**(Part 1), 2052 (1993)
160. Bajaj, K.K.: Mater. Sci. Eng. B **79**, 203 (2001)
161. Bimberg, D., Grundmann, M., Ledentsov, N.N.: Quantum Dot Heterostructures. Wiley, New York (1999)
162. Wang, Zh.M. (ed.): Lecture notes in nanoscale science and technology: Self-Assembled Quantum Dots. Springer, New York (2008)

Towards a Controlled Growth of Self-assembled Nanostructures: Shaping, Ordering, and Localization in Ge/Si Heteroepitaxy

L. Persichetti, A. Capasso, A. Sgarlata, M. Fanfoni, N. Motta, and A. Balzarotti

Abstract The strain-induced self-assembly of suitable semiconductor pairs is an attractive natural route to nanofabrication. To bring to fruition their full potential for actual applications, individual nanostructures need to be combined into ordered patterns in which the location of each single unit is coupled with others and the surrounding environment. Within the Ge/Si model system, we analyze a number of examples of bottom-up strategies in which the shape, positioning, and actual growth mode of epitaxial nanostructures are tailored by manipulating the intrinsic physical processes of heteroepitaxy. The possibility of controlling elastic interactions and, hence, the configuration of self-assembled quantum dots by modulating surface orientation with the miscut angle is discussed. We focus on the use of atomic steps and step bunching as natural templates for nanodot clustering. Then, we consider several different patterning techniques which allow to harness the natural self-organization dynamics of the system, such as: scanning tunneling nanolithography, focused ion beam and nanoindentation patterning. By analyzing the evolution of the dot assembly by scanning probe microscopy, we follow the pathway which leads to lateral ordering, discussing the thermodynamic and kinetic effects involved in selective nucleation on patterned substrates.

L. Persichetti • A. Sgarlata • M. Fanfoni • A. Balzarotti
Dipartimento di Fisica, Università di Roma Tor Vergata,
Via della Ricerca Scientifica 1, 00133, Rome, Italy

A. Capasso • N. Motta (✉)
School of Engineering Systems, Queensland University of Technology,
GPO BOX 2434, Brisbane, QLD 4001, Australia
e-mail: n.motta@qut.edu.au

S. Bellucci (ed.), *Self-Assembly of Nanostructures: The INFN Lectures, Vol. III*,
Lecture Notes in Nanoscale Science and Technology 12,
DOI 10.1007/978-1-4614-0742-3_4,

1 Introduction

During the last few decades, the field of nanoscience and nanotechnology has undergone a revolution that parallels the extraordinary advances of surface science instruments and techniques. The *nano* era is touted to have begun in the early 1980s with the invention of scanning probe microscopy which allowed one to monitor, measure, and manipulate matter at the nanoscale level. Dramatic new insights have come from the application of this and other new experimental tools. The ability to precisely control atoms and build molecules at extremely small length scales is leading to unprecedented breakthroughs in electronics [1, 2], photonics [3–7], medicine [8, 9], and energy production [10, 11].

Such a diverse range of potential applications arises from the exploitation of the unique phenomena that matter exhibits at the nanoscale. Here, at the cross-road between quantum and classical realms, the electronic, optical, chemical, magnetic and, mechanical properties of matter display a strong size and shape dependence because of quantized effects due to the reduced dimensions and surface effects arising from the increased surface-to-volume ratio. In the nanometer regime, the quantum manifestation of nature becomes relevant and measurable because the overall dimensions of objects are comparable to the characteristic length scale for fundamental excitations in materials, such as the electron wave function, the coherence length of phonons and magnons, or the correlation length for collective phenomena such as ferromagnetism and superconductivity. As the characteristic length scale decreases upon entering the nanoscale regime, the time scale of physical phenomena is concomitantly shifted. For example, the rate of kinetic processes is generally increased in nanomaterials due to the preponderance of surfaces. The latter effect leads to the increased effectiveness of catalytic nanomaterials [12–15] and nanosensors [16, 17].

To bring to fruition their full potential for actual applications, individual nanostructures need to be combined into ordered patterns in which the location of each single unit is coupled with others and the surrounding environment. The required length scale of the pattern depends on the ultimate application. For instance, the issue that precisely governs the absolute positioning of few nanostructures is the most stringent requirement for devices based on spatial coherence such as single electron transistors [18, 19], cellular automata [20, 21], or quantum computing devices [22, 23], whereas, different goals, i.e., high packing density and size uniformity, are the focus of the task for other applications such as light emitting diodes (LED) and lasers [6, 7].

The variety of methods that aims to create structures at the nanoscale can be broadly classified into two approaches. The *top-down* routes attempt to manipulate progressively smaller pieces of matter by photolithography and related techniques. The manufacturing of electronic circuitry in modern semiconductor technology is a typical example of a top-down approach in which the traditional workshop of externally controlled microfabrication tools is used to make small-scale components out of a macroscopic silicon wafer. It is generally agreed that the conventional photolithographic techniques are rapidly nearing the upper limits of their capabilities due to severe technical and economical limitations for dimensions smaller than 10–20 nm [24].

The capability to manufacture nanostructures in the sub-10 nm range is a great advantage of the *bottom-up* approach. This strategy exploits the spontaneous tendency of nano- or subnano-scale objects (namely, atoms, or molecules) to form kinetically or thermodynamically preferred structures and patterns without external direction. An attractive *natural* route for nanofabrication is the strain-induced self-assembly of suitable semiconductor pairs. A semiconductor film which is epitaxially grown onto a surface of another semiconductor with the same symmetry and crystal structure, but different lattice parameter, is inherently unstable due to the strain accumulation at the heterojunction interface. The lattice mismatch between the constituents is at the origin of excess elastic strain energy which tends to be relieved by surface roughness, occurring with a typical nanometer length scale in the form of coherently strained islands (usually referred to as *quantum dots*) on top of a two-dimensional wetting-layer (WL). This growth mode is known as Stranski–Krastanov (SK) growth [25–28]. Early attempts to grow perfectly smooth films for quantum wells had focused on suppressing these morphological instabilities. However, after the pioneering studies of Ekimov and Onushchenko [29] and Brus [30] demonstrated the amazing electronic and optical properties deriving from exciton confinement in QDs, the SK heteroepitaxy has been intensely pursued as an attractive natural route to build up and engineer nanostructures [31–35] because it offers an economical parallel process with the additional advantage of being compatible with the existing Si technology. Nevertheless, the long-range coherence of the process is still usually poor and perhaps the most challenging task of current research is to harness it.

This article surveys some recent research, mostly based on the authors' and their collaborators' works, which attempts to address this issue.

In Sect. 2, we provide a number of examples of self-organization strategies in which the shape, positioning, and actual growth mode of epitaxial nanostructures are manipulated by taking advantage of the physical processes that define the dynamics and thermodynamics of nanostructures on surfaces [36–42]. In this context, we focus on the role of elastic interactions in determining the configuration and the evolution of self-assembled quantum dots. Since the elastic forces in solid materials act at the nanometer range [43–45], these interactions are particularly suited for bottom-up fabrication of semiconductor quantum dots [31, 46–49]. Here, we discuss the possibility of controlling the elastic environment of interacting islands by modulating substrate orientation with the miscut angle.

Substrate miscut provides a way to introduce atomic steps in excess on the surface. Alternatively, the step density can be controlled through current-induced step bunching [50]. Either way, steps lend themselves to function as natural templates for overlayers growth, thus exerting a range of roles in the self-assembly and nanoscale evolution of materials' surfaces [51].

To design complex architectures of nanostructures, spontaneous self-assembling mechanisms may be integrated with pre-patterned surface templates. In Sect. 3, we discuss these hybrid bottom-up/top-down approaches. Several lithographic techniques are examined, such as scanning tunneling microscopy (STM) nanolithography, nanoindentation patterning, and focused ion beam (FIB) lithography.

In the course of the article, we consider Ge/Si as a prototype model system for semiconductor heteroepitaxy. Besides being promising for microelectronic applications [52–61], this system offers a fascinating and instructive example of self-assembled nanostructure formation [62–67].

2 Surface Steps as Natural Nanotemplates for Ge Growth

2.1 Continuum Description of Stepped Surfaces

On the nanometer scale, a seemingly smooth crystalline surface is actually modulated by a periodic arrangement of steps. These line boundaries at which the surface changes height, by one or more atomic units, are the key building blocks in surface growth processes [68]. Therefore, any self-organization scheme is affected by steps and should, in principle, try to benefit from their presence. To this purpose, one has to master the problem of describing the evolution of steps within a range of length scales from nanometers to microns, since the properties of solid-state devices generally lie in this size regime. A mesoscopic approach, known as continuum step model (CSM), has been proven to be extremely powerful in this case [51]. In a CSM framework, short-wavelength fluctuations at the atomic scale, such as kinks, are integrated out by approximating step profiles with continuum functions $x(y)$, where x is the coordinate perpendicular to the mean orientation of the steps and y is the coordinate parallel to it, as denoted in Fig. 1a. For this configuration, the free energy functional, or continuum step Hamiltonian, which controls fluctuations of coarse-grained steps can be written as [40, 51, 69]

$$H[x(y)] = \int \mathrm{d}y \left[\beta[\theta(y)] \sqrt{1 + \left(\frac{\mathrm{d}x}{\mathrm{d}y}\right)^2} + V[x(y)] \right]. \tag{1}$$

Here, the free energy per unit length of a step, β, is expressed in terms of the local orientation θ relative to the mean direction of the step and $V[x(y)]$ is the step–step interaction potential. By expanding $\beta(\theta)$ and the radical in (1) to the second order in $\theta \approx \tan\theta = \partial x / \partial y$, one has

$$\sqrt{1 + \left(\frac{\mathrm{d}x}{\mathrm{d}y}\right)^2} \cong 1 + \frac{1}{2}\left(\frac{\mathrm{d}x}{\mathrm{d}y}\right)^2 \cong 1 + \frac{1}{2}\theta^2, \tag{2}$$

and

$$\beta(\theta) \cong \beta(0) + \beta'(0)\theta + 1/2\beta''(0)\theta^2. \tag{3}$$

For close-packed steps $\beta'(0) = 0$ and the effective Hamiltonian becomes thereof

$$H[x(y)] = \int \mathrm{d}y \left[\beta(0) + \frac{1}{2}\tilde{\beta}(0)\left(\frac{\mathrm{d}x}{\mathrm{d}y}\right)^2 + V[x(y)] \right], \quad \tilde{\beta}(0) = \beta(0) + \beta''(0), \tag{4}$$

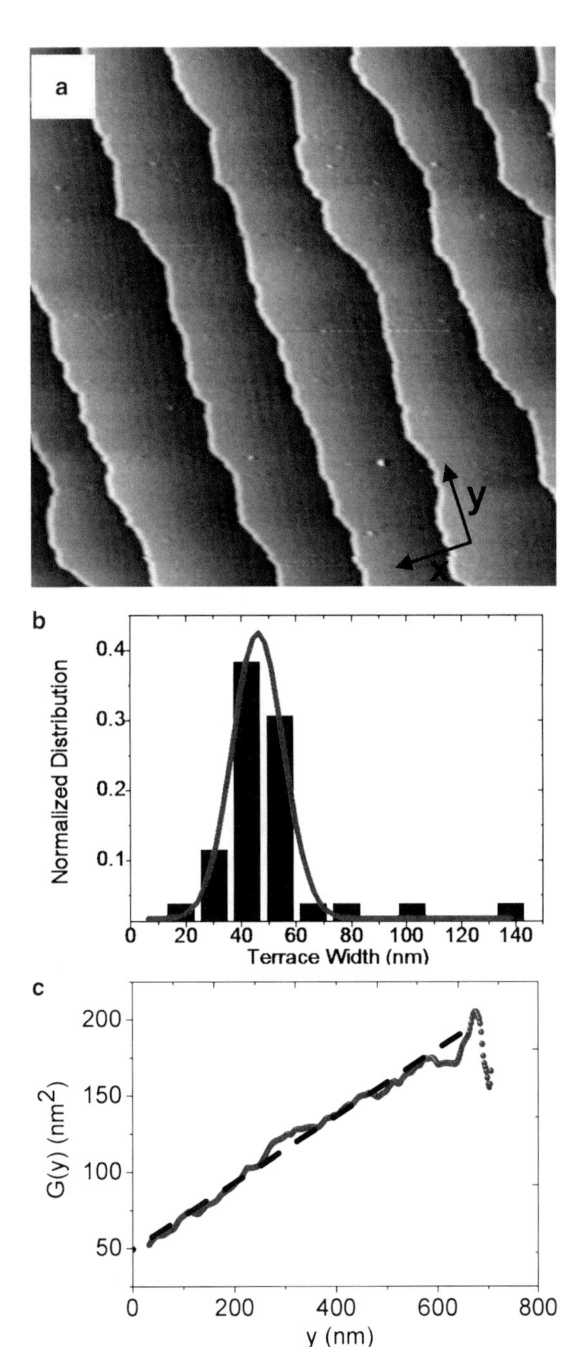

Fig. 1 (**a**) STM image ($325 \times 325\ \text{nm}^2$) of the clean Si(111) surface. (**b**) Terrace width distribution measured on the flat Si(111) surface. The average terrace width corresponds to $\langle l \rangle = (46 \pm 2)$ nm. The best Gaussian fit to the data, shown as *solid line*, has a variance of $\langle \sigma^2 \rangle = (0.078 \pm 0.017)\langle l \rangle^2$. This results in an interaction straight $A = (0.68 \pm 0.07)$ eV A. (**c**) Spatial correlation function G (y) averaged over 50 steps and a total step length of about 700 nm. The *straight dashed line* is a fit to the data from which the step stiffness reported in the text is evaluated

where $\tilde{\beta}$ is the *step stiffness* [70]. By recognizing y as a time-like variable, an interesting analogy can be drawn between the 2D configuration of a step and the word line of a particle in 1D [71]. Since steps cannot cross to avoid overhangs, these particles are described as fermions [72]. In this perspective, step stiffness translates to the inertia of massive particles.

From the noncrossing condition, a steric or entropic repulsive interaction arises. Besides, an explicit energetic interaction is caused by the lattice distortion at the step edges. The interaction among steps can be treated within a mean-field picture (Gruber–Mullins approximation (GM) [73]) in which a step is considered to wander in the mean potential created by all the other steps. Moreover, a strictly one-dimensional interaction, occurring between segments of neighboring steps at constant y, is assumed.

Firstly, we describe the case in which the energetic interaction is negligible with respect to the entropic repulsion. In GM approximation, the step becomes a particle in a potential well of width $2\langle L\rangle$, where $\langle L\rangle$ is the average step spacing. The terrace-width distribution (TWD) and the entropic repulsion energy can be obtained by the ground-state properties of the specular fermion description

$$E_0 = \frac{(\pi kT)^2}{8\tilde{\beta}\langle L\rangle^2},$$
$$P(s) = \sin^2\left(\frac{\pi s}{2}\right), \tag{5}$$

where we defined the normalized step distance $s = L/\langle L\rangle$. As it will be shown in the next paragraph, this simple model describes very well the step–step interactions on vicinal surfaces, where terrace widths become comparable to thermal step fluctuations.

For explicit energetic repulsion, the mean-field potential is instead [43]

$$V(s) = \frac{A}{\langle L\rangle^2}\sum_{n=1}^{\infty}\left[\frac{1}{(n+s)^2} + \frac{1}{(n-s)^2}\right] \approx \frac{A}{\langle L\rangle^2}\left(\sum_{n=1}^{\infty}\frac{2}{n^2}\right) + \frac{\pi^4}{15}\frac{A}{\langle L\rangle^4}s^2, \tag{6}$$

where the constant A is the *interaction strength* and in the final form we make use of the harmonic approximation around $s = 0$. The step distribution function is then described by the familiar Gaussian ground-state wave function

$$P_{\mathrm{G}}(s) = \frac{1}{\sigma\sqrt{2\pi}}\mathrm{e}^{\left[-\frac{(s-1)^2}{2\sigma^2}\right]}, \tag{7}$$

with the variance σ^2 which is given by

$$\sigma^2 = \frac{k_{\mathrm{B}}T}{\sqrt{48A\tilde{\beta}}}, \tag{8}$$

when only nearest-neighbor interactions are taken into account [69].

With the CSM approach, the structure and the evolution of steps can be described in terms of empirical parameters, step stiffness and interaction strength, that are directly accessible to experiment. Whereas the interaction strength is derived by fitting the experimental distribution of step–step spacings to (7) (Fig. 1b), the step stiffness is extracted from measurements of the mean-square wandering along the step edge in STM images:

$$G(y) = \left\langle [x(y) - x(0)]^2 \right\rangle. \tag{9}$$

Operationally, the analysis of STM images proceeds as follows: the profile of each step is digitalized and then fitted to a proper $x(y)$ continuous function. The correlation function $G(y)$ is calculated thereof and averaged over a significant statistical ensemble of steps. For distances smaller than the correlation length, the wandering of a step edge can be considered as a simple random walker moving on either sides of neighboring steps while advancing in the y direction. Pursuing this analogy, the differential displacement $G(y)$ is expected to increase linearly with "time" y:

$$G(y) = \left\langle [x(y) - x(0)]^2 \right\rangle \frac{kT}{\tilde{\beta}} y. \tag{10}$$

In Fig. 1a, a typical STM image of the flat Si(111) surface with monoatomic steps is displayed. The related correlation function $G(y)$ (Fig. 1c) shows that a linear behavior is effectively found over a y range larger than the average step–step distance (approximately 46 nm), while for larger y values, the linear dependence ceases, in qualitative agreement with theory [74]. From the slope of $G(y)$, the step stiffness $\tilde{\beta} = (43.2 \pm 0.4)$ meV/A is derived at the freeze-in temperature of $T = 1{,}103$ K [75]. This value compares well with previous reflection electron microscopy data from which a step stiffness $\tilde{\beta} \approx 46$ meV/A is calculated [76, 77].

2.2 *Vicinal Surfaces*

A regular array of steps can be produced by cutting a single crystal few degrees away from a low-index plane. If the cleavage plane is slightly misoriented from a crystallographic plane, the surface breaks up into a staircase of terraces limited by steps, and it is referred to as *vicinal*. The relevant parameters of the step array depend on the misorientation or *miscut* angle θ of the vicinal surface from the low-index plane (Fig. 2). The possibility of tuning the step–step separation by changing the miscut angle paves the way to the formation of lateral superlattices on the surface [78–84]. Nonetheless, the thermodynamic properties of steps are also strongly modified by vicinality and, hence, their behavior should be carefully assessed before building at the nanoscale.

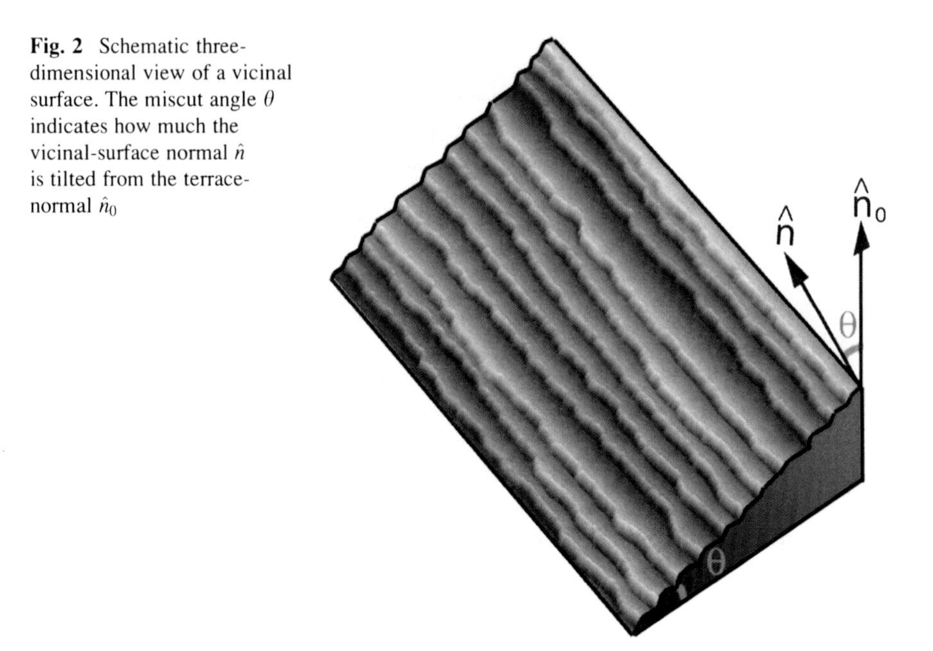

Fig. 2 Schematic three-dimensional view of a vicinal surface. The miscut angle θ indicates how much the vicinal-surface normal $\hat{n}$ is tilted from the terrace-normal $\hat{n}_0$

2.3 STM Probing of Miscut-Dependent Energetics of Steps

A wide range of atomistic and continuum models has been developed to explain the energetics of steps on vicinal Si(001) surfaces [85–90]. Here, we describe an experimental study of step stiffness and step–step interaction for substrates misoriented toward the [110] direction from 0.2° to 8° (for experimental details, see [75]). Following the methodology outlined in Sect. 2.1, we assess the dependence of these empirical parameters on the miscut angle and we show how to interpolate between continuum mesoscopic models and atomistic calculations of vicinal surfaces.

Before proceeding with the work-up, a brief digression on the structural properties of flat and vicinal Si(001) surfaces is in order. To minimize the surface energy of unreconstructed Si(001), neighboring silicon atoms on the surface dimerize along the $\langle 110 \rangle$ directions to form a (2×1) structure [91, 92]. The (2×1) reconstruction is sketched schematically in Fig. 3a. High-resolution STM images show that the orientation of the dimer rows rotates by 90° for two adjacent terraces (Fig. 3b, c). Following the notation of Chadi [93], we define S_A and S_B steps as those in which the upper-terrace-dimerization direction is perpendicular or parallel to the step ledge, respectively. Since the formation energy of kinks is higher in a S_A step than in a S_B step, the profiles of A-steps appear smooth while the B-steps can be jagged (Fig. 3c). On vicinal surfaces with low miscut angle, the two types of steps alternate but, as the miscut angle is increased, an evolution to a surface made up mostly of double-height B-type steps (D_B) occurs [89, 94, 95] (Fig. 4). This transition from single to double steps has been characterized both experimentally

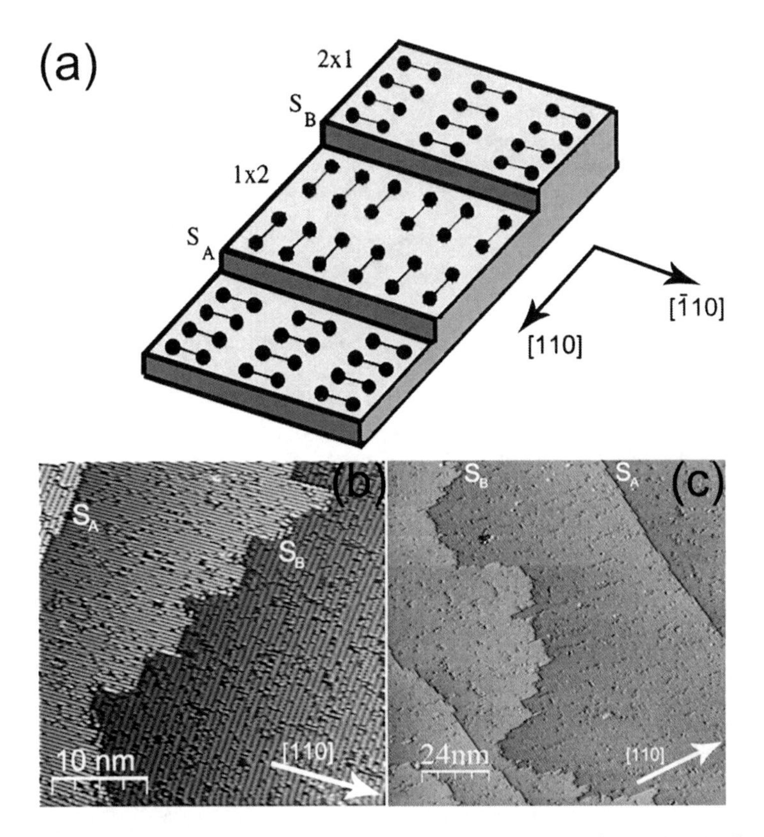

Fig. 3 (**a**) Schematic representation of the alternating (2 × 1) and (1 × 2) domains leading to S_A and S_B steps on the flat Si(001) surface. (**b, c**) High-resolution STM images of the flat Si(001) surface. S_A and S_B steps are indicated

[96, 97] and theoretically [89, 90]. In particular, Pehlke and Tersoff described this transition as an equilibrium between two phases $S_A + S_B$ and D_B, leading to a reaction $S_A + S_B \rightarrow D_B$ that proceeds rightwards as the miscut angle gets higher. By means of atomistic calculations based on the Stilling–Weber potential, they predict that during this transition, a mixed phase of single- and double-height steps should be observed, since it is lower in energy than any other combination of pure phases. STM measurements provide convincing confirmation of this model. For a given miscut angle θ, the terrace width $\langle L \rangle$ of a single-stepped surface is half that of a double-stepped one and the average terrace width in the mixed phase can be written as

$$\langle L \rangle = \frac{[n_S(\theta) + 2n_D(\theta)]a}{\sqrt{8}\tan\theta}, \tag{11}$$

where $a = 0.384$ nm is the surface lattice constant of Si(001) and n_S and n_D are the relative densities of single $S_A + S_B$ and double D_B steps [97]. In Fig. 5a, the average step spacings measured on STM images are compared with (11) for a mixed phase (upper curve) and for the limiting case of a pure single height phase (lower curve).

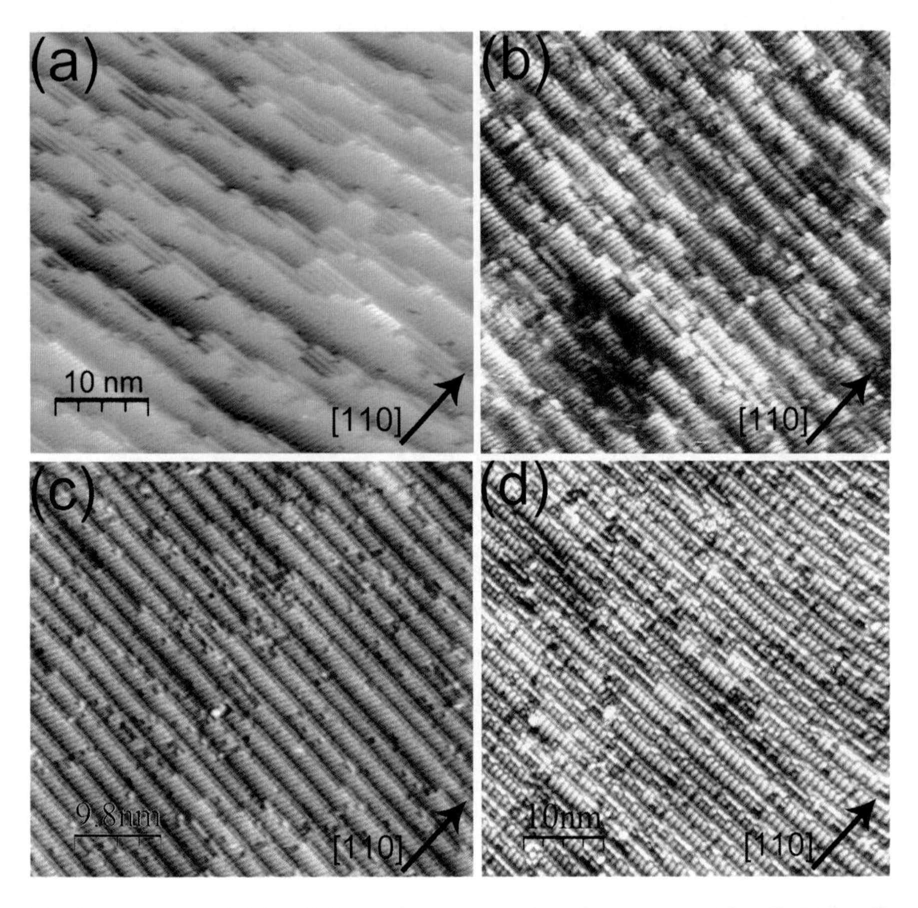

Fig. 4 STM images of vicinal Si(001) surfaces at increasing miscut angles: (**a**) $\theta = 2°$, (**b**) $\theta = 4°$, (**c**) $\theta = 6°$, (**d**) $\theta = 8°$

Confirming the predictions of atomistic calculations, we find that the experimental behavior is well described by a binary mixture of single- and double-height steps, while the pure single-height phase decreases more rapidly with increasing miscut angle. From visual inspection of Fig. 4, it can be clearly seen that steps are stiffer at high miscuts. This is quantitatively accounted for by a gradual increase in step stiffness measured at different miscut angles (Fig. 5b). The rise of $\tilde{\beta}(\theta)$ at large misorientations is due to the drop of terrace widths and to the appearance of D_B steps for which bending costs more energy. The thermodynamic values of the step stiffness can be related to the atomistic description of steps, using simple models of atomic interactions, such as lattice models. According to the terrace-step-kink (TSK) model, the dependence of $\tilde{\beta}$ on the kink energy ε, taking $a_{\perp} = a_{\parallel} = a$, is given by [98]

$$\tilde{\beta} = \frac{2k_B T}{a} \sinh^2\left(\frac{\varepsilon}{2k_B T}\right). \tag{12}$$

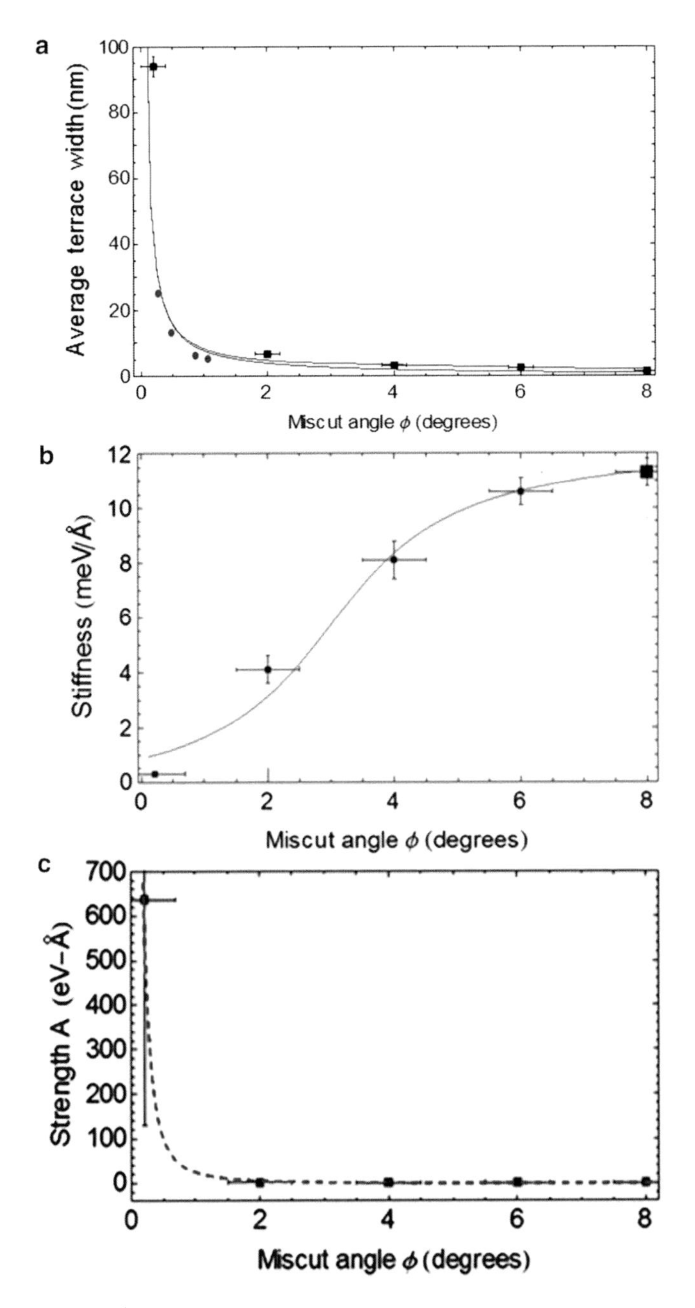

Fig. 5 (**a**) Measured average terrace widths of mixed ($S_A + S_B$) and double D_B steps as a function of miscut angle for Si(001). The *continuous lines* are obtained from (11). The *lower curve* refers to the phase only. *Full dots* are data points taken from [97]. (**b**) Step stiffness measured at different miscut angles and calculated fitting function of the TSK model. The highlighted point is the extrapolated value of the step stiffness at 8°. (**c**) Step–step interaction strength on vicinal Si(001) surface. The *squares* are experimental data obtained in the two limit regimes: (1) energetic interactions for $\theta<2°$, (2) entropic interactions above

Table 1 Experimental parameters of continuum step model for vicinal Si(001) surfaces

θ	$\langle L \rangle$ (nm)	A (eV Å)	$\tilde{\beta}$ (meV/Å)
0.2°	94 ± 3	636 ± 504	0.30 ± 0.04
2°	6.8 ± 0.2	(22 ± 17) 1.3 ± 0.2	4.1 ± 0.5
4°	3.23 ± 0.09	(29 ± 20) 0.54 ± 0.03	8.1 ± 0.7
6°	2.52 ± 0.05	(41 ± 24) 0.52 ± 0.05	10.6 ± 0.5
8°	1.36 ± 0.02	0.47 ± 0.02	11.3 ± 0.5[a]

The value of the interaction strength A for the quasi flat surface is extracted from the Gaussian TWD; at larger miscut angles, where the harmonic-well approximation breaks down, the values of A evaluated from the Gaussian TWD (*in brackets*) are compared with their estimates in entropic regime (*not in brackets*)

[a] Extrapolated from the fitting curve shown in Fig. 5b

By expressing ε in terms of the step densities n_{S} and n_{D}, one can fit the stiffness data into (12) and obtain for $T = 767$ K the kink energies $\varepsilon_{\mathrm{S}} = 20$ meV/atom and $\varepsilon_{\mathrm{D}} = 70$ meV/atom for the $S_{\mathrm{A}} + S_{\mathrm{B}}$ and D_{B} steps, respectively. The latter energies can be compared to the values $\varepsilon_{\mathrm{SA}} = 28 \pm 2$ meV/atom, $\varepsilon_{\mathrm{SB}} = 90 \pm 10$ meV/atom, measured [97] at 873 K. The similarity between S_{B} and D_{B} energies is not fortuitous since at high temperatures the meandering weakens the distinction between single and double steps. The same fitting curve is used to extrapolate the step stiffness at 8°, where the continuum elastic approximation is no longer reliable because the amplitude of step fluctuations becomes the order of step separation.

If one could construct an estimate of the interaction strength $A(\theta)$ from microscopic models, this would be a direct way of quantifying the range of validity of the harmonic-well approximation (8), i.e., above a certain miscut angle, the step–step interaction should be dominated by entropy and noncrossing conditions. To this purpose, the average ledge energy of a binary solution of single and double steps has been determined from atomistic calculations [89] of equilibrium energy of each type of steps as

$$E(\theta) = E_{\mathrm{S}}(\theta) n_{\mathrm{S}}(\theta) + E_{\mathrm{D}}(\theta) n_{\mathrm{D}}(\theta), \quad (13)$$

where E_{S} and E_{D} are the ledge energies computed using the Stilling–Weber potential for a pair of single and double steps, respectively. Then, the interaction strength $A(\theta) = E(\theta)\langle L \rangle^2$ has been calculated as a function of θ. The result is plotted in Fig. 5c as a dashed line. Comparing these atomistic estimates of A with results of mesoscopic models reported in Table 1, it can be seen that the harmonic-well approximation is in agreement with atomistic calculations only for the quasi-flat surface. At larger angles, it overestimates the atomistic data, although the overall functional behavior is preserved. For $\theta \geq 2°$, the estimates of A in the entropic regime (5) are instead in excellent agreement with atomistic theory, as shown in Fig. 5c.

Ultimately, the possibility of relating atomic scale properties to a large scale behavior of steps is extremely useful for bridging different length scales in the study of surface. Continuum step models are particularly suitable for this purpose, since they require only a small number of experimentally measured parameters.

2.4 Experimental Investigation and Continuum Modeling of Thermodynamic Stability of Ge Islands on Si(001) Surface

Strain-driven self-assembly in heteroepitaxial systems has been studied extensively as an attractive route to fabricate semiconductor quantum dots. Controlling the mechanisms of heteroepitaxial self-assembly well enough to induce predictable island properties still remains a significant challenge. In this section, we aim to illustrate how substrate vicinality allows a fine shaping of semiconductor nanostructures in the Ge/Si(001) model system. The complex miscut-dependent nature of vicinal surfaces introduces the concept of asymmetry into the basic phenomena leading to the formation and evolution of self-assembled quantum dots [99–101]. Besides its important implications for the growth process, asymmetry has a technological relevance, since it is potentially able to split degeneracy of quantum dot states and provide optical anisotropy [102, 103].

Before discussing the effect of vicinality, we summarize hereafter the surface evolution of Ge islands on the flat Si(001) surface. Due to the 4.2% lattice mismatch between Ge and Si, the growth proceeds in a Stranski–Krastanov fashion [104]. A pseudomorphic WL is formed up to a critical thickness of about three to four monolayers (ML). Then, the strain energy stored in the film is partially relaxed through the formation of unfaceted mounds or *prepyramids* (Fig. 6a) [105, 106]. As the Ge deposition increases, the prepyramids evolve into square-based pyramids bounded by {105} facets (Fig. 6b). The latter facets can be clearly distinguished as four spots in the surface orientation map (SOM) of the STM image (inset of Fig. 6b). In this two-dimensional histogram, each spot represents the local normal orientation relative to the (001) plane (located at the center of the plot), while the intensity is proportional to the amount of surface with that orientation. When the growth continues, a morphological transition to multifaceted islands (the so-called *domes*) [107–109] occurs (Fig. 6c). The domes have a perfect fourfold symmetry like the pyramids, but more complex morphology, including four {105}, four {113} and eight {15 3 23} facets [110], as shown by SOM plot (inset of Fig. 6c). Finally, dome islands may eventually dislocate and larger islands named *superdomes* with interfacial misfit dislocations may appear (Fig. 6d) [111–114]. Following this evolution in terms of height-to-base aspect ratio (r), a progressive increase in aspect ratio occurs with increasing island volume V. As a result, the small prepyramids ($r < 0.1$) and pyramids ($r = 0.1$) are very shallow. Instead, the large domes are markedly steeper, being 0.22 in r. The essential features of these morphological changes can be understood within a simple thermodynamical model [108]. It is well known that the total energy gain associated with the formation of a three-dimensional island of volume V on the WL is[1]

[1] In the following simplified approach, we neglect the contribution of edge-energy terms on island stability.

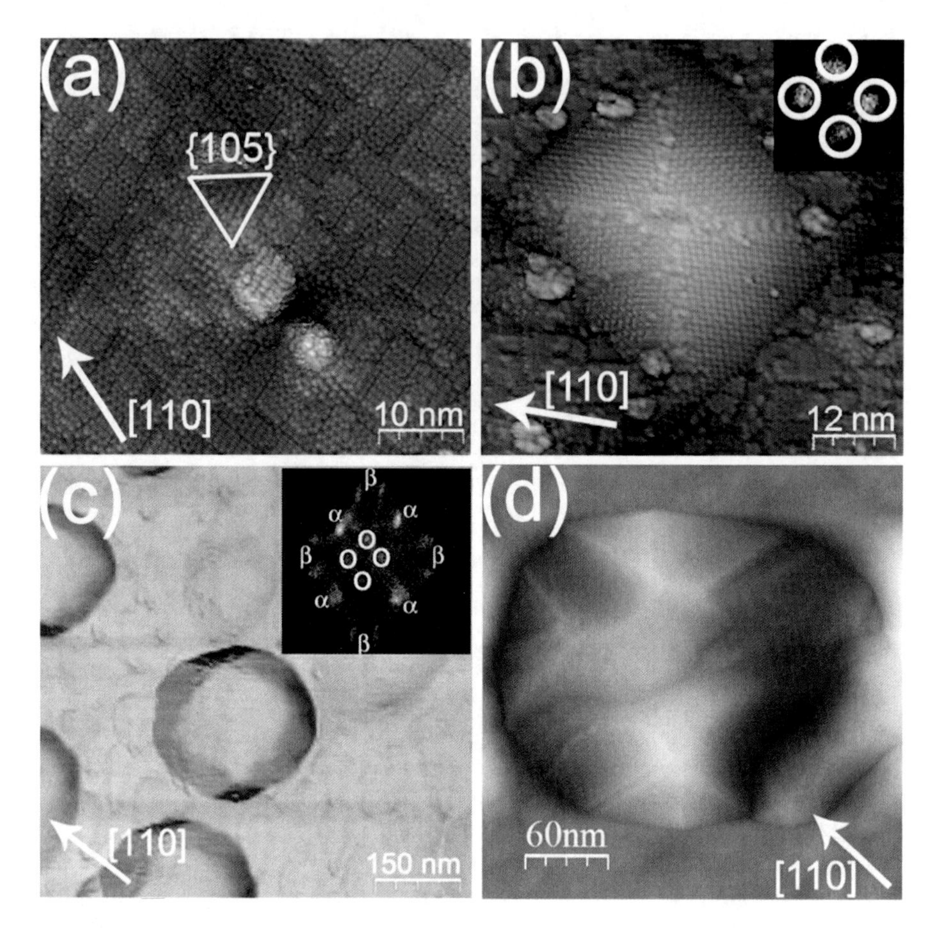

Fig. 6 Shape evolution of Ge islands on the flat Si(001) surface: (**a**) pre-pyramid, (**b**) {105}-pyramid, (**c**) multifaceted dome, and (**d**) dislocated superdome. In the *insets* the facet plots are shown. The spots of the various facets are labeled as follows: {105} by ▯; {113} by α; {15 3 23} by β

$$E_{\mathrm{tot}} = e_{\mathrm{relax}} V + e_{\mathrm{surf}} V^{2/3}, \tag{14}$$

where the first term represents the elastic strain relief caused by the three-dimensional geometry while the second one accounts for the formation of island facets. By using finite element (FE) simulation, the solutions of the elastic problem can be found for realistic three-dimensional island shapes. The strain tensor, ε_{ij}, in the system (Ge island + Si substrate) is determined using the FE method to solve the set of the equations of linear elasticity for an elastic body at equilibrium which experiences a misfit strain in absence of volumetric and surface forces

$$\begin{cases} \sigma_{ij,j} = 0, \\ \varepsilon_{ij} = (u_{i,j} + u_{j,i})/2, \\ \sigma_{ij} = C_{ijkl}(\varepsilon_{kl} - \varepsilon_0 \delta_{kl}), \end{cases} \tag{15}$$

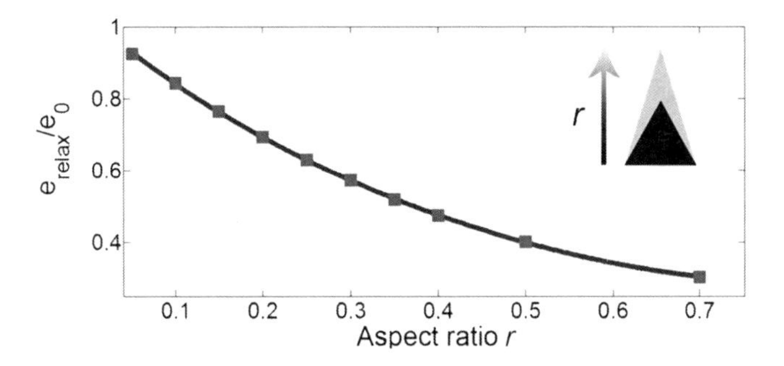

Fig. 7 Elastic relaxation vs. aspect ratio for Ge pyramids on Si(001)

where σ_{ij} are the stresses, C_{ijkl} is the anisotropic stiffness tensor, and ε_0 is the mismatch strain. The strain energy density $\rho = \frac{1}{2}C_{ijkl}(\varepsilon_{ij} - \varepsilon_0\delta_{ij})(\varepsilon_{kl} - \varepsilon_0\delta_{kl})$ is calculated from the solution of (15) and e_{relax} is obtained straightforward by the difference between the residual strain energy stored in the system after relaxation and the energy in an equivalent volume of a fully strained epitaxial Ge film (e_0). The surface term, e_{surf}, is the extra surface energy per unit area due to the presence of the island:

$$e_{\text{surf}} = \left[\sum_{i=1}^{N} \gamma_i S_i - \gamma_0 S_0\right] V^{-2/3}, \tag{16}$$

where γ_i and S_i denote the surface energy and surface area of the ith surface of the island. S_0 is the base area and γ_0 is the energy per unit area of the WL. As shown in Fig. 7, the formation of islands with a large aspect ratio allows for a better bulk strain relaxation. Nonetheless, it also involves a larger surface energy cost. Thus, in the small-volume limit where the surface term is dominant, shallow islands result to be energetically favored. As the island volume grows, the volumetric term becomes increasingly important, leading to steeper morphologies. To illustrate the point, we plotted in Fig. 8a the dependence of E_{tot} vs. V for different Ge pyramids. For concreteness, we consider the case of identical facet energies $\gamma_i = \gamma_0 = 65$ meV/A^2. Even in this over-simplified example, the sequence of shape changes toward higher aspect ratios is confirmed and the {105} pyramids ($r = 0.1$) have the lower energy for small island volumes, consistently with the experiment. While instructive and intuitively enlightening, the above description neglects some important processes which take place in real systems. Actually, SiGe alloying offers a competitive and simultaneous route to three-dimensional islanding in the reduction of the strain energy. The experimental characterization [42, 115–119] and theoretical modeling [120, 121] of intermixing and the study of the fascinating phenomena related to it [120–123] have been the subject of intense research work which goes well beyond the scope of this lecture. Furthermore, our analysis is focused on the onset of SK growth where the effect of intermixing is limited.

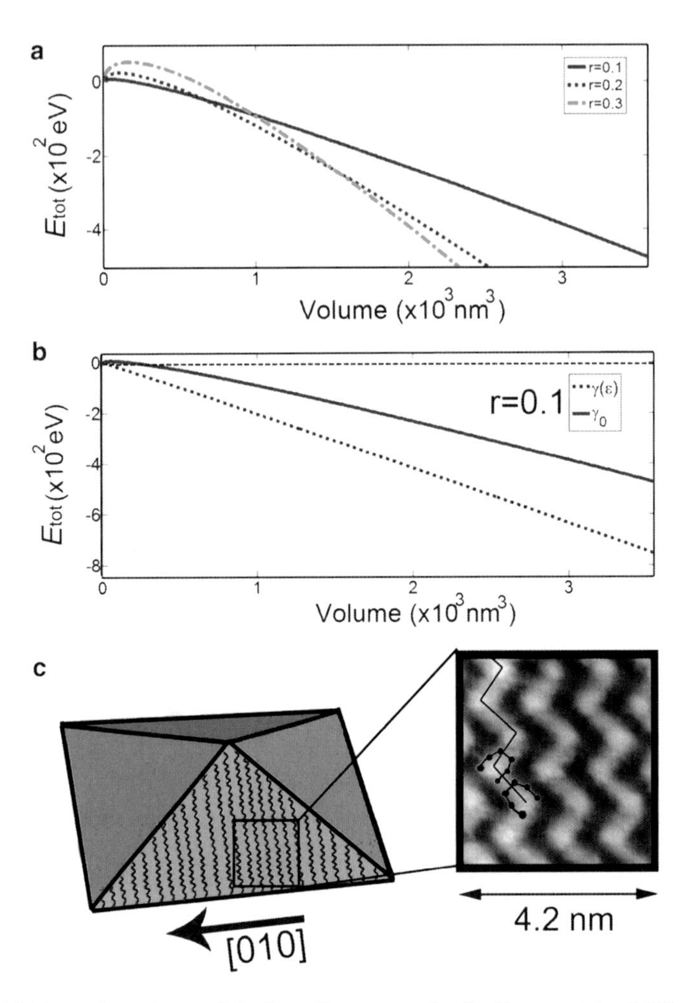

Fig. 8 (**a**) Volume dependence of the formation energy E_{tot} for Ge pyramids of different aspect ratios r calculated using (14) for the case of identical facet energies $\gamma_i = \gamma_0 = 65\ \text{meV/A}^2$. (**b**) Formation energy of a {105} Ge pyramid calculated assuming a strain-independent surface energy (*continuous line*) in comparison to the curve obtained after taking into account the dependence of $\gamma_{\{105\}}$ on strain (*dashed line*). (**c**) Schematics showing the orientation of the rows of the RS reconstruction for Ge pyramids on the flat Si(001) surface. In the blow up, a STM image of the Ge(105) RS reconstruction is displayed

Instead, what is more important for the thermodynamic stability of {105} Ge pyramids is the peculiar surface reconstruction of the {105} facets [124, 125]. On the Ge(105) surface, atoms form ordered arrays of *U-shaped* structures which are organized into zigzag rows orthogonal to the [010] direction [rebonded-step (RS) reconstruction] [126, 127] (Fig. 8c). This structure was found to be strongly stabilized under compressive strain of Ge/Si epitaxy [126]. Hence, to obtain more realistic results, it is important to take into account the strain-dependent correction to the surface energies:

$$e_{\mathrm{surf}} = \left[\sum_{i=1}^{N} \gamma_i(\varepsilon_i) S_i - \gamma_0 S_0 \right] V^{-2/3}. \tag{17}$$

This is done by interpolating published ab initio results [126] for the dependence of the Ge(105) surface energy on strain with the in-plane components of the strain averaged over each facet, taken from FE calculations. In Fig. 8b, the corrected stability curve of {105} pyramids (dashed line) is compared with the nominal one (continuous line). The substantial energy gain associated with the {105} faceting under compressive strain flattens the activation energy for island formation, describing the barrierless nucleation process which is actually observed experimentally [106, 128].

2.5 *Tuning Ge Island Shape with Substrate Vicinality*

Several experimental studies have observed extended {105} faceting on Si(001) misoriented substrates [100, 129–134]. Therefore, the {105} energetics appears to be crucial in determining the morphological evolution of Ge islands even in the case of growth on vicinal Si(001) surfaces. Whereas on the flat surface {105} Ge pyramids have a perfect fourfold symmetry and almost square base with each side oriented along the $\langle 010 \rangle$ directions (Fig. 6b), they progressively elongate along the $[\bar{1}10]$ miscut direction, as the substrate misorientation gets higher (Fig. 9). This shape transition is accompanied with the increase of surface area of the facets along the step-down direction at the expense of the other two facets. We now show that there is a strict correlation between the morphological evolution and the energetic factors which govern the {105} faceting at atomic scale. Consider a simple geometric model of a {105} pyramid grown on a vicinal surface (Fig. 10a). The $[\bar{5}51]$ intersection line of adjacent {105} facets forms an 8.05° angle with the (001) plane. To allow {105} faceting, this angle must never change, producing the observed elongation toward the miscut direction. By using elementary geometry, it is straightforward to calculate the expected miscut-dependent asymmetry in terms of the ratio between the longest and the shortest island side (continuous line in Fig. 10b). The values of the same ratio measured on STM nanotopographies are plotted as circles for comparison. The match with the experiment is impressive. Moreover, this simple geometric model sheds light on the special situation for miscuts around 8°. When the miscut angle equals the aforementioned 8.05° angle, a pyramidal shape can no longer form, since the $[\bar{5}51]$ line should run parallel to the substrate orientation; thus, the system rearranges itself into an elongated prism of triangular cross section bounded by two adjacent {105} facets named *nanoripples* [131, 134–136] (Fig. 11b). Their morphology can be easily imaged as the result of

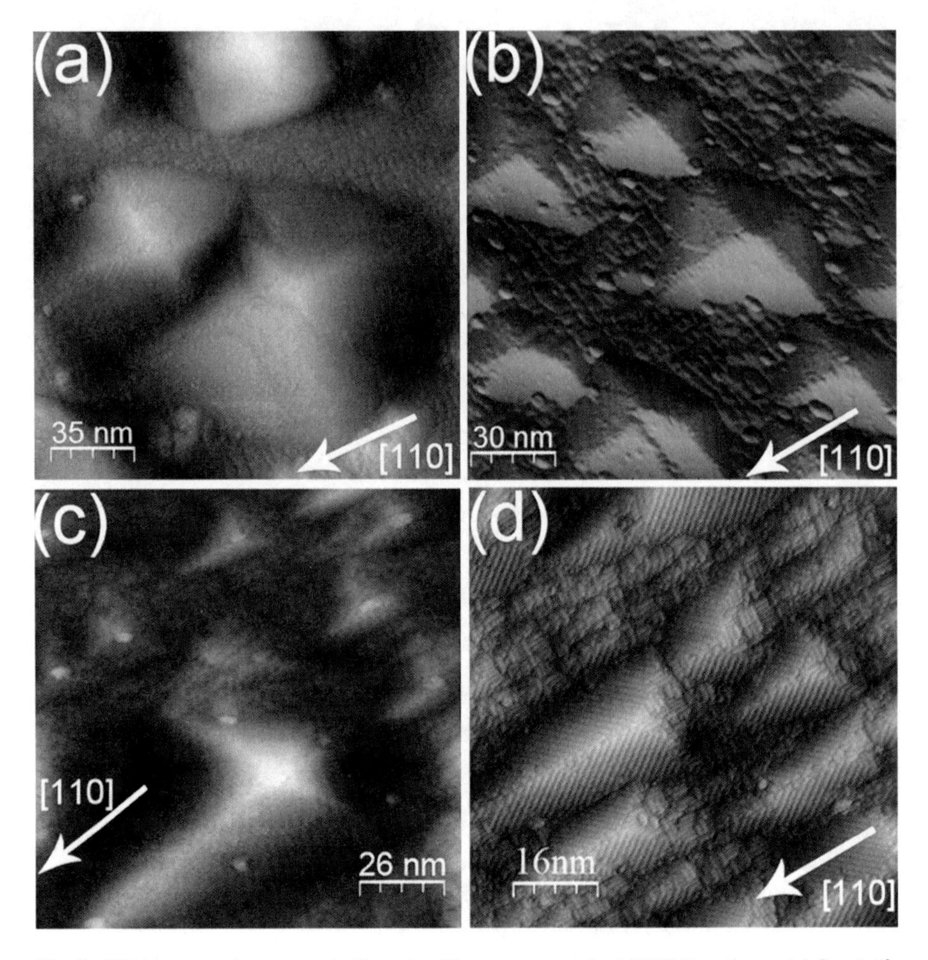

Fig. 9 STM images of asymmetric Ge pyramids grown on vicinal Si(001) surfaces: (**a**) $\theta = 1.5°$, (**b**) $\theta = 2°$, (**c**) $\theta = 4°$, (**d**) $\theta = 6°$

cutting a {105} pyramid along the [110] direction with a plane tilted by 8° from the (001) surface, as schematically displayed in Fig. 11a. From the sketch, it is clear that due to geometric constraints, the down side of the ripple cannot be bounded by {105} facets. This is precisely what is observed on the STM images in Fig. 11c, d, which display the opposite ends of the island along the miscut direction. On one side (Fig. 11c), the ripple is not closed by any real facet but gradually lowers in height and width as the number of the stacked {105} layers decreases near the end of the island.

From what we have described above, it is clear that the significant free energy gain associated to {105} faceting reduces the problem of dot shape to a matter of geometry. This offers an additional degree of control over the shape and symmetry of self-assembled nanostructures with significant novel potential for applications in nanotechnology. We emphasize that the possibility of predicting the shape of the

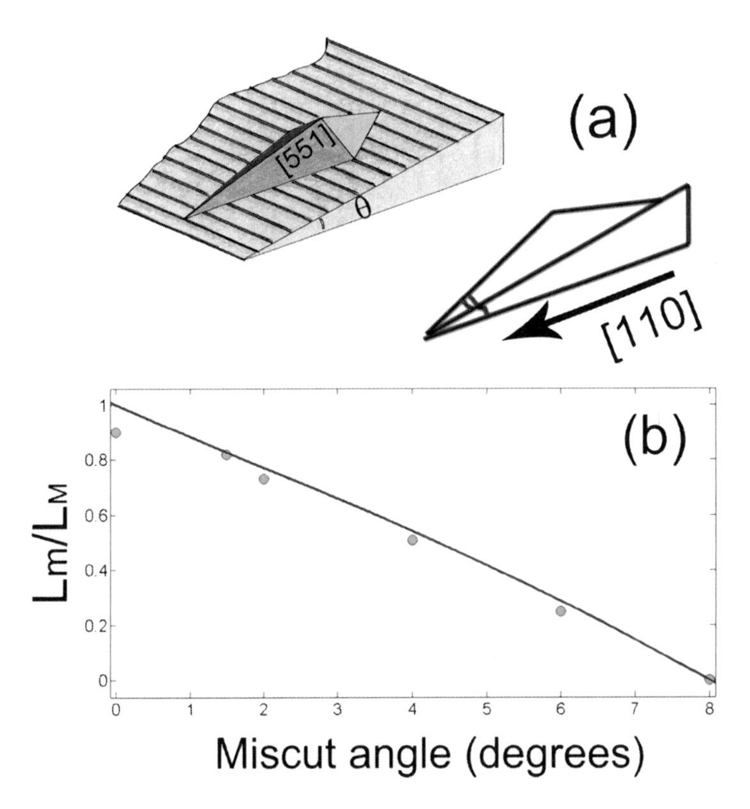

Fig. 10 (**a**) Schematic representation of an island lying on a vicinal surface. (**b**) (L_m/L_M) ratio as a function of the miscut angle. The *filled dots* are the experimental values measured from STM images, while the *continuous line* represents the calculated ratio for an ideal {105} pyramid

dots with such a geometric approach is due to the fact that the structure of the dot facets is not altered at atomic scale. In Sect. 2.4, we have seen that the rows of the RS reconstruction of Ge(105) facets are orthogonal to the ⟨010⟩ directions (Fig. 8c) and, hence, to the pyramid edge (Fig. 12a). Since a vicinal surface consists of arrays of (001) terraces separated by steps, in order to ensure a good matching to the WL, the rows must be kept orthogonal to the ⟨001⟩ directions. As a consequence, they form an angle $<90°$ with the edge of islands grown on vicinal substrates (Fig. 12b, c).

Figure 13 shows FE calculations of the elastic energy per unit volume for all the Ge islands imaged on flat and vicinal Si(001) substrates. It can be seen that the morphological transition to rippled structures is accompanied by an effective strain relief. Evidently, the energetics of ripples and thus their relative stability with respect to domes are markedly different from that of pyramids, because strain relaxation can only occur in the lateral direction and it is geometrically suppressed along the prisms [135]. As shown by a recent experimental study [136], this results in a peculiar morphological evolution which differs significantly from that active on the flat surface.

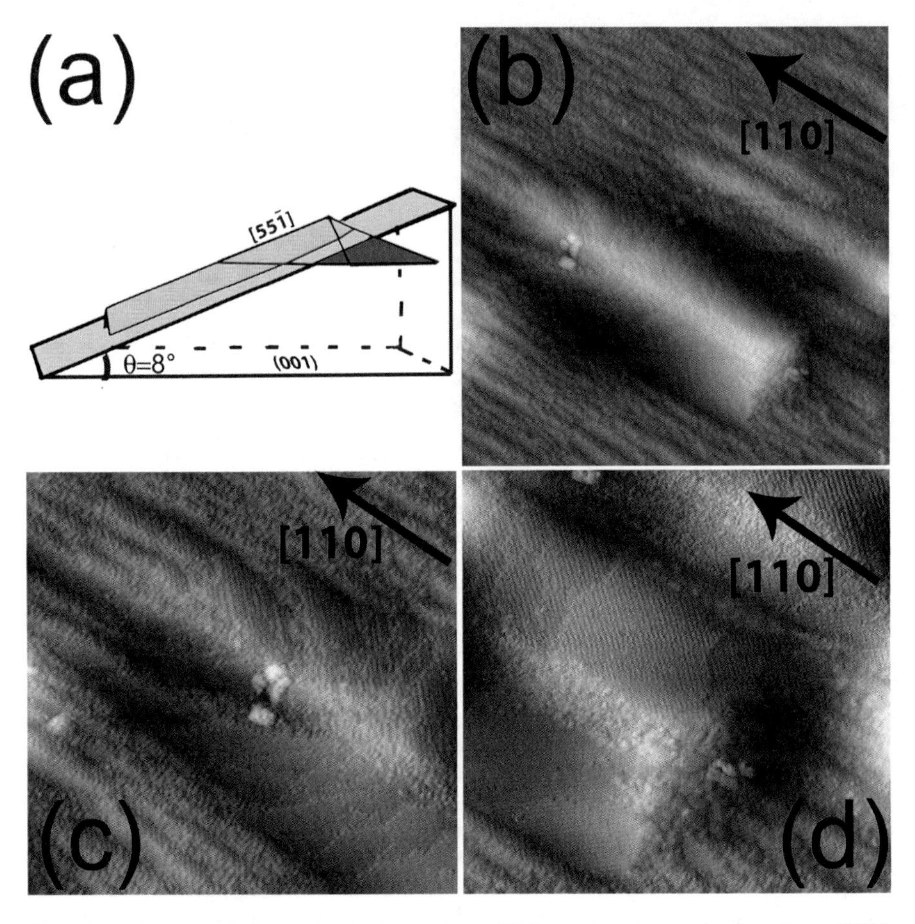

Fig. 11 (**a**) The morphology of a ripple can be easily imaged as the result of cutting a {105} pyramid along the [110] direction with a plane tilted by 8° from the (001) surface. (**b–d**) STM images of Ge ripples grown on the 8°-miscut Si(001) surface

As islands grow and transform, island morphologies are no more just skewed versions of the symmetric shape on the flat substrate. As predicted by Spencer and Tersoff [99], experiments have shown that *topologically* asymmetric islands appear in the shape sequence [136]. The SOM plots displayed in Fig. 14 reveal that the domes grown on substrates with a high misorientation angle have a different set of facets on one side than on the other. Thus, asymmetry is an intrinsic feature of the growth on vicinal surfaces.

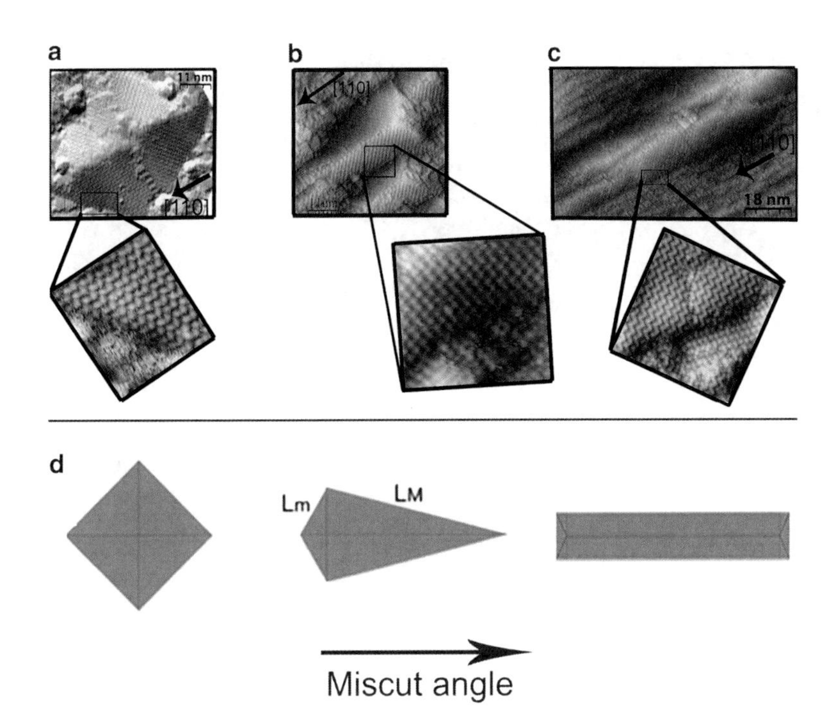

Fig. 12 STM images of Ge islands on flat and vicinal Si(001) surfaces: (**a**) square-base pyramid on the flat Si(001) surface, (**b**) asymmetric pyramid on the 6°-miscut Si(001) surface, (**c**) ripple on the 8°-miscut Si(001) surface. The corresponding blow ups show the RS reconstruction near the island edge. (**d**) Morphological evolution of Ge islands as a function of the miscut angle

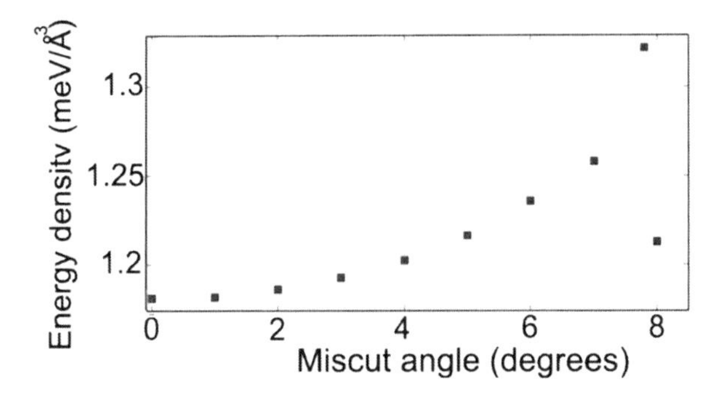

Fig. 13 Elastic energy per unit volume computed by FE simulations for the different Ge island shapes observed on vicinal Si(001) surfaces

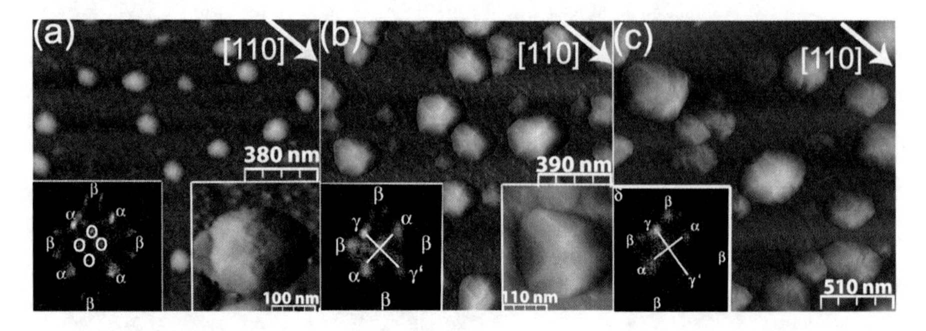

Fig. 14 Morphology of Ge domes: (**a**) on the flat, (**b**) on the 8°-miscut and (**c**) on the 10°-miscut Si(001) surface. In the *insets* the corresponding surface orientation maps are shown. The spots of the different facets are labeled as follows: {105} by ○, {113} by α; {15 3 23} by β; {111} by δ; the new facets along the miscut direction on vicinal substrates are indicated by γ and γ'

2.6 Tailoring the Elastic Interaction Potential Between Ge Islands with Substrate Vicinality

Engineering the growth of strained epitaxial films is a fascinating perspective in modern nanoscale science. The challenge is to control the strain-relief mechanisms which govern the growth process to direct them toward desired pathways. This can be done artificially, e.g., exploiting patterning surface features, as will be shown in Sect. 3. Here, we introduce a conceptually different approach based on tailoring the elastic interaction potential between Ge nanostructures with substrate vicinality [101, 137]. In the previous section, we have shown that a fine shaping of Ge islands is possible on Si(001) by changing the miscut angle. This offers a direct way to alter the elastic-interaction potential among islands which is greatly influenced by the detailed island's shape. The resulting effect depends on the intensity of the elastic field and, hence, on the island size. For small volume islands (pyramids and ripples), the symmetry breaking of the elastic field induced by vicinality modifies the local spatial ordering of islands. For large multifaceted domes, the modified elastic pattern is able to force the growth of Ge toward completely different pathways than on the flat substrate.

Misfit islands interact repulsively through their mutual strain fields in the substrate [44, 49, 138]. Their mutual interaction energy U is the extra energy density needed to create an island in a certain location when another island already exists nearby and it is given by

$$U = U(\mathbf{r}) - U(\infty), \tag{18}$$

where $U(\mathbf{r})$ is the total strain energy (per unit volume) stored in the substrate and in the islands for the relative position of the island pair defined by $\mathbf{r}$.

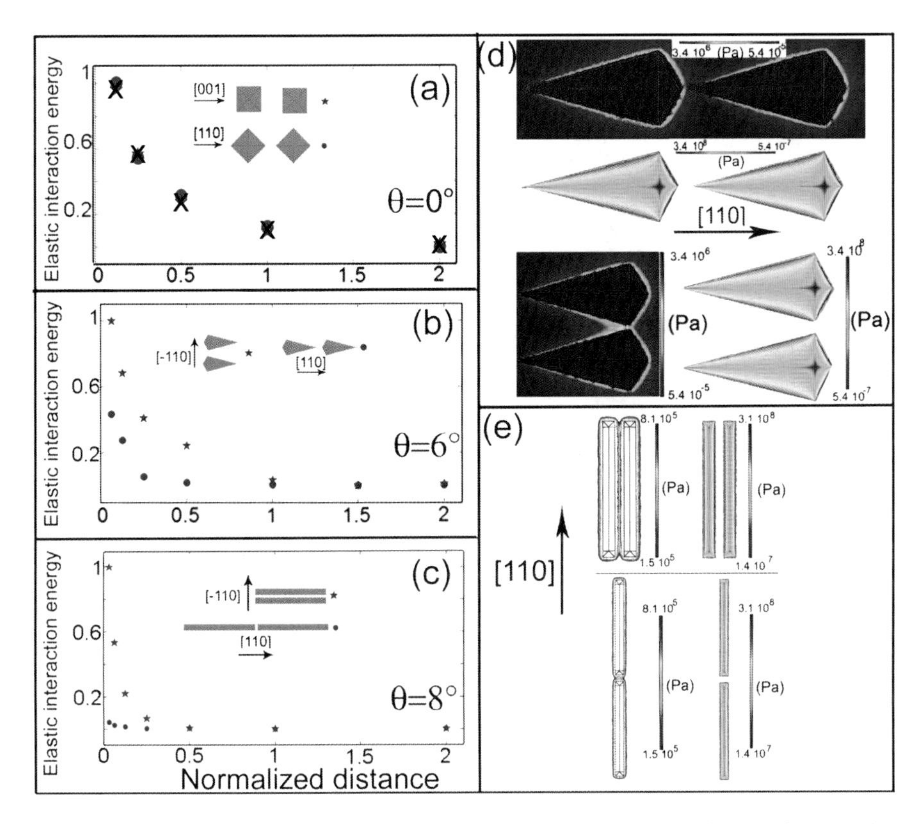

Fig. 15 (**a–c**) Elastic interaction energy for different configurations of an island pair (**a**) on the flat, (**b**) on the 6°-miscut, and (**c**) on the 8°-miscut Si(001) surfaces (*vertical axis* in arbitrary units, *horizontal axis* in units of the average island side). Elastic energy density maps of an island pair on (**d**) 6°-miscut and (**e**) 8°-miscut surfaces. Each plot is displayed with two different scales giving the elastic relaxation within the islands and on the substrate around them

Figure 15a shows FE calculations of the elastic interaction energy for square-based Ge pyramids on the flat Si(001) surface. For the two relevant configuration of an island pair, the interaction energy is almost isotropic. This is not the case of the Ge islands grown on vicinal substrates for which elastic interactions have a strong directional dependence (Fig. 15b, c). When the misorientation angle is increased, the lowest-energy configuration is achieved by aligning the pair along the [110] miscut direction. The latter configuration allows for a larger elastic relaxation of the substrate in between the islands, as shown by the energy maps displayed in Fig. 15d, e. The anisotropy of strain field profiles inside and around each island modifies the local spatial organization of Ge nanostructures. By measuring the spatial distribution of nearest-neighbor distances (SDNN) on different vicinal substrates, it is found that the local arrangement of islands becomes anisotropic with increasing miscut angle (Fig. 16). The SDNN is calculated from a systematic analysis of STM images. First the centers of mass of all islands are identified. Then, for each island, the nearest neighbor is found by calculating

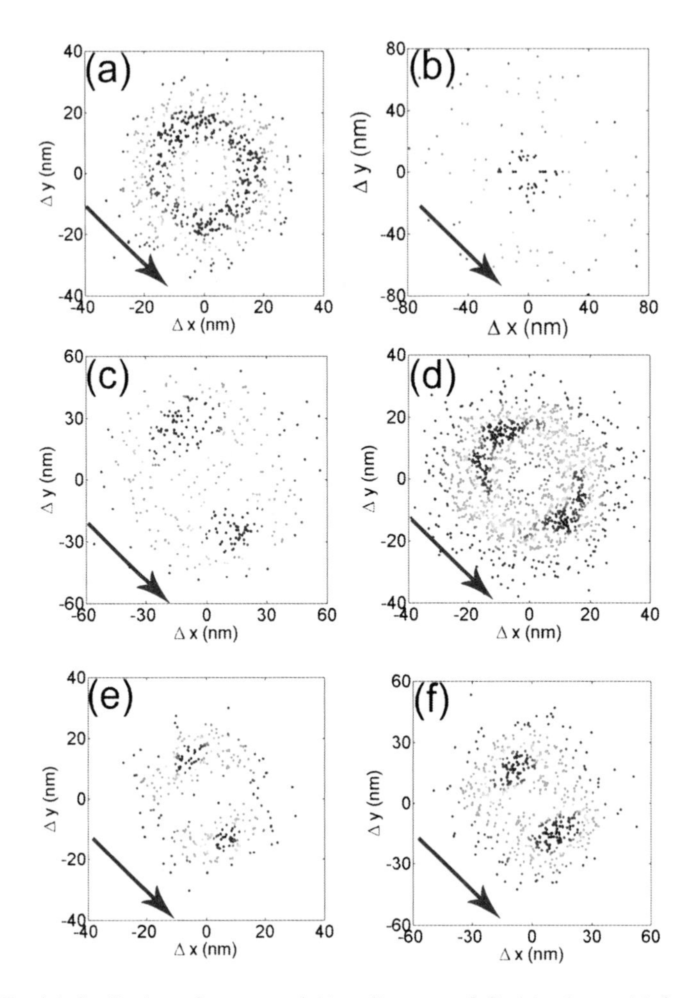

Fig. 16 Spatial distribution of nearest-neighbor distances of Ge islands on: (**a**) flat, (**b**) 1.5°-miscut, (**c**) 2°-miscut, (**d**) 4°-miscut, (**e**) 6°-miscut, and (**f**) 8°-miscut Si(001) samples. The arrows indicate the [110] direction

the distances between the corresponding centers of mass. Each panel in Fig. 16 shows the position of the nearest neighbors measured on the related vicinal substrate. It can be seen that the relative density of nearest neighbors (given by the color scale) is almost isotropic for flat substrates, whereas it is markedly increased along the [110] direction at high miscuts. Thus, the morphological anisotropy of islands at high misorientation angles breaks the isotropy of elastic potential, producing directions of reduced elastic interaction energy. As long as the volume of islands is small (e.g., for pyramids and ripples), the effect of elastic anisotropy on Ge/Si heteroepitaxy is limited to short-range local ordering interactions. Nonetheless, the much more intense elastic interactions between Ge domes can also be tuned with substrate vicinality. Figure 17a shows the island's

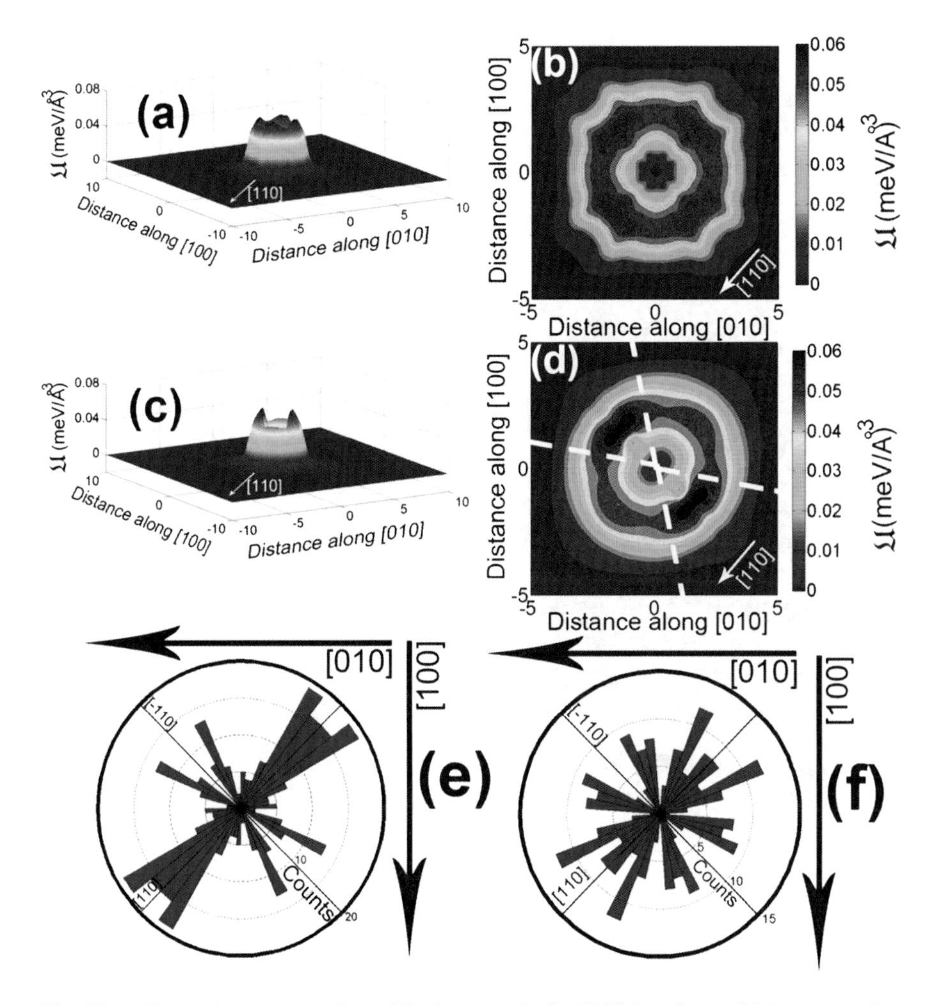

Fig. 17 (**a**) Interaction energy surface of Ge domes on the flat Si(001) surface and (**b**) corresponding contour plot. (**c**) Interaction energy surface of Ge domes on 10°-miscut Si(001) surface and (**d**) corresponding contour plot (The region of reduced interaction energy around the miscut direction is *highlighted*). Angular distribution of impingement directions measured (**e**) on 8°-miscut Si(001) substrates and (**f**) on 10°-miscut Si(001) substrates

interaction energy calculated for a pair of Ge domes grown on the flat Si(001) surface; the corresponding contour plot is reported in Fig. 17b. The interaction potential reflects the fourfold symmetry of the island and results in an energetic barrier to island coalescence with local minima around the ⟨001⟩ directions. The shape of the interaction energy surface is strongly modified for the domes on the vicinal substrate (Fig. 17c, d). The breaking of the island's symmetry induced by substrate vicinality produces directions along which islands can get into contact with low-elastic repulsion. Specifically, elastically soft configurations are achieved for islands interacting within an angular window of approximately ±60° about the

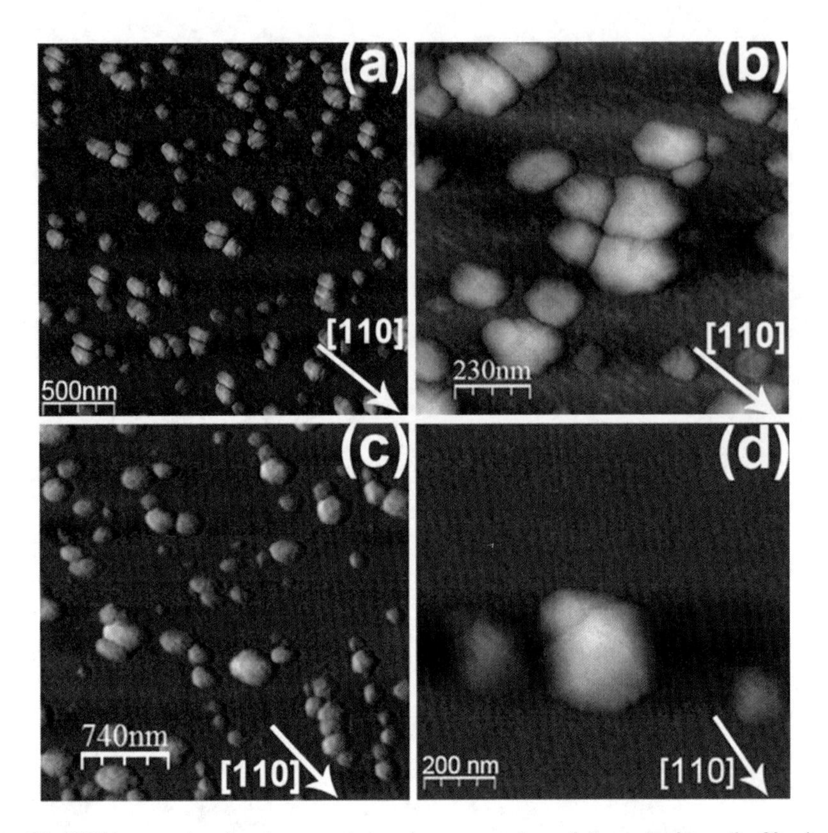

Fig. 18 STM images showing the extended coalescence regime of domes (**a, b**) on the 8°-miscut and (**c, d**) on the 10°-miscut Si(001) surfaces

[110] miscut direction (Fig. 17d). This modified elastic pattern orients Ge/Si heteroepitaxy towards an extended coalescence regime in which the impingement directions are dictated by the shape of the elastic potential. STM images clearly show that extensive coalescence occurs on 8° and 10°-miscut Si(001) surfaces (Fig. 18). Moreover, along the elastically soft directions around [110], the number of impingements is impressively higher, as shown by a statistical analysis of the distribution of impingement directions of domes grown on highly misoriented substrates (Fig. 17e, f). This indicates that the elastic interaction anisotropy is the main driving force for the observed growth evolution of Ge on vicinal surfaces. The experimental evidence that the symmetry breaking of the elastic field can be used to effectively direct the pathway of Ge heteroepitaxy lays the groundwork for new self-assembling strategies designed to suit the natural shape of the elastic interactions among nanostructures. Finally, this analysis is readily applicable to other heteroepitaxial systems for which an elastic field is a common key parameter.

2.7 *Self-organized Ordering of Ge Islands on Step-Bunched Si(111) Surfaces*

To modulate the arrangement of atomic steps on Si surfaces, a complementary approach to surface vicinality is electric-current-induced step bunching. Since the seminal paper of Latyshev et al. [139], it has been shown that the periodicity of steps on Si(111) substrates can be controlled by varying the direction of the heating current during high-temperature flash-annealing treatments. When the sample is heated with direct current, uniform step trains can rearrange into closely spaced bunches separated by large terraces. For $T > 1{,}220°C$, step bunching is induced by a direct current perpendicular to the step edges in the step-down direction. The process is fully reversible, since a regular array of steps is obtained again by reversing the current direction. The mechanism of step rearrangement for step-down heating currents has been described within a generalized Burton–Cabrera–Frank (BCF) which takes account of electromigration of adatoms [140]. Similarly to the effect of Schwoebel barrier, the step-down electromigration decreases the contribution of the upper terrace to the motion of a step during sublimation by pushing adatoms back to the step and favoring their reattachment to the step edges. As a result, the step bunch together, until the repulsive interaction between them compensates for the tendency to bunching induced by the electric force. Although rather tempting, the model based on BCF theory cannot explain the occurrence of step-bunching instabilities for step-up current which is experimentally observed in different temperature ranges [139, 141, 142]. To address this point, various step-bunching mechanisms have been proposed [143–146]. However, from a theoretical point of view, the problem is not yet fully understood and it still represents a challenge. A detailed discussion of electromigration affected sublimation goes well beyond the scope of this lecture and will therefore not be discussed any further. Instead, this section will be focused on possible application of step bunches as natural templates for the nucleation of low-dimensional nanostructures. Since the periodicity of step bunches can be reasonably controlled and it is well-known that growing islands preferentially decorate surface irregularities, step-bunched surfaces can be used to influence the location and the spatial ordering of nanodots [132, 147–149].

Here, we focus on the influence of bunching instabilities on the SK growth of Ge. We show that, with respect to regular substrates, an evident self-ordering of Ge islands is found on step-bunched surfaces.

Si(111) surfaces with miscut angle $<0.5°$ were flashed for 30–60 s at about 1,250°C by direct current heating, keeping the pressure below 5×10^{-8} Pa. The regular (R) and the step-bunched (SB) surfaces were obtained with a current flow oriented in the step-up and the step-down directions, respectively. In Fig. 19a, a STM image of a typical R surface is displayed. It consists of a staircase of equally spaced bilayer steps of $\approx$65 nm width and 0.31 nm height (Fig. 19b). By contrast, an example of SB surface obtained by step-down heating is reported in Fig. 19c. In this case, the terraces have an average width of 1,350 nm and are separated by bunches

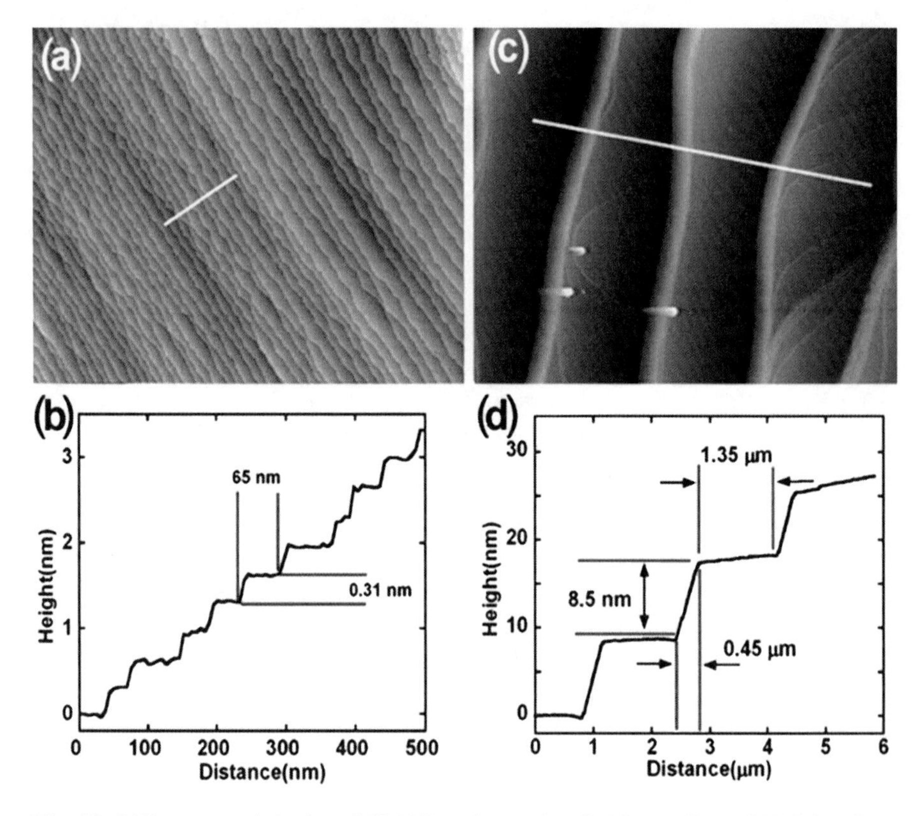

Fig. 19 Different morphologies of Si(111) surfaces after flashing at $T = 1{,}250^\circ C$ by direct current heating. (**a**) STM image ($1{,}700 \times 1{,}700 \times 10\ nm^3$) of a R surface obtained by current flowing in the step-up direction. (**b**) Height profile taken along the *white line* in (**a**). (**c**) STM image ($7{,}000 \times 7{,}000 \times 36\ nm^3$) of a SB surface obtained by current flowing in the step-down direction. (**d**) Height profile taken along the *white line* in (**c**)

about 8.5 nm high, which correspond to 27 atomic steps (Fig. 19d). Both experimental results [150] and growth simulations [151] indicate that mesas on Si(111) substrates act as nucleation sites for Ge islands. Thus, in our case, we expect preferential nucleation on step bunches. This is indeed observed on SB substrates after deposition of 8 ML of Ge at $T = 450^\circ C$ (Fig. 20a). Ge islands first nucleate and evolve along step edges and subsequently on flat terraces. Islands grown on step bunches undergo complete ripening and form continuous ribbons. When the evolution on the step edges is complete, nucleation takes place at the center of terraces. The capture of adatoms toward the high-density, early-nucleated dots at the step bunches is responsible for the islands depletion alongside the long ribbons, which in turn forces further dots nucleation to occur in the central part of the terraces. By increasing Ge coverage, step bunches appear fully decorated and the size of the islands on terraces increases (Fig. 20b). On flat terraces, 1D areas crowded with islands that run parallel to the step edges are defined. The whole process is quite

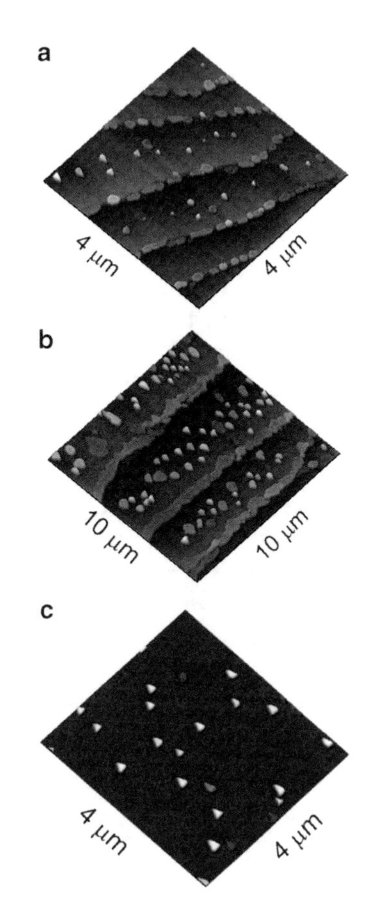

Fig. 20 (**a**) STM image of a SB surface after deposition of 8 ML of Ge. (**b**) STM image of a SB surface after deposition of 17 ML of Ge. (**c**) STM image of a R surface after deposition of 8 ML of Ge

analogous to a lithographic process (i.e., a definition of different areas on a surface), with the peculiarity of being totally self-driven. The lateral ordering on SB surfaces is highlighted by the comparison with island distribution on R substrates on which Ge dots appear randomly distributed (Fig. 20c).

The present results show that controlled positioning of Ge islands can be achieved without lithographic patterning by exploiting natural bunching instabilities of Si(111) surfaces. Besides being undoubtedly fascinating for the microscopic physics underlying, spontaneous step bunching provides a natural template for the nucleation of low-dimensional nanostructures, thus opening interesting perspectives for self-ordering strategies complementary to artificial lithography.

3 Nucleation and Growth of Ge Islands on Patterned Substrates

3.1 FIB Patterning

A focused ion beam (FIB) is a multi-purpose instrument similar to a scanning electron microscope, which uses a high energetic beam of Ga^+ ions to scan a sample surface. Due to the Ga ions' cross section and to the interaction of such ions with the matter, a FIB provides functions far beyond the sole imaging, being capable of material removal, nanostructuring and nanofabrication. It was originally conceived and devoted to integrated circuits' processing, in applications such as failure analysis [152], circuit maintenance [153], and mask repair [154]; however, due to its flexibility, this instrument has been applied in a wide range of other different fields [155].

In the present context, we will discuss the surface modification at the nanoscale that FIB structuring enables focusing on FIB patterning. By this technique, it is possible to engrave a sample surface with defined outlines, in which any feature (i.e., shape, size, and depth) is set up by the control software. Bitmap files can also be used as templates for the region to be patterned, allowing any kind of grayscale picture to be engraved on the substrate. Compared to photolithography, FIB milling exhibits many advantages, as a finer resolution, a higher aspect ratio and the possibility to work on almost any material without the need for a mask or an etching stage, at the expense of a lower throughput.

In the last decade, FIB patterning has been extensively applied as an artificial means for controlling the self-assembly mechanism of metal and semiconductor dots, with a great deal of effort being pursued especially on Ge quantum dots over Si surfaces [156, 157]. To this extent, we have studied the assembly and the lateral localization of Ge dots and islands on Si(001) and SiO_2/Si(001) nano-structured surfaces. When sputtering a surface by FIB, the process of material removal and the consequent morphologic modification are often regarded as the only output of the exposure to the ion beams. Taking in consideration nanoscale patterns milled in silicon, the typical ion beam energy is around 30 keV, with currents ranging from 1 to 100 pA. In a simplified model, the beam's current density is the key parameter that determines the interaction and thus the effect of the ions impinging on a surface. At low-dose irradiation (10^{14}–10^{16} ions/cm^2), the disruption of the Si lattice causes a crystalline-to-amorphous transformation, which can lead to a pronounced swelling of the processed area. Ga ions are implanted underneath the surface at a mean depth of about 26 nm. Conversely, when the current density is increased (10^{17}–10^{20} ions/cm^2), sputtering becomes the predominant effect producing an effective removal of material. For 30 keV Ga ions on silicon, a constant sputter yield of approx. 2.4 atoms/ion is assumed for doses higher than $\sim 4 \times 10^{17}$ cm^{-2}.

Implantation and deposition occur at the same time during FIB processing and, although their respective contribution varies with the beam energy, they can greatly affect the surface properties and must be then taken into account. Above all, these phenomena have been deemed responsible for alterations of the strain and of the chemistry of the surface, ultimately determining the nucleation and assembly of deposited materials, as Ge dots. For these reasons, several research groups have taken different approaches to obtain a high degree of order in self-assembled nanostructures, by exploiting all these heterogeneous effects that ion beams induce on a surface. Kammler et al. have employed the surfactant effect of the implanted Ga^+ ions to achieve a good control over the Ge QDs' position [158]. Other groups induced a complex modification of a Si surface's properties in terms of strain, chemistry, and topography by a two-step process combining a low-dose FIB implantation and a thermal annealing, obtaining a precise one-island-per-feature nucleation over a wide range of growth condition [159].

In our case, we have performed specific investigations focused on the development of a cleaning procedure that could guarantee desorption of implanted Ga atoms along with total removal of crystalline defects, without modifying the holes' shape and size. Among the many procedures proposed, a chemical recipe is the most effective to this purpose [160]. It consists of two ultrasonic baths in H_2SO_4: H_2O_2 and HCl:H_2O_2:H_2O, followed by a 20′ annealing at 850°C in UHV. The first bath in acid solution aims at removing all organic contaminants still on the surface, while the second continuously creates and etches oxide layers on the wafer, thus eliminating any metal impurity. The heating stage provides a thorough recovery of the crystalline structure of the Si lattice.

A very accurate AFM characterization has been performed after each stage of this cleaning procedure, demonstrating that the pattern morphology, in terms of shape and size, is always preserved. Cleaned samples have been analyzed by Auger spectroscopy (Fig. 21a), for evaluating the content of Ga and undesired species after the patterning. The Ga content on the surface is below the detection limit of 10^{16} atoms/cm^2. Moreover, Rutherford backscattering spectroscopy (RBS) measurements (Fig. 21b) have been carried out on cleaned samples after the deposition of Ge, and they reveal an amount of 5.2×10^{15} Ge atoms (consistent with the actual deposition of 8 ML), without measurable amounts of contaminants such as Ga, K, Ca, or transition metals, which are detrimental in very large scale integration (VLSI).

Our investigations show that both FIB and nanoindentation can be methods for arranging preferential nucleation sites on semiconductor surfaces. Although the key idea of directing island nucleation toward pre-defined sites might appear rather straightforward, the effect of a pattern on the ordering of Ge dots is still under debate. It has been shown that preferred nucleation locations can vary significantly as a function of the interplay between strain and surface energy. As a matter of fact, many parameters must be considered and tuned to create the desired configuration of Ge QDs by artificial means.

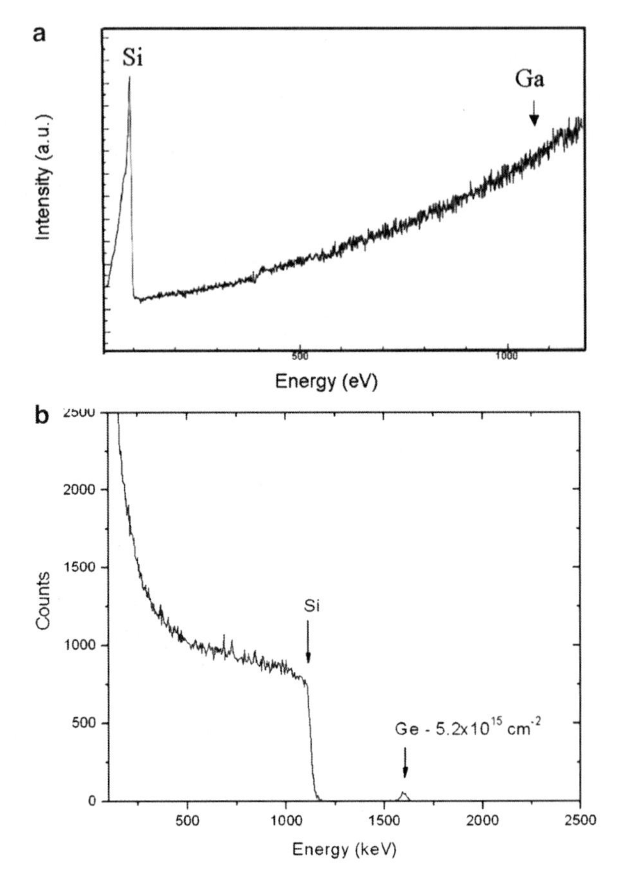

Fig. 21 Investigations of the composition of a Si surface cleaned after FIB patterning: (**a**) Auger spectroscopy; (**b**) Rutherford backscattering spectroscopy

3.1.1 Si(001)

One of the first experiments to order quantum dots by nanopatterning has been successfully performed by placing InAs dots [161] on GaAs substrates via a lithographic etching of an array of holes placed at regular distances. Kim provided a similar result by ordering Ge dots on a Si(001) substrate covered by an oxide layer [162].

Si(001) is indeed a very important substrate for all electronic devices, and it is a test bed for growth of QDs. Parameters such as hole depth and diameter as well as distance are to be controlled in order to provide a suitable data set for case studies. We have recently studied the nucleation of Ge islands on clean Si(100) substrate patterned by FIB and found that they nucleate selectively inside the holes creating an ordered array, whereas a random distribution is observed in nonpatterned regions. It is believed that stepped regions of the hole walls can supply primary nucleation sites at the nanoscale. An island growing in a pit benefits also of an enhanced elastic relaxation [163, 164]; thereby, the nucleation will be even more favorable from a thermodynamic point of view. The deposited

Ge atoms accumulate inside the holes, while the nucleation is suppressed around them. A full understanding of the assembly on patterned Si substrates is yet to be achieved because of the many variables involved in the epitaxial growth. Temperature, flux, presence of buffer layers, and deposition methods are some of the principal variables that determine the surface energy anisotropy as well as the total energy gain associated to the formation of an island. In the case of physical vapor deposition (PVD) and molecular beam epitaxy (MBE), the main parameter that affects the assembly on patterned substrates is the growth temperature, which in turn controls the diffusion kinetics of the deposited material. Also the evaporation rate introduces variation in the kinetics, leading in some cases to very different results.

The driving force for preferential nucleation on patterned substrates is the difference of the chemical potential of a patterned surface as compared to a planar surface which can be expressed by [165]

$$\mu = \Omega\gamma k(x, y) + \Omega E_{\mathrm{el}}(x, y), \tag{19}$$

where the first term $\Omega\gamma k(x, y)$ describes the change of the surface energy γ with the surface curvature $k(x, y)$ and the second term $\Omega E_{\mathrm{el}}(x, y)$ accounts for the change of the local strain energy $E_{\mathrm{el}}(x, y)$ induced by the holes. The first term is lowered on concave surfaces (inside the holes), while the second is lowered for convex ones (e.g., hole edges) where the compressed Ge bonds can be relaxed more easily. From a thermodynamic point of view, the balance between these two competitive effects determines the equilibrium location of islands. However, depending on the growth temperature, the organization of islands may also be markedly influenced by kinetics. Broadly speaking, we can distinguish two ranges of temperatures: 550–700°C referred hereafter as low temperature and 700–850°C as high temperature. At low temperature, the growth is kinetically limited by the low mobility of Ge atoms which have not enough energy to overcome the diffusion barrier of the pits. Thus, the deposited Ge tends to be gathered inside the pits, especially on the sidewalls, which act as preferential nucleation sites. Due to a higher strain relaxation of the bottom of the pits, Ge islands are attracted there in a second stage, drawing atoms from the dots originally nucleated on the sidewalls. These islands are subject to a ripening process at the expenses of the small dots, cleaning the area inside and around the holes.

Conversely, when the deposition temperature is high, Ge atoms are able to move out of the pits and assemble on the terraces around them. Therefore, even though the pits still act as nucleation sites, they do not represent the final location for the Ge islands. In fact, the atoms possess now enough energy to escape the diffusion barrier and assemble on the terraces, which have been found to be the sites of minimum total energy [160].

Pascale et al. [166] report a random mixed phase presenting dots on terraces and within the holes at 650°C. The authors suggest the existence of a saddle point between the metastable and the equilibrium states at this temperature, resulting from limited out-diffusion of Ge from the pits.

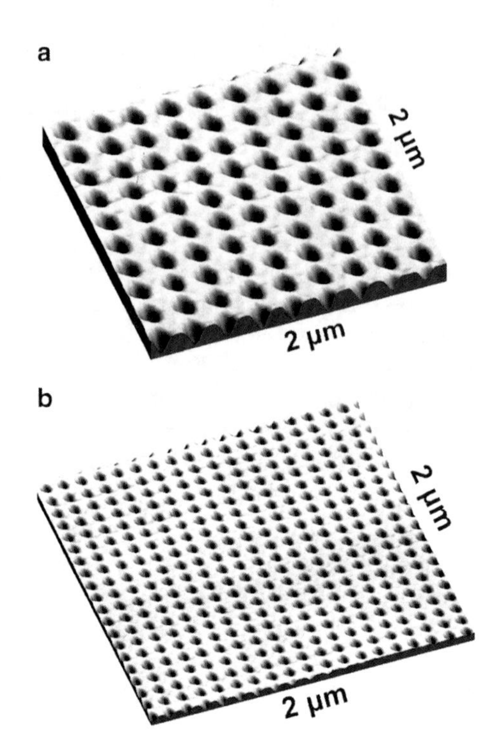

Fig. 22 3D AFM images of FIB patterned Si: (**a**) 200 nm pitch, 70 nm hole diameter, 30 nm depth; (**b**) 100 nm pitch, 50 nm hole diameter, 20 nm depth

Square arrays of holes have been engraved in selected areas of a Si substrates, by a dual beam FIB (Quanta 3D-FEI), which employs a liquid metal ion source to generate Ga^+ ion beams ($I = 1$–100 pA, $V = 30$ kV). We prepared several substrates with pitch distance ranging between 50 and 700 nm and depths between 2 and 40 nm. As an example, in Fig. 22, two AFM images of arrays with pitch of 200 nm (a) and 100 nm (b) are reported. Low-current ion beams have been used (1–10 pA), in order to achieve the maximum resolution and the minimum collateral damage due to ion bombardment. In the following, we discuss the effect of Ge evaporation on the 500 nm array.

After the cleaning procedure followed by an additional annealing at 600°C for 200 min and a 1 ML Si buffer layer [one monolayer (ML) $= 6.78 \times 10^{14}$ atoms/cm^2] deposition, 6 ML Ge [one monolayer (ML) $= 6.29 \times 10^{14}$ atoms/cm^2] have been evaporated keeping the substrate at 600°C. Ex situ AFM characterization shows that a regular distribution of Ge large islands nucleates following the underlying template array pattern (Fig. 23a), while in the nonpatterned area small dots nucleate randomly (Fig. 23b).

To quantify the long- and short-range order of the pattern we evaluated the pair distribution function (PDF) of each assembly of dots, starting from the images shown in Fig. 23. For our scope, we define the two-dimensional PDF as:

$$g(r) = N(r)/2\pi r \Delta r \rho, \tag{20}$$

where $N(r)$ is the number of dots found in the circular crown of area $2\pi r \Delta r$ at distance r from a reference dot, Δr is the radial width of the circular crowns

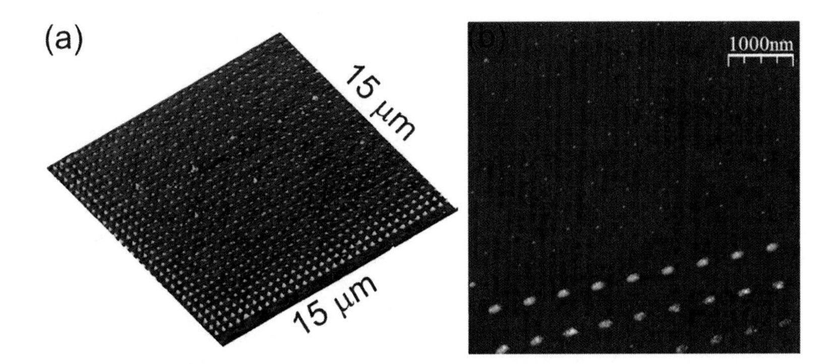

Fig. 23 AFM images FIB patterned Si(001) substrate: (**a**) patterned area at 6.0 ML coverage (z-scale = 25 nm); (**b**) distribution of islands near the edge of the pattern

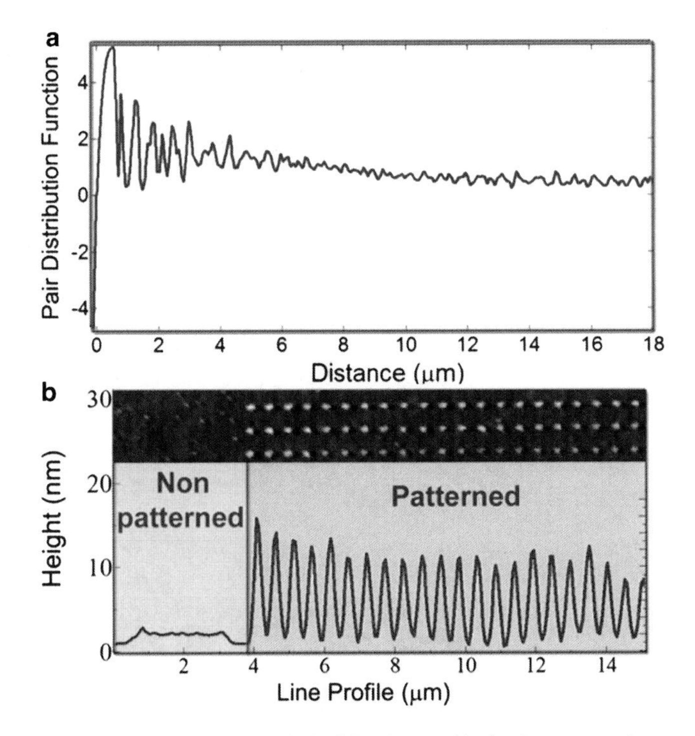

Fig. 24 (**a**) Pair distribution function of the island assembly in the patterned area; (**b**) height profile of the patterned and the unpatterned area of the AFM image in the *inset*

considered (i.e., the bin size of the distance distribution), and ρ is the mean dot density in the image. The PDF has an asymptotic value of 1, and deviations from unity are interpreted as density fluctuations at a given distance r from a dot, within a bin size Δr. To gain quantitative insights into the controlled self-assembly, PDF and surface profilometry are shown in Fig. 24a, b, respectively. The profile shows a

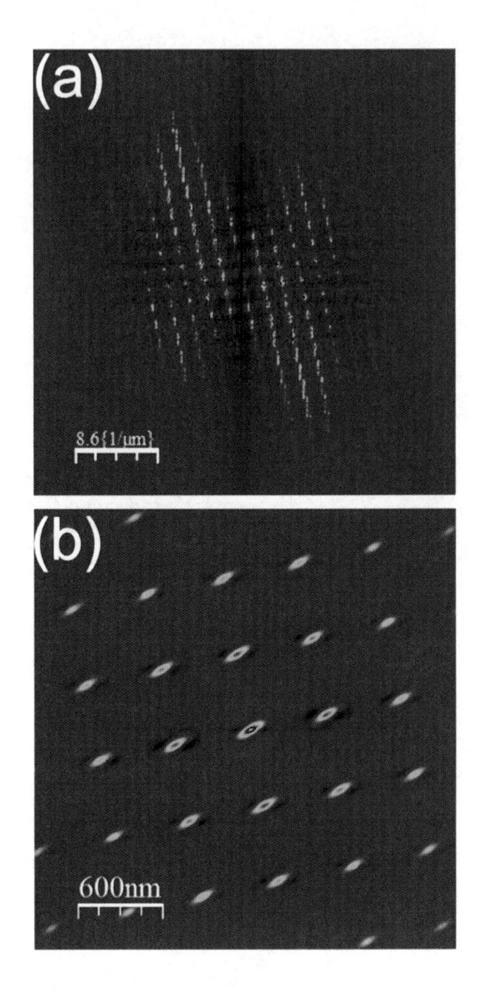

Fig. 25 Analysis of the patterned area in Fig. 23: (**a**) fast Fourier transform; (**b**) auto-correlation function

regular sequence of peaks with periodicity corresponding to the pitch spacing (500 nm). The long-range correlation of the island assembly is further evidenced by the presence of several peaks in the PDF. As expected for an ordered configuration, besides the nearest-neighbor peak at the pit distance, other peaks corresponding to the neighboring shells of a square lattice are present. The perfect 2D regularity is confirmed by using fast Fourier transform (FFT) and auto-correlation function (ACF) analysis (Fig. 25).

3.1.2 SiO_2/Si(100)

Growth on Flat SiO_2 Substrates

Before growing Ge on a patterned oxide, we have investigated the growth of Ge nanocrystals embedded in a SiO_2 layer to study the nucleation dynamics on this surface. The mechanism of Ge islanding on SiO_2 is different from island formation

Fig. 26 (**a**) STM image of a clean SiO_2/Si(100) surface. (**b**) STM image of 4 ML of Ge deposited on a SiO_2/Si(100) surface at room temperature. (**c**) STM image of the sample shown in (**b**) after 30 min annealing at 500°C

on Si. Because of the limited surface diffusion of Ge on this oxide, there is no formation of a wetting layer but a Volmer–Weber growth of Ge droplets [167–170]. This growth mode can be exploited in order to obtain, at the same time, high density, size reduction, and electrical insulation of Ge dots, providing applications of such dots as memory cell devices [156, 171, 172]. By using Si or Ge inclusions to replace the floating gate, one obtains an increased storage capability together with faster erase/write times and a reduction of the dissipation power [170, 173]. Due to the smaller band gap and the valence band offset at the interface with the substrate, Ge nanocrystals enable a further improvement of the memory characteristics with respect to Si inclusions. In optoelectronics the interest has been motivated by the observation of visible photoluminescence for Ge nanostructures in a SiO_2 matrix [174–176] which has been ascribed to quantum confinement phenomena and the presence of the Ge/SiO_2 interface [177].

For Ge/SiO_2, where surface energy should be almost isotropic, several analyses [163, 171] have shown that the total free energy gain favors the formation of islands inside the holes of the pattern where the nucleation barrier is found to be smaller. This is in line with our experimental observations.

We prepared clean SiO_2 surfaces by rapid thermal annealing of Si(001) substrates at $T = 950°C$ in UHV. STM images confirmed that the shape of the Si (001) surface underneath is exactly preserved, i.e., a perfect sequence of terraces 70 nm wide and monoatomic steps. On the terraces we observed small isolated brighter protrusions, with a density of about 1×10^{13} cm^{-2}, forming a rough substrate with root mean square (RMS) roughness equal to 0.06 nm (Fig. 26a).

On such oxidized surfaces, the deposition of 4 ML at room temperature (evaporation by electron gun at 0.1 nm/min, $P = 10^{-10}$ torr) leads to the formation of Ge amorphous clusters, which are randomly distributed but closely interconnected (Fig. 26b). Ge dots are then formed by a 30 min annealing at $T = 500°C$ (Fig. 26c). The dots segregate and become crystalline, with an average size of 5 nm and a low aspect ratio. The density decreases around 3×10^{12} cm^{-2}. They

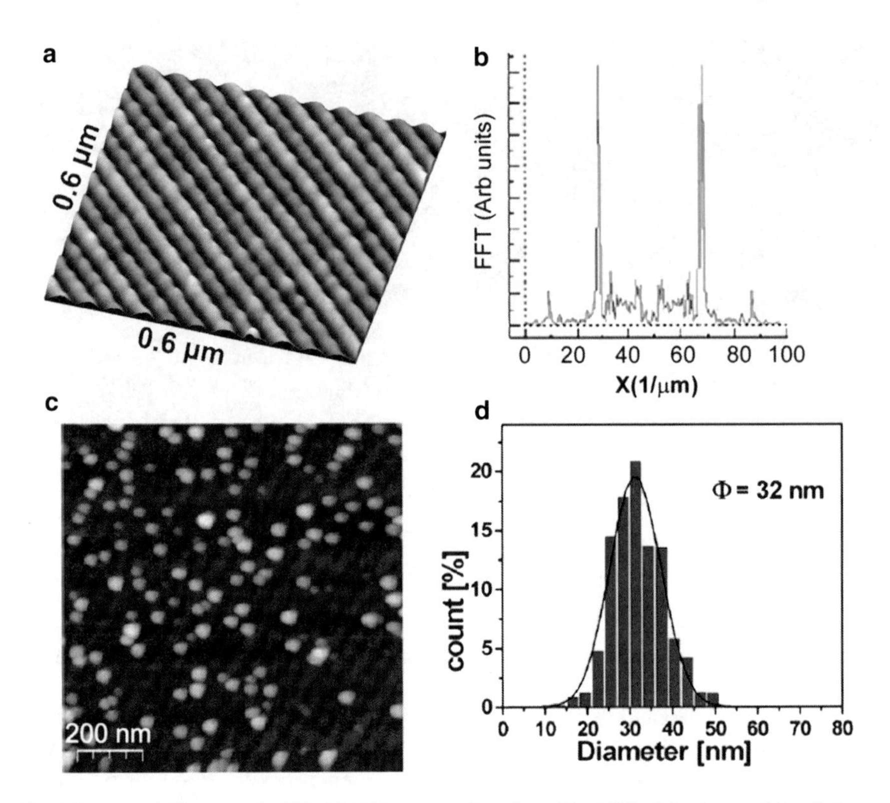

Fig. 27 (**a**) AFM image of a SiO_2/Si FIB patterned surface. The FFT of the image (**b**) enlightens the average distance of 51 nm between holes. The average depth of the holes is 4 nm. (**c**) STM image of the surface after deposition of 5.2 ML at 600°C. (**d**) Corresponding size distribution

appear almost flat with the onset of (001) and (113) faceting. Such morphology is attributed to the thermodynamic equilibrium shape of Ge [108], as the (001) and (113) are the minimum surface energy facets.

Growth on Patterned SiO_2 Substrates

Dense arrays of holes (4×10^{10} cm^{-2}, diameter ~30 nm) with a pitch of ~50 nm are produced by FIB milling selected areas of a Si substrate (Fig. 27a, b). After the cleaning treatment, a 5 nm thick layer of clean SiO_2 was grown on this substrate. The deposition of 5.2 ML of Ge at $T = 600$°C leads to the formation of randomly nucleated islands on the patterned SiO_2 surface (Fig. 27c). Some islands seem to nucleate inside the holes, others nearby. Some regions are free of islands. Nevertheless, a mean diameter of 32 nm is extracted from the islands size distribution with a full width at half maximum (FWHM) of 10 nm (Fig. 27d). At this growth temperature, the surface topography does not play a crucial role and does not affect the lateral ordering.

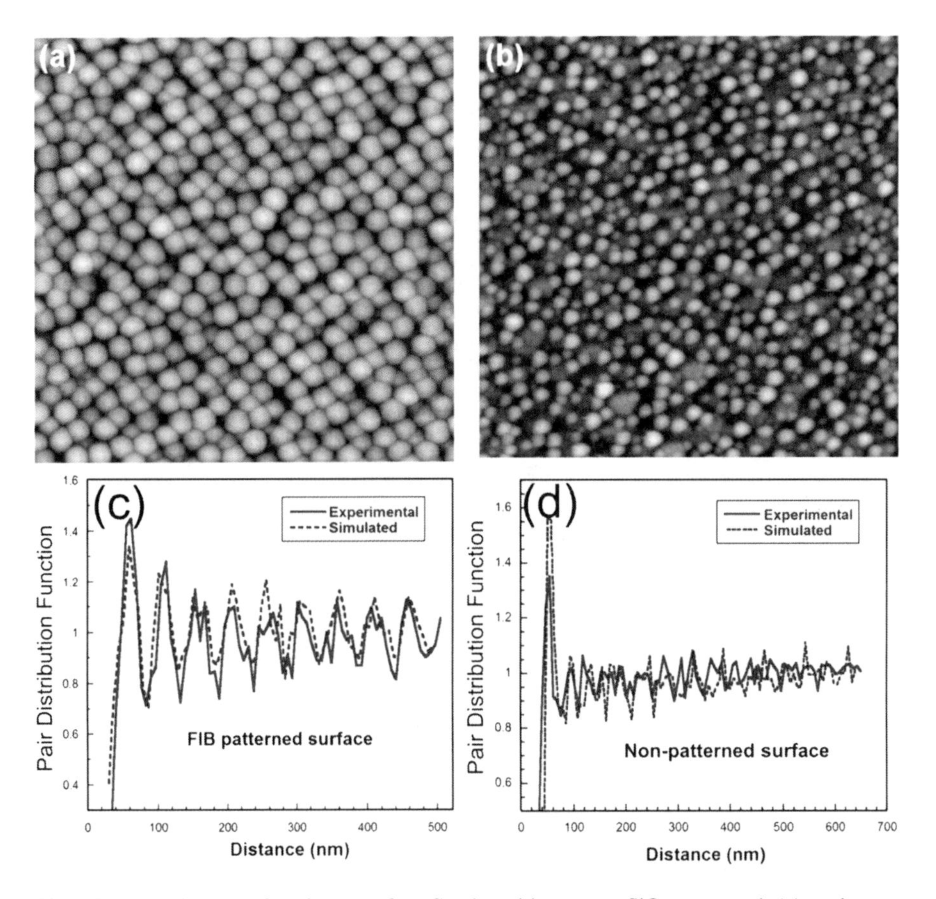

Fig. 28 AFM images (2 × 2 μm) after Ge deposition on a SiO_2 patterned (**a**) and non-patterned region (**b**). (**c**) Plots of two PDFs for islands on a FIB patterned surface. The *solid curve* is obtained by analyzing the image reported in (**a**), and the *dashed curve* is the PDF of an equivalent simulated dot pattern. Both curves show evidences of long-range order. (**d**) PDF of experimental and simulated images of Ge QDs on nonpatterned Si oxide surface. The first peak is a consequence of the short range order existing within the dots pattern, while the noisy signal at higher distances demonstrates that no long-range order is present

Island ordering is fostered by a two stage growth process: (1) the deposition of a few MLs of Ge producing an amorphous layer and (2) a thermal annealing at 500°C for 20 min, in order to form Ge nanocrystals (NCs).

During post-growth annealing, the amorphous film dewets to form nanocrystalline Ge droplets (Fig. 28a). The PDF analysis indicates that a full two-dimensional correlation in the dot arrangement with the same periodicity of the patterned substrate is finally achieved after this annealing cycle. The PDF related to the patterned area of the sample (Fig. 28c) shows the presence of several peaks with a value well above 1, each of them corresponding to a set of dots shells. The first peak occurs at a nearest-neighbor distance of 57 ± 3 nm, which is slightly larger than the pattern pitch, while the inflection point is located at 40 nm, i.e., the same value as the islands' average diameter (some islands touch). We model the experimental PDF using a two-dimensional square lattice of points with $a = 50$ nm, much like the holes pattern prepared with FIB. To account for the finite dots size and uncertainty on the nucleation site, we position a single dot at random in a 30 nm diameter circle centered on each lattice point. The PDF extracted from this simulated pattern is strikingly similar to the one obtained from experimental data. This confirms that, when Ge nucleates in the absence of strain, one dot forms at each hole of the FIB pattern, in our experimental conditions and presumably as long as the diffusion length is larger than the pitch distance. These results suggest that patterning favors first of all the formation of nanostructures with sizes matching the pitch of the pattern.

No lateral order is observed on nonpatterned areas, which are homogeneously covered with Ge dots of average diameter 30 nm with a seemingly random spatial distribution (Fig. 28b). Despite the random location of the dots, on nonpatterned regions some degree of order is expected as well, owing to an island's growth mechanism involving an adatom diffusion process. As a matter of fact, the presence of a capture zone around each nucleated dot decreases the nucleation probability around each stable nucleus and hinders the formation of two very close islands. As a consequence, a peak is found in the PDF at a nearest-neighbor distance of 49 ± 3 nm, while only noise contributes to the signal at higher distances (Fig. 28d). The trend of the PDF clearly points out an effect of short-range ordering, totally hidden at visual inspection but easily unveiled through the use of our software routine.

3.2 *Nanoindentation Patterning*

Indentation is commonly used in hardness tests to determine the resistance of a material to deformation [178, 179]; however, this technique has been recently refined in order to manipulate a sample surface on the nanoscale, renamed as nanoindentation [180]. In indentation hardness testing, a diamond indenter of specific geometry is impressed into the surface of the test specimen using a known applied force (commonly called a "load") of 1–1,000 gf. Microindentation tests typically have forces of 2 N (roughly 200 gf) and produce indentations of about 50 μm. Microhardness testing can be used to observe changes in hardness on the microscopic scale. Nanoindentation tests are a variety of indentation hardness tests that can be applied to smaller volumes. In nanoindentation, small loads and tip sizes are used, so that the indentation area may only be a few square micrometers or even nanometers [180].

Indenters with a geometry known to high precision (usually a Berkovich tip, which has three-sided pyramid geometry) are generally employed. During the course of the instrumented indentation process, a record of the depth of penetration is made, and then the area of the indent is determined using the known geometry of the indentation tip. While indenting, various parameters, such as load and depth of penetration, can be measured. A record of these values can be plotted on a graph to create a load–displacement curve. These curves can be used to extract mechanical properties of the material.

In the last few decades, there has been great interest in the mechanical properties of Si, particularly in regards to the phase transformations that occur during nanoindentation [181].

New phases in the nanoindented holes in Si could affect the crystal structure of Ge QDs grown by epitaxy inside the indents. In addition, the shape of the holes made by an indenter could lead to the growth of dots or island with specific facet inclination and then could provide a high degree of control on the deposition outputs.

Recently, nanoindentation patterning has been successfully used to guide the assembly of InAs quantum dots on GaAs(001) [182]. This maskless technique is highly controllable yet rather simple, and potentially allows patterning of large areas via localized dislocations without the need for chemical processing. In this section, we describe an effective pathway toward ordered configuration of Ge islands on substrates patterned by nanoindentation. By the analysis of the evolution of the dot assembly, we gain experimental insight into the initial stage of the ordering process induced by nanomechanical patterning.

Nanoindentation of Si(001) substrates has been performed at room temperature using a Hysitron Triboindenter system fitted with a Berkovich diamond tip. A maximum indentation load of 5 mN was used to form inverted pyramidal impressions with lateral size of 500 nm and depth of 30 nm. Two dimensionally ordered square arrays of indents were patterned with pitches ranging from 1 to 10 μm. As an example of the patterned morphology prior to Ge deposition, we show AFM images of the samples with 1 and 2 μm of pitch in Fig. 29. The pit shape is an inverted (upright) pyramid with {105} sidewalls forming an angle of about 11° with the (001) plane and has a mean depth of 30 nm (Fig. 29c). Once the patterns were made and fully characterized by SEM and AFM, the samples were transferred to the UHV chamber for a cleaning stage and Ge deposition. The substrates were annealed in UHV ($p < 4 \times 10^{-9}$ Pa) at 600°C for several hours and then flashed for 1 min at 1,200°C, by passing a direct current of few amperes through the samples [147]. This standard cleaning procedure is largely used to provide an atomically flat surface as a template for self-assembly. After the flashes 7 Ge ML were deposited by physical vapor deposition at 580°C under a constant flux of 0.16 ML/min.

After the initial stages of Ge deposition and dot assembly (6 ML of Ge coverage), Ge dots nucleate both inside and between the indents: in particular, they gather at the corners and on the slope of pit sidewalls which, consequently, assume smooth, rounded shapes. Further analyses of these data reveal details of the nucleation mechanism. The island density at various length scales from the center of an indent is shown in Fig. 30a. A sharp peak in density is observed at a distance corresponding to the average half-side length of a hole (250 nm), indicating enhanced nucleation inside the indents. Outside the pit region the density first decreases and then flattens far away from the indent. To stress the pattern contribution, we have measured the dot density on a nonpatterned

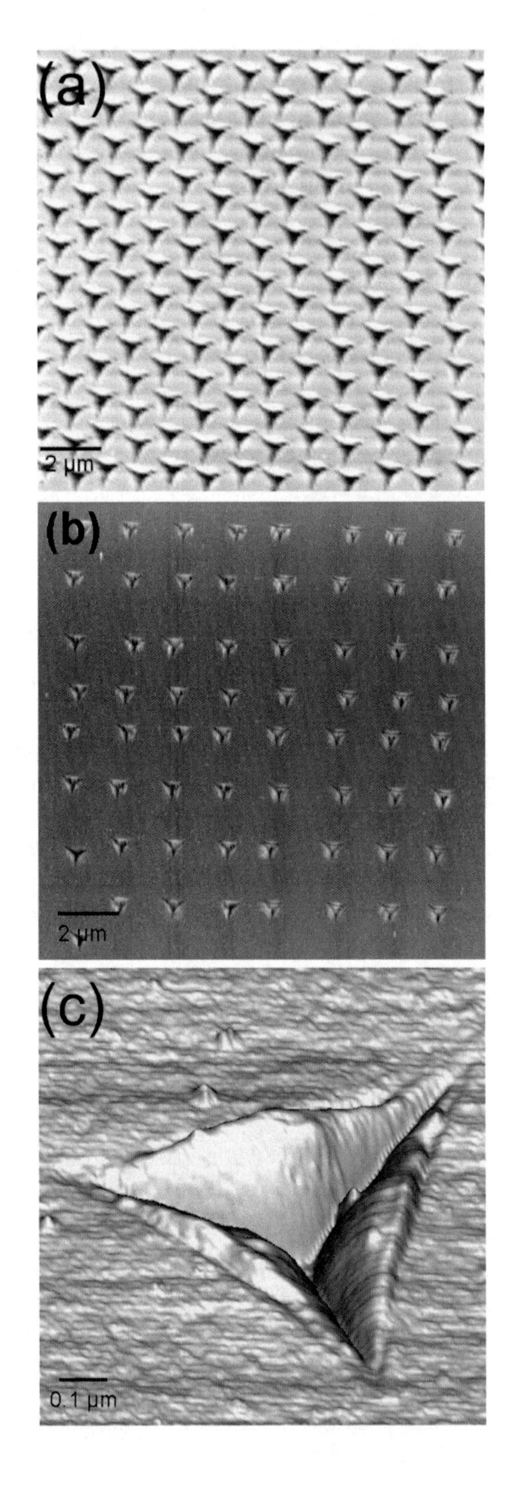

Fig. 29 AFM images of nanoindented Si substrates prior to Ge deposition: (**a**) 1 μm-pitch; (**b**) 2 μm-pitch; (**c**) 3D view ($0.7 \times 0.7 \times 0.03\ \mu m^2$) of a single pit

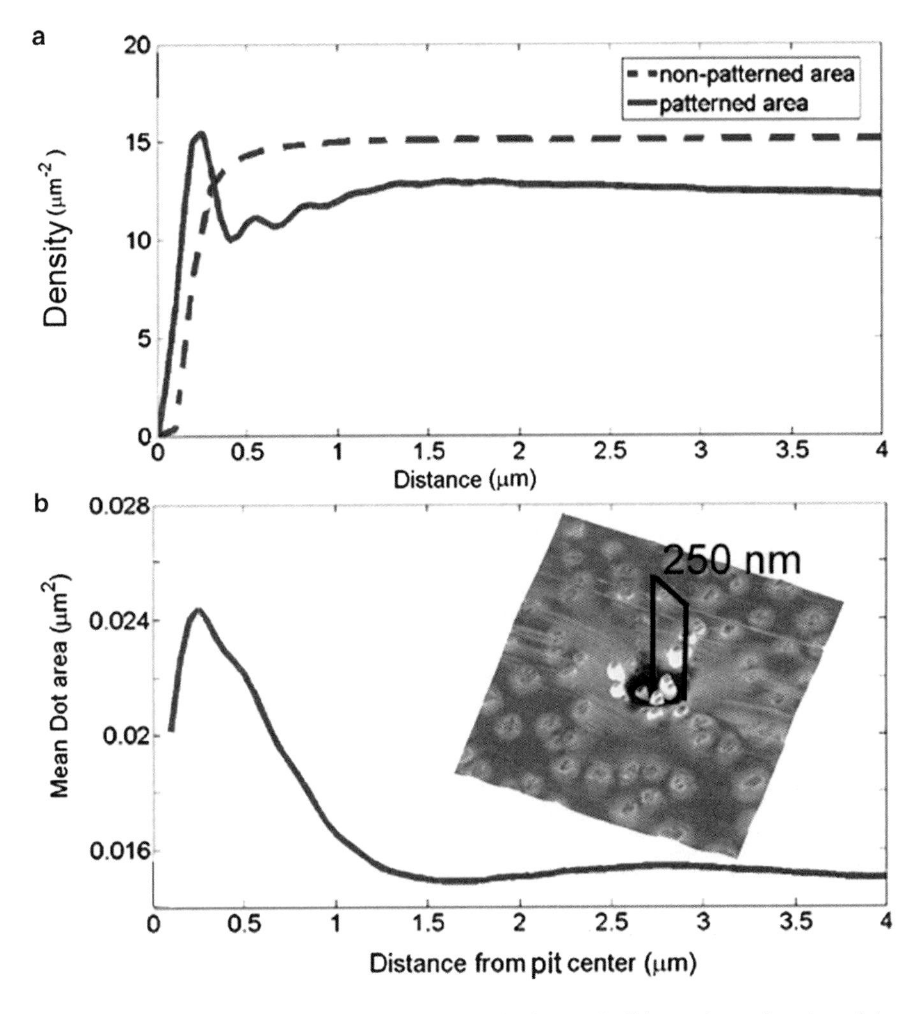

Fig. 30 (**a**) Island density plotted: on the patterned substrate (*solid curve*) as a function of the distance from the pit center; on the non-patterned surface (*dashed curve*) with respect to the center of mass of the dots. (**b**) Mean dot area as a function of the distance from the pit center. In the *inset* an AFM image ($2 \times 2\ \mu m^2$) showing the dot distribution around a pit is displayed

area of the same sample, using as reference points for the density estimate the center of mass of each dot (dashed curve in Fig. 30a). In this case, the density is smooth, demonstrating that the indents strongly modify the island assembly. Furthermore, the pattern affects the dot size which is strongly correlated to the distance from the pit center. In Fig. 30b, the mean dot area is higher inside the indents than outside and reaches a maximum at the indent boundary. In line with theoretical predictions [163], our experimental results suggest that inside the indents, the island formation energy is smaller than on the flat surface, due to the enhanced strain relaxation.

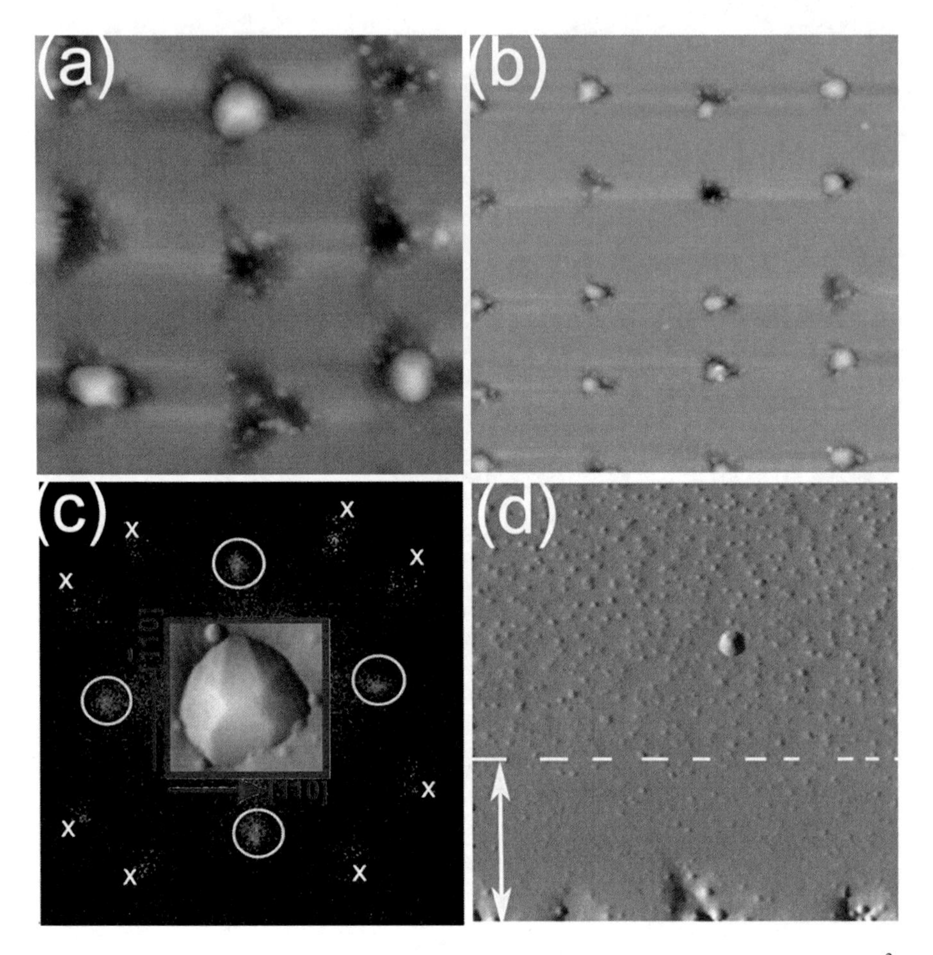

Fig. 31 AFM images of the surface morphology at 7 ML of Ge coverage: (**a**) (3 × 3 × 0.2 μm^2) 1 μm-pitch array; (**b**) (5 × 5 μm^2) 2 μm-pitch array. (**c**) Surface orientation map of the dome-shaped islands; in the *inset* (0.6 × 0.6 × 0.1 μm^2) a single Ge dome is displayed. (**d**) (3.2 × 3.2 × 0.060 μm^2) non-patterned area near the pattern edge

After the first stages of Ge dot nucleation, the system keeps evolving at a larger coverage of 7 ML. The resulting surface morphology for 1 and 2 μm pitch distance is shown in Fig. 31a, b. Two main features stand out from simple inspection of AFM images: (1) the disappearance of the small dots on the surface from between the indents; (2) the formation of larger faceted Ge islands inside some of the indents. The surface orientation map of these faceted islands, reported in Fig. 31c, shows two types of dominant facets, the {113} and {15 3 23} facets marked by circles and crosses, respectively. These are the same facets which are usually measured for multifaceted islands grown on the flat surface (the so-called domes).

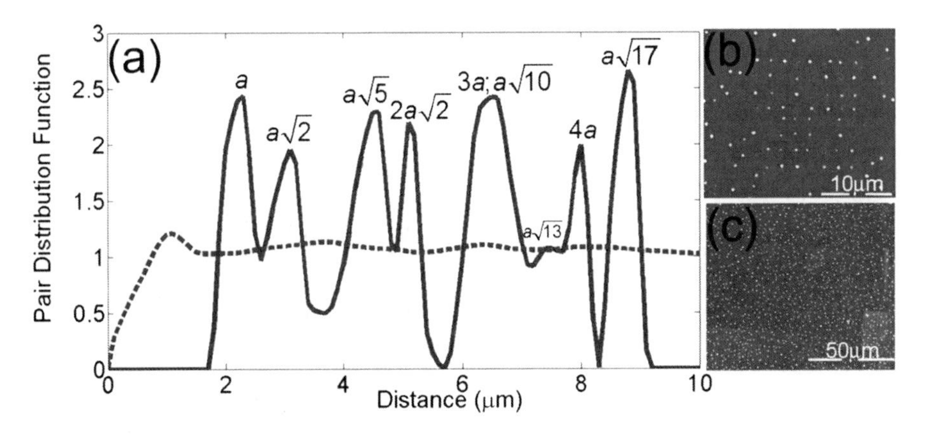

Fig. 32 (**a**) Plots of the PDF on the 2 μm-pitch array (*solid curve*) and on a non-patterned area (*dashed curve*). Near the PDF peaks the radius of the relative coordination shell is indicated ($a = 2$ μm). The SEM images used for evaluating the PDFs of the patterned and non-patterned areas are shown in (**b**) and (**c**), respectively

The island surface appears to be not marked by dislocations. We suggest that, after flash-annealing and subsequent coverage by about 3 ML of wetting layer, the indented-induced dislocations are partially healed in the region close to the surface. Nevertheless, the dislocation strain field, which guides nucleation, is largely unaffected during the Ge growth at 853 K, since we observe an effective enhancement of the long-range order on the patterned regions. Figure 32a shows the PDF of Ge islands on the 2 μm pitch pattern (solid curve) along with the PDF measured on a nonpatterned area (dashed curve). On the patterned templates (Fig. 32b), the PDF shows well-defined peaks corresponding to the coordination shells of the square lattice; while on the unpatterned region (Fig. 32c) the single smooth peak of the nearest neighbor shell is solely present (Fig. 32a).

We argue that island ordering is due to the same mechanism which drives the morphological change in island shapes. Ge islands that form at the indented sites are trapped because of the lower strain energy induced by the relaxation. During subsequent Ge deposition at $T = 600°C$, Ge diffuses on the Si(001) surface allowing these trapped islands to ripen at the expense of Ge islands surrounding the indents. In fact, the volume of the large islands which form at the indented sites is compatible with the total volume of the small dots in the capture zone of an indent. This is natural if the shrinking is due to coarsening, since material would flow preferentially to the closest indent. Comparing the average volume of the large islands (~3×10^6 nm^3) with the volume of the small islands before ripening (~2×10^4 nm^3), we find that the critical value for the first-order transition to dome-shaped islands ($V_C = 1.3 \times 10^6$ nm^3) [183] is exactly in between. This suggests that the coarsening process is driven by the discontinuous change in island chemical potential at the shape transition ("anomalous coarsening") [108]. When the island volume becomes larger than the critical volume V_C, anomalous coarsening activates, producing the changes of the island shape seen in Fig. 31. Further

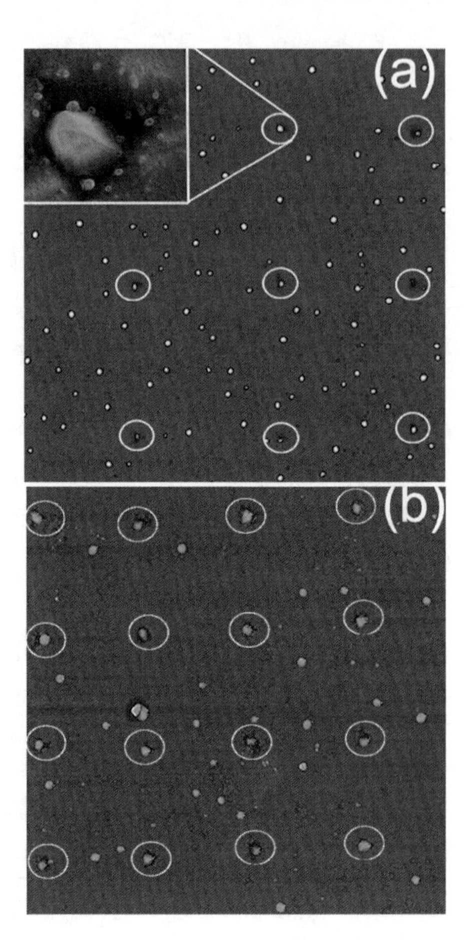

Fig. 33 Surface morphology and analysis of the island spatial organization at 7 ML of Ge coverage. (**a**) AFM image (30 × 30 × 0.30 μm^2) of the 10 μm-pitch array; in the *inset* (1.25 μm^2) a dot grown inside a pit is displayed. (**b**) AFM image (20 × 20 × 0.22 μm^2) of the 5 μm-pitch array; the *white circles* mark the pit position

evidence of coarsening is given by the depleted regions that are all around the patterns, at a distance of about 1 μm from the pattern borders (Fig. 31d). These regions of reduced islands density, which are caused by the material that migrated toward the pits, have a length consistent with the migration length of Ge in similar experimental conditions (≈1 μm) [147, 184]. As expected, at island volumes lower than V_C, the boundary of the pattern, as well as the area in between the indents, are not depleted.

Since island equilibration is triggered by surface diffusion, optimum ordering occurs when the pattern pitch is comparable to the Ge diffusion length. Our analysis of the pattern periodicity confirms that site-selective nucleation is observed only for pitch distance 2 μm, which is roughly twice the migration length of Ge adatoms in our experimental conditions. Conversely, some dots between the pits are still present on the arrays with a pitch of 5 and 10 μm (Fig. 33a, b), which is consistently larger than the Ge migration length.

Island ordering is therefore strictly driven (and limited) by surface diffusion. In line with previous results [182], we suggest that the origin of the diffusion field toward the nanostamped features is the local tensile strain induced by indentation which lowers the lattice mismatch between film and substrate. The ability to direct the spatial organization of epitaxial nanostructures in large areas without the need of lithographic masks and chemical cleaning treatments makes this patterning technique particularly easy to implement and, thus, rather attractive. However, nanoindentation is intrinsically limited in spatial resolution and in the minimum size/depth of the pits that can be produced. In case a finer nanostructuring of the surface is needed, a different patterning method has to be used.

These results demonstrate that the pathway which leads to lateral ordering is determined by the peculiar growth process on Si(001) and that it can be effectively modulated by changing the pattern parameters.

4 STM Patterning

A scanning tunneling microscope is well suited for substrate patterning at atomic resolution, as the tip position can be controlled with the precision of a fraction of angstrom, and atoms on the surface can be attracted to the tip or discharged from it by applying a small voltage pulse.

Substrate patterning by STM has been pioneered by Snow and Campbell [185], who oxidized a H-passivated Si(100) substrate by applying a negative tip (4 V) bias by STM in air and applied a standard liquid etch procedure to create a 120 nm deep pattern in the unexposed areas. They used also AFM in a similar manner [186] with a Ti-coated tip obtaining a highly reproducible pattern.

Successive improvements and modifications of this technique led researchers to create nanoscale patterns in different semiconductors in ultra-high vacuum with very high resolution. Kohmoto and coworkers used the STM tip to create small defects in GaAs(100) [187] by releasing a tiny amount of material at selected locations with a 7 V pulse, giving origin to small holes in the successive GaAs epitaxial layers. These holes were then used to attract InAs clusters in the final epitaxial step.

In a recent paper [188], we demonstrated the use of the STM tip in UHV to dig tiny holes in a Si(100) reconstructed surface by a Z pulse, obtaining a regular array pits that kept their shape also after a long 600°C annealing. The Si(001) substrate (p type, $\rho = 0.1$–0.5 cm) was firstly annealed by dc heating at 1,250°C in ultra-high-vacuum obtaining a (2×1) reconstructed surface. The pits obtained by Z pulse had a diameter in the range 8–15 nm and depth of 1–2 ML. We have also attempted to create nanostructures by current pulses (10 nA for 20 ms) or voltage pulses (10 V for 20 ms). The results are shown in Fig. 34: the current pulse gives origin to clusters of material while the voltage pulse digs irregular cluster of holes, probably connected to the shape of the tip. Lateral tips are very common in STM probes, and it is very likely that material is released from each of them in the current pulse or material is attracted from the surface by the voltage pulse. The cleanest holes are generated by Z pulse, where the only tip interacting with the surface is the one at the very apex.

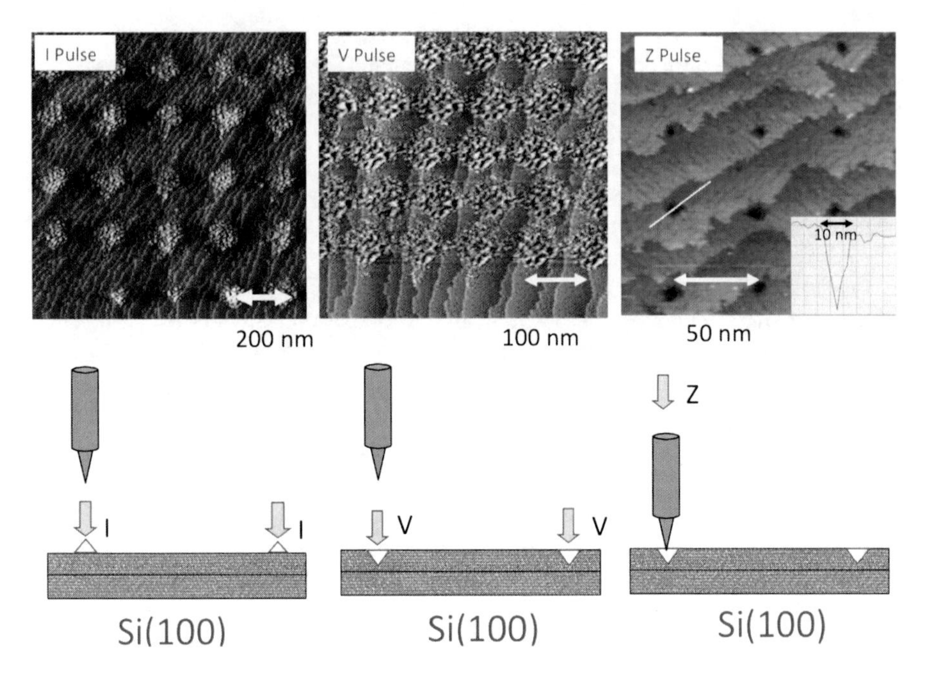

Fig. 34 Nanopatterning on Si(100) by STM. *Left*: by current pulse ($\Delta I = 10$ nA, $\Delta t = 20$ ms); *middle*: by voltage pulse ($\Delta V = 10$ V, $\Delta t = 20$ ms); *right*: by Z pulse ($\Delta Z = 1.5$ nm, $\Delta t = 20$ ms). The *inset* shows the cross section of a pit along the *white line*

On the nanopatterned surface, the growth of Ge by physical vapor deposition of $(2.6 \pm 0.3) \times 10^{-3}$ ML/s was recorded in real-time by STM [188]. The Ge coverage was estimated from the increasing area of terraces between two successive images during the layer-by-layer growth.

Figure 35a–d display four images extracted from an STM movie of Ge growth. At the beginning a step flow process occurs, meaning that the WL forms on the step edge enlarging the terraces.

An accurate data analysis confirms that Ge atoms do not fill the pit, but rather stop at the step edges around. The pit diameter increases, jumping every 1.15 ML of Ge coverage. This corresponds to the formation of a new layer at an equilibrium distance of 1.4 ± 0.3 nm with respect to the edge of the layer below, matching the size of a $p(2 \times 2)$ or $c(4 \times 2)$ reconstructed unit cell [188].

The evolution of a specific hut from WL to pyramid is then followed confirming the existence of a pre-pyramid stage, which evolves with the progressive insertion of {105} facets.

The 2D–3D transition takes place between 3 and 4 ML of Ge coverage. In Fig. 36a–d we show four representative images selected among the many taken during the growth. The arrows indicate the position of the pits, which drive the nucleation of the Ge huts. In the top-right region of Fig. 36a, b, it is possible to see the formation of a pre-pyramid that transforms into a pyramidal hut

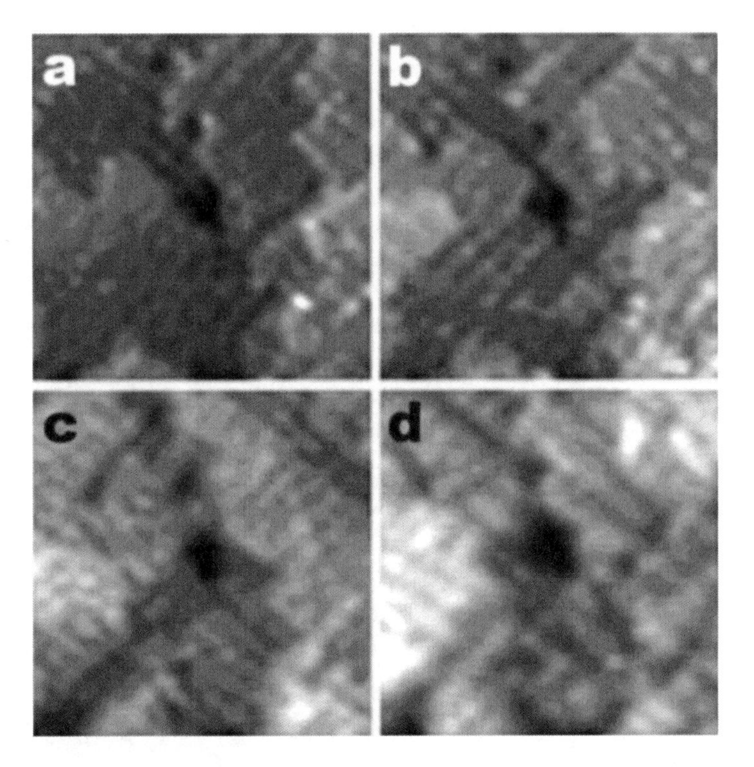

Fig. 35 Real time STM images acquired during Ge deposition on Si(100) around a pit. (**a**) 0.87 ML; (**b**) 0.99 ML; (**c**) 1.24 ML; (**d**) 1.48 ML

in the later stage of the growth Fig. 36c, d. By analyzing the evolution of the 3D line profile and volume of the above cluster, we have been able to determine the transition between platelet growth and pyramid formation, occurring at a height of 0.8 nm. At $\theta = 3.79$ ML, a complete square base pyramid is observed which grows by developing its four {105} facets [104, 124]. The results suggest that arrays of intentionally produced pits drive the nucleation process at selected sites. This remarkable result can be explained by evaluating the relaxation energy originating from elastic interactions between islands and pits. To this purpose we model our system (Fig. 37) by a stepped island (with steps of equal spacing s) aside a square pit four-layers deep. By taking into account repulsive or attractive interactions between the steps [189, 190]

$$\Delta G_{\mathrm{r}} = A \sum_{i \neq j} \mathrm{sign}(i)\mathrm{sign}(j)(s_{ij}) \ln\left(\frac{s_{ij}}{a}\right), \tag{21}$$

where A is a constant corresponding to the energy of the epilayer proportional to the step height, sign $= \pm 1$ counts the attractive (+) or repulsive (−) interaction between two steps with the same or opposite orientation, respectively, s_{ij}

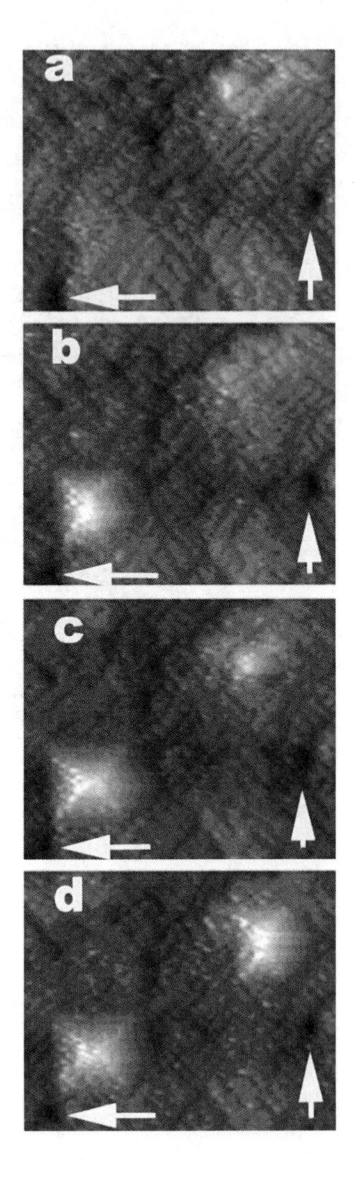

Fig. 36 Sequence of STM images at increasing Ge coverage, showing the formation of mounds that evolves into square pyramids close to the pits (indicated by *arrows*) It is interesting to note that the pit is always close to one of the corners of the pyramid

is the distance between the steps i and j, and a is a cutoff length of the order of the lattice parameter of Ge. The value of $\Delta G_r/A$ is increasing as a function of s/a for a one-layer high island, so in this case the growth is unlikely; for three-layer and four-layer high pre-pyramids, $\Delta G_r/A$ is decreasing and thus they have high probability to grow at values of $s/a > 3$ and $s/a > 2$, respectively. For a square-base Ge pre-pyramid, a width of 16 nm and an aspect ratio of 0.017 are calculated consistent with the experiment.

The ultimate STM patterning would be to move single atoms, and to create atomic precise drawings on a surface. This technique has been developed soon

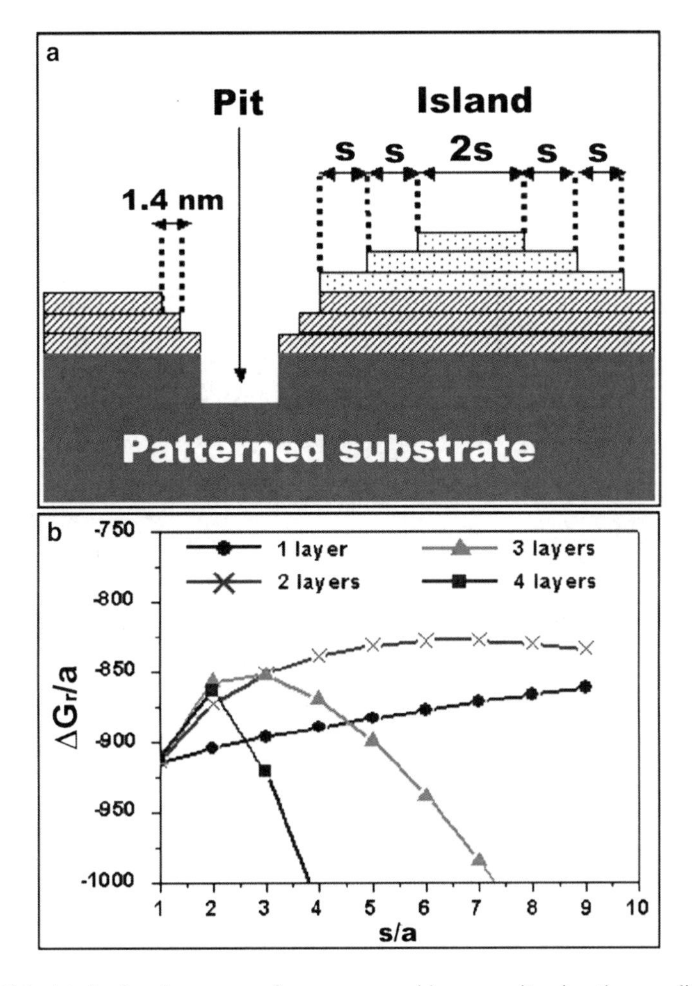

Fig. 37 Calculated relaxation energy for a pre-pyramid near a pit using the one-dimensional elastic interaction model described by (1) of the text. (**a**) Cross-sectional schematic representation of a pit and a three-layer high island considered for the calculation; (**b**) evolution of the normalized relaxation energy, $\Delta G_r/A$, as a function of the step width s for an island one- to four-layer high

after the invention of the STM by demonstrating the nanoscale oxidation and H-passivation of a Si(100) 2×1 surface [191]. Outstanding results have been achieved more recently by groups attempting to create nanostructures suitable for quantum computing applications [192–195], by careful removal of single H atoms on a Si(100):H surface, with the STM exposing reactive Si dangling bond sites, followed by selective deposition of P atoms in the exposed areas. Hydrogen termination can be achieved by chemical methods or by heating the sample and exposing it to a beam of atomic hydrogen for a few minutes. Tip desorption is obtained by using high positive voltage (4–7 V) moderate current (1–2 nA), and a scanning speed

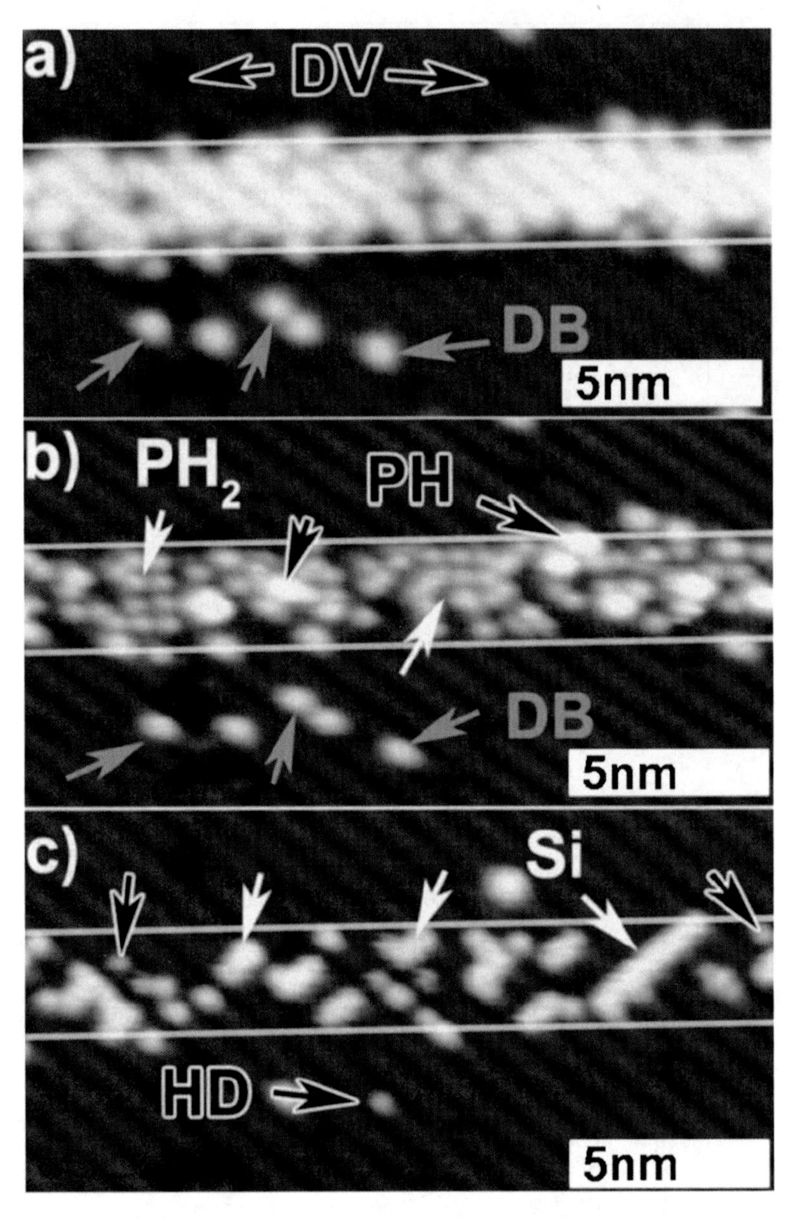

Fig. 38 STM-based fabrication of P dopant nanostructures. Each STM image shows the same region of the surface. (**a**) STM lithography of a 4 nm wide dangling bond wire; (**b**) phosphine dosing; (**c**) phosphorus incorporation annealing (5 s at 350°C) (from [194], Courtesy of M. Simmons)

of 100 nm/s in the selected areas. P dosing is obtained by exposing the surface to PH_3. Incorporation of P in the matrix is obtained by a further annealing (Fig. 38). This process, precise and very clean, has been used to realize quantum devices [196].

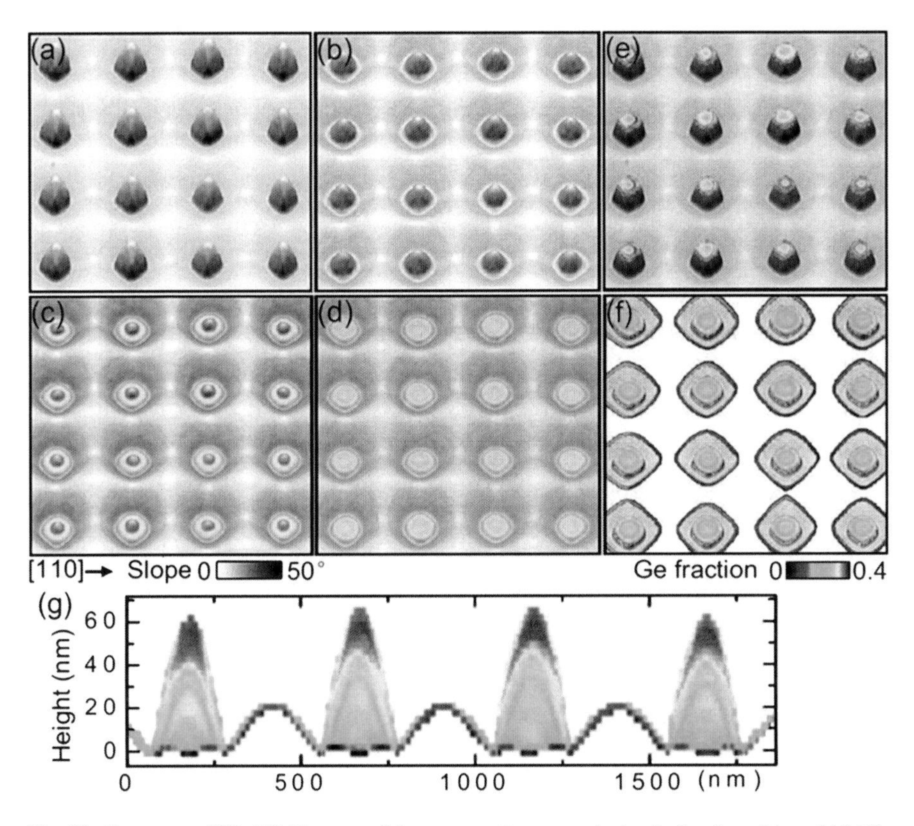

Fig. 39 Sequence of 3D AFM images of the same surface area obtained after deposition of 12 ML Ge on patterned Si(001) with a period of 500 nm at 720°C (**a**) and after progressive selective etching in NHH solution (**b–d**). Horizontal cross-cuts of the islands shown in (**a**) with in-plane Ge compositions at heights of 36 nm (**e**) and 10 nm (**f**) with respect to the level of the island bases. (**g**) Vertical cross-cut of the Ge content across one island row (from [201] Courtesy of J.J. Zhang)

More recently, by using a careful preparation of the surface, this process has been extended to Ge(100), with the successful fabrication of a P nanowire [197–199].

4.1 Lithographic Patterning

Lithography has always been the basis of the patterning in microelectronic and nanoelectronics. In spite of the fact that, following the Moore law, technology is attaining the ultimate limit for top down approach, there is still much interest in creating a pattern by a combination of optical or electron beam lithography and self-assembly. Results recently achieved in the growth of ordered Ge islands on Si demonstrate that, with careful preparation, it is possible to obtain perfect nanostructures with an impressive control on shape and ordering. Zhang et al. [200, 201] realized arrays of holes on Si(001) with variable spacing (500–1,000 nm). A Si buffer layer is used to correct the pit morphology and it is successively covered by a layer of Ge (3–7 ML) at 700°C. The pattern gives origin to a very regular array of square-based islands with perfect lateral ordering. At higher depositions these islands transform into "domes" and then into "barns" [201]. An accurate study of the Ge composition, by using different grades of Ge etching, shows that the Ge concentration at the base decreases with the increasing amount of deposited Ge (Fig. 39).

5 Conclusions

In this article, we reviewed some recent research findings related to the self-organization of Ge nanostructures at Si surfaces. The first part of our contribution examined the possibility to control the positioning and the actual growth mode of epitaxial nanostructures by manipulating the intrinsic mechanisms of the SK process. For example, by tailoring the elastic environment of interacting islands with substrate misorientation, it is possible to engineer the growth of strained epitaxial films and direct it toward desired pathways. By modulating the substrate miscut, a fine tuning of island shape is obtained as well. The location of Ge nanostructures can be influenced by atomic steps and step bunching which serve as natural templates for nanodot clustering.

In the second part, our analysis points to patterning techniques which allow one to harness the natural self-organization dynamics of the system. The key idea of growth on patterned substrates is the correspondence between specific surface features and preferential sites for island nucleation. Here, selective nucleation has been observed and discussed for patterned features obtained by different techniques. STM nanolithography enables one to produce in a controlled way surface defects as shallow as those occurring naturally on surfaces and, hence, it is particularly useful to study the influence of elastic strain fields of defects in heterogeneous nucleation. FIB patterning of Si and SiO_2 surfaces affects the nucleation sites of Ge nanostructures deposited, increasing both ordering and homogeneity of the islands. The ability to direct spatial organization of epitaxial nanostructures in large areas without the need of chemical processing makes nanoindentation patterning particularly easy to implement and, thus, rather attractive.

Acknowledgments The authors are in debt to S. Ruffell for providing nanoindented samples for this study and E. Placidi for his help with AFM measurements. We acknowledge the financial support of the Queensland Government through the NIRAP project "Solar Powered Nanosensors."

References

1. Kastner, M.A.: Proceedings of the 23rd International Conference on Physics of Semiconductors 1, 27 (1996)
2. Chen, M., Porod, W.: Design of gate-confined quantum-dot structures in the few-electron regime. J. Appl. Phys. **78**, 1050 (1995)
3. Kirstaedter, N., Ledentsov, N.N., Grundmann, M., Bimberg, D., Ustinov, V.M., Ruvimov, S. S., Maximov, M.V., Kopev, P.S., Alferov, Z.I., Richter, U., Werner, P., Gosele, U., Heydenreich, J.: Low threshold, large T0 injection laser emission from (InGa)As quantum dots. Electron. Lett. **30**, 1416 (1994)
4. Bimberg, D., Kirstaedter, N., Ledentsov, N.N., Alferov, Zh.I., Kopev, P.S., Ustinov, V.M.: InGaAs-GaAs quantum-dot lasers. IEEE J. Sel. Top. Quantum Electron. **3**, 196 (1997)
5. Chang, W., Chen, W., Chang, H., Hsieh, T., Chyi, J., Hsu, T.: Efficient single-photon sources based on low-density quantum dots in photonic-crystal nanocavities. Phys. Rev. Lett. **96**, 117401 (2006)
6. Fafard, S., Hinzer, K., Raymond, S., Dion, M., McCaffrey, J., Feng, Y., Charbonneau, S.: Red-emitting semiconductor quantum dot lasers. Science **274**, 1350 (1996)
7. Yoffe, A.D.: Semiconductor quantum dots and related systems: electronic, optical, luminescence and related properties of low dimensional systems. Adv. Phys. **50**, 1 (2001)
8. Freitasjr, R.: What is nanomedicine? Nanomedicine: nanotechnology. Biol. Med. **1**, 2 (2005)
9. Wagner, V., Dullaart, A., Bock, A.-K., Zweck, A.: The emerging nanomedicine landscape. Nat. Biotechnol. **24**, 1211 (2006)
10. Steele, B., Heinzel, A.: Materials for fuel-cell technologies. Nature **414**, 345 (2001)
11. Dresselhaus, M.S., Thomas, I.L.: Alternative energy technologies. Nature **414**, 332 (2001)
12. Bell, A.: The impact of nanoscience on heterogeneous catalysis. Science **299**, 1688 (2003)
13. Grunes, J., Zhu, J., Somorjai, G.: Catalysis and nanoscience. Chem. Commun., 2257 (2003)
14. Somorjai, G., Tao, F., Park, J.: The nanoscience revolution: merging of colloid science, catalysis and nanoelectronics. Top. Catal. **47**, 1 (2008)
15. Gellman, A., Shukla, N.: Nanocatalysis: more than speed. Nat. Mater. **8**, 87 (2009)
16. Kalantar-zadeh, K., Fry, B.: Inorganic nanotechnology enabled sensors. In: Nanotechnology-Enabled Sensors, pp. 283. Springer, New York (2008)
17. Bogue, R.: Nanosensors: a review of recent progress. Sens. Rev. **28**, 12 (2008)
18. Fulton, T.A., Dolan, G.J.: Observation of single-electron charging effects in small tunnel junctions. Phys. Rev. Lett. **59**, 109 (1987)
19. Kastner, M.A.: The single electron transistor and artificial atoms. Ann. Phys. **9**, 885 (2000)
20. Lent, C.: Bypassing the transistor paradigm. Science **288**, 1597 (2000)
21. Cole, T.: Quantum-dot cellular automata. Progr. Quant. Electron. **25**, 165 (2001)
22. DiVincenzo, D.: Quantum computation. Science **270**, 255 (1995)
23. Kane, B.E.: Silicon-based quantum computation. Fortschr. Phys. **48**, 1023 (2000)
24. Thompson, S., Parthasarathy, S.: Moore's law: the future of Si microelectronics. Mater. Today **9**, 20 (2006)
25. Krastanov, L., Stranski, I.: Zur theorie der orientierten ausscheidung von Ionenkristallen aufeinander. Sitz. Ber. Akad. Wiss. Wien, Mat. Nat. 146 (1938)
26. Markov, I.V.: Crystal Growth for Beginners: Fundamentals of Nucleation, Crystal Growth, and Epitaxy. World Scientific, Singapore (1995)

27. Pimpinelli, A., Villain, J.: Physics of Crystal Growth. Cambridge University Press, Cambridge (1998)
28. Venables, J.: Introduction to Surface and Thin Film Processes. Cambridge University Press, Cambridge (2000)
29. Onushchenko, A.A., Ekimov, A.I.: Quantum size effect in three dimensional microscopic semiconductor crystals. JETP Lett. **34**, 345 (1981)
30. Rossetti, R., Nakahara, S., Brus, L.E.: Quantum size effects in the redox potentials, resonance Raman spectra, and electronic spectra of CdS crystallites in aqueous solution. J. Chem. Phys. **79**, 1086 (1983)
31. Shchukin, V., Bimberg, D.: Spontaneous ordering of nanostructures on crystal surfaces. Rev. Mod. Phys. **71**, 1125 (1999)
32. Liu, F.: Self-organized nanoscale structures in Si/Ge films. Surf. Sci. **386**, 169 (1997)
33. Moriarty, P.: Nanostructured materials. Rep. Progr. Phys. **64**, 297 (2001)
34. Arthur, J.: Molecular beam epitaxy. Surf. Sci. **500**, 189 (2002)
35. Rosei, F.: Nanostructured surfaces: challenges and frontiers in nanotechnology. J. Phys. Condens. Mat. **16**, S1373 (2004)
36. Venables, J.: Atomic processes in crystal growth. Surf. Sci. **299–300**, 798 (1994)
37. Cammarata, R.: Surface and interface stress effects in thin films. Prog. Surf. Sci. **46**, 1 (1994)
38. Mulheran, P.A., Blackman, J.A.: The origins of island size scaling in heterogeneous film growth. Phil. Mag. Lett. **72**, 55 (1995)
39. Williams, S., Ribeiro, G., Kamins, T., Ohlberg, D.: Thermodynamics of the size and shape of nanocrystals: epitaxial Ge on Si(001). Annu. Rev. Phys. Chem. **51**, 527 (2000)
40. Giesen, M.: Step and island dynamics at solid/vacuum and solid/liquid interfaces. Prog. Surf. Sci. **68**, 1 (2001)
41. Berbezier, I., Ronda, A., Portavoce, A.: SiGe nanostructures: new insights into growth processes. J. Phys. Condens. Mat. **14**, 8283 (2002)
42. Ratto, F., Costantini, G., Rastelli, A., Schmidt, O., Kern, K., Rosei, F.: Alloying of self-organized semiconductor 3D islands. J. Exp. Nanosci. **1**, 279 (2006)
43. Parshin, A.Y., Marchenko, V.I.: Elastic properties of crystal surfaces. Sov. J. Exp. Theor. Phys. **52**, 129 (1980)
44. Rickman, J., Srolovitz, D.: Defect interactions on solid surfaces. Surf. Sci. **284**, 211 (1993)
45. Liu, F.: Modeling and simulation of strain-mediated nanostructure formation on surface. In: Schommers, W., Rieth, M. (eds.) Handbook of Theoretical and Computational Nanotechnology, p. 577. American Scientific Publishers, Los Angeles (2006)
46. Alerhand, O.L., Vanderbilt, D., Meade, R., Joannopoulos, J.D.: Spontaneous formation of stress domains on crystal surfaces. Phys. Rev. Lett. **61**, 1973 (1988)
47. Floro, J.A., Lucadamo, G.A., Chason, E., Freund, L.B., Sinclair, M., Twesten, R.D., Hwang, R.Q.: SiGe island shape transitions induced by elastic repulsion. Phys. Rev. Lett. **80**, 4717 (1998)
48. Shenoy, V., Freund, L.: A continuum description of the energetics and evolution of stepped surfaces in strained nanostructures. J. Mech. Phys. Solids. **50**, 1817 (2002)
49. Ramasubramaniam, A., Shenoy, V.B.: Three-dimensional simulations of self-assembly of hut-shaped Si/Ge quantum dots. J. Appl. Phys. **95**, 7813 (2004)
50. Yagi, K.: Step bunching, step wandering and faceting: self-organization at Si surfaces. Surf. Sci. Rep. **43**, 45 (2001)
51. Jeong, H.: Steps on surfaces: experiment and theory. Surf. Sci. Rep. **34**, 171 (1999)
52. Brunner, K., Abstreiter, G.: Ordering and electronic properties of self-assembled Si/Ge quantum dots. Jpn. J. Appl. Phys. **40**, 1860 (2001)
53. Drucker, J.: Self-assembling Ge(Si)/Si(100) quantum dots. IEEE J. Quantum Electron. **38**, 975 (2002)
54. Berbezier, I., Ronda, A.: Si/SiGe heterostructures for advanced microelectronic devices. Phase Transit. **81**, 751 (2008)

55. Rowell, N.L., Lockwood, D.J., Berbezier, I., Szkutnik, P.D., Ronda, A.: Photoluminescence efficiency of self-assembled Ge nanocrystals. J. Electrochem. Soc. **156**, H913 (2009)
56. Singha, R.K., Manna, S., Das, S., Dhar, A., Ray, S.K.: Room temperature infrared photoresponse of self assembled Ge/Si(001) quantum dots grown by molecular beam epitaxy. Appl. Phys. Lett. **96**, 233113 (2010)
57. Oehme, M., Karmous, A., Sarlija, M., Werner, J., Kasper, E., Schulze, J.: Ge quantum dot tunneling diode with room temperature negative differential resistance. Appl. Phys. Lett. **97**, 012101 (2010)
58. Rezaev, R.O., Kiravittaya, S., Fomin, V.M., Rastelli, A., Schmidt, O.G.: Engineering self-assembled SiGe islands for robust electron confinement in Si. Phys. Rev. B **82**, 153306 (2010)
59. Lin, Z., Brunkov, P., Bassani, F., Bremond, G.: Electrical study of trapped charges in nanoscale Ge islands by Kelvin probe force microscopy for nonvolatile memory applications. Appl. Phys. Lett. **97**, 263112 (2010)
60. Simmons, C.B., Koh, T., Shaji, N., Thalakulam, M., Klein, L.J., Qin, H., Luo, H., Savage, D. E., Lagally, M.G., Rimberg, A.J., Joynt, R., Blick, R., Friesen, M., Coppersmith, S.N., Eriksson, M.A.: Pauli spin blockade and lifetime-enhanced transport in a Si/SiGe double quantum dot. Phys. Rev. B **82**, 245312 (2010)
61. Bera, C., Soulier, M., Navone, C., Roux, G., Simon, J., Volz, S., Mingo, N.: Thermoelectric properties of nanostructured Si1-xGex and potential for further improvement. J. Appl. Phys. **108**, 124306 (2010)
62. Brunner, K.: Si/Ge nanostructures. Rep. Progr. Phys. **65**, 27 (2002)
63. Stangl, J., Holy, V., Bauer, G.: Structural properties of self-organized semiconductor nanostructures. Rev. Mod. Phys. **76**, 725 (2004)
64. Voigtlander, B.: Fundamental processes in Si/Si and Ge/Si epitaxy studied by scanning tunneling microscopy during growth. Surf. Sci. Rep. **43**, 127 (2001)
65. Teichert, C.: Self-organization of nanostructures in semiconductor heteroepitaxy. Phys. Rep. **365**, 335 (2002)
66. Baribeau, J.M., Wu, X., Rowell, N.L., Lockwood, D.J.: Ge dots and nanostructures grown epitaxially on Si. J. Phys. Condens. Mat. **18**, R139 (2006)
67. Berbezier, I., Ronda, A.: SiGe nanostructures. Surf. Sci. Rep. **64**, 47 (2009)
68. Misbah, C., Louis, O., Saito, Y.: Crystal surfaces in and out of equilibrium: a modern view. Rev. Mod. Phys. **82**, 981 (2010)
69. Bartelt, N., Einstein, T., Williams, E.: The influence of step-step interactions on step wandering. Surf. Sci. **240**, L591 (1990)
70. Fisher, M., Fisher, D., Weeks, J.: Agreement of capillary-wave theory with exact results for the interface profile of the two-dimensional Ising model. Phys. Rev. Lett. **48**, 368 (1982)
71. de Gennes, P.G.: Soluble model for fibrous structures with steric constraints. J. Chem. Phys. **48**, 2257 (1968)
72. Zandvliet, H.: An experimentalists view on the analogy between step edges and quantum mechanical particles. Solid State Commun. **94**, 677 (1995)
73. Gruber, E., Mullins, W.: On the theory of anisotropy of crystalline surface tension. J. Phys. Chem. Solids. **28**, 875 (1967)
74. Ihle, T., Misbah, C., Louis, P.: Equilibrium step dynamics on vicinal surfaces revisited. Phys. Rev. B **58**, 2289 (1998)
75. Persichetti, L., Sgarlata, A., Fanfoni, M., Bernardi, M., Balzarotti, A.: Step-step interaction on vicinal Si(001) surfaces studied by scanning tunneling microscopy. Phys. Rev. B **80**, 075315 (2009)
76. Alfonso, C., Bermond, J., Heyraud, J., Metois, J.: The meandering of steps and the terrace width distribution on clean Si(111): an in-situ experiment using reflection electron microscopy. Surf. Sci. **262**, 371 (1992)
77. Bartelt, N.C., Goldberg, J.L., Einstein, T.L., Williams, E., Heyraud, J.C., Metois, J.J.: Brownian motion of steps on Si(111). Phys. Rev. B **48**, 15453 (1993)

78. Baski, A.A., Erwin, S.C., Whitman, L.J.: The structure of silicon surfaces from (001) to (111). Surf. Sci. **392**, 69 (1997)
79. Viernow, J., Lin, J.L., Petrovykh, D.Y., Leibsle, F.M., Men, F.K., Himpsel, F.J.: Regular step arrays on silicon. Appl. Phys. Lett. **72**, 948 (1998)
80. Rauscher, H.: One-dimensional confinement of organic molecules via selective adsorption on CaF1 versus CaF2. Chem. Phys. Lett. **303**, 363 (1999)
81. Brunner, K.: Self-organized periodic arrays of SiGe wires and Ge islands on vicinal Si substrates. Physica E **7**, 881 (2000)
82. Himpsel, F.J., Altmann, K.N., Bennewitz, R., Crain, J.N., Kirakosian, A., Lin, J.L., McChesney, J.L.: One-dimensional electronic states at surfaces. J. Phys. Condens. Mat. **13**, 11097 (2001)
83. Men, F.K., Liu, F., Wang, P.J., Chen, C.H., Cheng, D.L., Lin, J.L., Himpsel, F.J.: Self-organized nanoscale pattern formation on vicinal Si(111) surfaces via a two-stage faceting transition. Phys. Rev. Lett. **88**, 096105 (2002)
84. Bennewitz, R., Crain, J.N., Kirakosian, A., Lin, J.L., McChesney, J.L., Petrovykh, D.Y., Himpsel, F.J.: Atomic scale memory at a silicon surface. Nanotechnology **13**, 499 (2002)
85. Zandvliet, H.J.W., Elswijk, H.B., van Loenen, E.J., Dijkkamp, D.: Equilibrium structure of monatomic steps on vicinal Si(001). Phys. Rev. B **45**, 5965 (1992)
86. Zandvliet, H.J.W., Elswijk, H.B.: Morphology of monatomic step edges on vicinal Si(001). Phys. Rev. B **48**, 14269 (1993)
87. Zandvliet, H.J.W., Poelsema, B., Elswijk, H.B.: Fluctuations of monatomic steps on Si(001). Phys. Rev. B **51**, 5465 (1995)
88. Zandvliet, H., Van Moere, R., Poelsema, B.: Free energy and stiffness of [110] and [010] steps on a (001) surface of a cubic lattice: revival of the solid-on-solid model. Phys. Rev. B **68**, 073404 (2003)
89. Poon, T., Yip, S., Ho, P., Abraham, F.: Equilibrium structures of Si(100) stepped surfaces. Phys. Rev. Lett. **65**, 2161 (1990)
90. Pehlke, E., Tersoff, J.: Nature of the step-height transition on vicinal Si(001) surfaces. Phys. Rev. Lett. **67**, 465 (1991)
91. Schlier, R.E., Farnsworth, H.E.: Structure and adsorption characteristics of clean surfaces of germanium and silicon. J. Chem. Phys. **30**, 917 (1959)
92. Tromp, R.M., Hamers, R.J., Demuth, J.E.: Si(001) dimer structure observed with scanning tunneling microscopy. Phys. Rev. Lett. **55**, 1303 (1985)
93. Chadi, D.J.: Stabilities of single-layer and bilayer steps on Si(001) surfaces. Phys. Rev. Lett. **59**, 1691 (1987)
94. Tong, X., Bennett, P.A.: Terrace-width-induced domain transition on vicinal Si(100) studied with microprobe diffraction. Phys. Rev. Lett. **67**, 101 (1991)
95. de Miguel, J.J., Aumann, C.E., Kariotis, R., Lagally, M.G.: Evolution of vicinal Si(001) from double- to single-atomic-height steps with temperature. Phys. Rev. Lett. **67**, 2830 (1991)
96. Wierenga, P.E., Kubby, J.A., Griffith, J.E.: Tunneling images of biatomic steps on Si(001). Phys. Rev. Lett. **59**, 2169 (1987)
97. Swartzentruber, B.S., Kitamura, N., Lagally, M.G., Webb, M.B.: Behavior of steps on Si (001) as a function of vicinality. Phys. Rev. B **47**, 13432 (1993)
98. Bartelt, N., Einstein, T., Williams, E.: The role of step collisions on diffraction from vicinal surfaces. Surf. Sci. **276**, 308 (1992)
99. Spencer, B.J., Tersoff, J.: Asymmetry and shape transitions of epitaxially strained islands on vicinal surfaces. Appl. Phys. Lett. **96**, 073114 (2010)
100. Persichetti, L., Sgarlata, A., Fanfoni, M., Balzarotti, A.: Shaping Ge islands on Si(001) surfaces with misorientation angle. Phys. Rev. Lett. **104**, 036104 (2010)
101. Persichetti, L., Sgarlata, A., Fanfoni, M., Balzarotti, A.: Breaking elastic field symmetry with substrate vicinality. Phys. Rev. Lett. **106**, 055503 (2011)

102. Nakaoka, T., Kako, S., Ishida, S., Nishioka, M., Arakawa, Y.: Optical anisotropy of self-assembled InGaAs quantum dots embedded in wall-shaped and air-bridge structures. Appl. Phys. Lett. **81**, 3954 (2002)
103. Kumar, J., Kapoor, S., Gupta, S., Sen, P.: Theoretical investigation of the effect of asymmetry on optical anisotropy and electronic structure of Stranski-Krastanov quantum dots. Phys. Rev. B **74**, 115326 (2006)
104. Mo, Y.W., Savage, D.E., Swartzentruber, B.S., Lagally, M.G.: Kinetic pathway in Stranski-Krastanov growth of Ge on Si(001). Phys. Rev. Lett. **65**, 1020 (1990)
105. Vailionis, A., Cho, B., Glass, G., Desjardins, P., Cahill, D., Greene, J.E.: Pathway for the strain-driven two-dimensional to three-dimensional transition during growth of Ge on Si (001). Phys. Rev. Lett. **85**, 3672 (2000)
106. Tersoff, J., Spencer, B.J., Rastelli, A., von Kanel, H.: Barrierless formation and faceting of SiGe islands on Si(001). Phys. Rev. Lett. **89**, 196104 (2002)
107. Medeiros-Ribeiro, G., Bratkovski, A., Kamins, T., Ohlberg, D., Williams, S.: Shape transition of germanium nanocrystals on a silicon (001) surface from pyramids to domes. Science **279**, 353 (1998)
108. Ross, F.M., Tersoff, J., Tromp, R.M.: Coarsening of self-assembled Ge quantum dots on Si (001). Phys. Rev. Lett. **80**, 984 (1998)
109. Montalenti, F., Raiteri, P., Migas, D.B., von Kanel, H., Rastelli, A., Manzano, C., Costantini, G., Denker, U., Schmidt, O.G., Kern, K., Miglio, L.: Atomic-scale pathway of the pyramid-to-dome transition during Ge growth on Si(001). Phys. Rev. Lett. **93**, 216102 (2004)
110. Rastelli, A., von Kanel, H.: Surface evolution of faceted islands. Surf. Sci. **515**, L493–L498 (2002)
111. LeGoues, F.K., Reuter, M.C., Tersoff, J., Hammar, M., Tromp, R.M.: Cyclic growth of strain-relaxed islands. Phys. Rev. Lett. **73**, 300 (1994)
112. Merdzhanova, T., Kiravittaya, S., Rastelli, A., Stoffel, M., Denker, U., Schmidt, O.G.: Dendrochronology of strain-relaxed islands. Phys. Rev. Lett. **96**, 226103 (2006)
113. Marzegalli, A., Zinovyev, V.A., Montalenti, F., Rastelli, A., Stoffel, M., Merdzhanova, T., Schmidt, O.G., Miglio, L.: Critical shape and size for dislocation nucleation in Si1-xGex islands on Si(001). Phys. Rev. Lett. **99**, 235505 (2007)
114. Gatti, R., Marzegalli, A., Zinovyev, V.A., Montalenti, F., Miglio, L.: Modeling the plastic relaxation onset in realistic SiGe islands on Si(001). Phys. Rev. B **78**, 184104 (2008)
115. Katsaros, G., Costantini, G., Stoffel, M., Esteban, R., Bittner, A.M., Rastelli, A., Denker, U., Schmidt, O.G., Kern, K.: Kinetic origin of island intermixing during the growth of Ge on Si (001). Phys. Rev. B **72**, 195320 (2005)
116. Ratto, F., Locatelli, A., Fontana, S., Kharrazi, S., Ashtaputre, S., Kulkarni, S.K., Heun, S., Rosei, F.: Chemical mapping of individual semiconductor nanostructures. Small **2**, 401 (2006)
117. Ratto, F., Rosei, F., Locatelli, A., Cherifi, S., Fontana, S., Heun, S., Szkutnik, P., Sgarlata, A., De Crescenzi, M., Motta, N.: Composition of GeSi islands in the growth of Ge on Si(111) by x-ray spectromicroscopy. J. Appl. Phys. **97**, 043516 (2005)
118. Medeiros-Ribeiro, G., Williams, S.: Thermodynamics of coherently-strained GexSi1-x nanocrystals on Si(001): alloy composition and island formation. Nano Lett. **7**, 223 (2007)
119. Montoro, L., Leite, M., Biggemann, D., Peternella, F., Batenburg, J., Medeiros-Ribeiro, G., Ramirez, A.: Revealing quantitative 3D chemical arrangement on Ge/Si nanostructures. J. Phys. Chem. C **113**, 9018 (2009)
120. Tu, Y., Tersoff, J.: Coarsening, mixing, and motion: the complex evolution of epitaxial islands. Phys. Rev. Lett. **98**, 096103 (2007)
121. Digiuni, D., Gatti, R., Montalenti, F.: Aspect-ratio-dependent driving force for nonuniform alloying in Stranski-Krastanow islands. Phys. Rev. B **80**, 155436 (2009)

122. Denker, U., Rastelli, A., Stoffel, M., Tersoff, J., Katsaros, G., Costantini, G., Kern, K., Phillipp, J., Jesson, D.E., Schmidt, O.G.: Lateral motion of SiGe islands driven by surface-mediated alloying. Phys. Rev. Lett. **94**, 216103 (2005)
123. Lang, C., Kodambaka, S., Ross, F.M., Cockayne, D.J.H.: Real time observation of GeSi/Si (001) island shrinkage due to surface alloying during Si capping. Phys. Rev. Lett. **97**, 226104 (2006)
124. Raiteri, P., Migas, D.B., Miglio, L., Rastelli, A., von Kanel, H.: Critical role of the surface reconstruction in the thermodynamic stability of 105 Ge pyramids on Si(001). Phys. Rev. Lett. **88**, 256103 (2002)
125. Fujikawa, Y., Akiyama, K., Nagao, T., Sakurai, T., Lagally, M.G., Hashimoto, T., Morikawa, Y., Terakura, K.: Origin of the stability of Ge(105) on Si: a new structure model and surface strain relaxation. Phys. Rev. Lett. **88**, 176101 (2002)
126. Migas, D., Cereda, S., Montalenti, F., Miglio, L.: Electronic and elastic contributions in the enhanced stability of Ge(105) under compressive strain. Surf. Sci. **556**, 121 (2004)
127. Cereda, S., Montalenti, F., Miglio, L.: Atomistic modeling of step formation and step bunching at the Ge(105) surface. Surf. Sci. **591**, 23 (2005)
128. Sutter, P., Lagally, M.G.: Nucleationless three-dimensional island formation in low-misfit heteroepitaxy. Phys. Rev. Lett. **84**, 4637 (2000)
129. Zhu, J., Brunner, K., Abstreiter, G.: Observation of 105 facetted Ge pyramids inclined towards vicinal Si(001) surfaces. Appl. Phys. Lett. **72**, 424 (1998)
130. Teichert, C., Bean, J.C., Lagally, M.G.: Self-organized nanostructures in Si1-xGex films on Si(001). Appl. Phys. A **67**, 675 (1998)
131. Ronda, A.: Self-patterned Si surfaces as templates for Ge islands ordering. Physica E **23**, 370 (2004)
132. Lichtenberger, H., Muhlberger, M., Schaffler, F.: Ordering of Si0.55Ge0.45 islands on vicinal Si(001) substrates: interplay between kinetic step bunching and strain-driven island growth. Appl. Phys. Lett. **86**, 131919 (2005)
133. Sutter, P., Sutter, E., Vescan, L.: Barrierless self-assembly of Ge quantum dots on Si(001) substrates with high local vicinality. Appl. Phys. Lett. **87**, 161916 (2005)
134. Szkutnik, P.D., Sgarlata, A., Balzarotti, A., Motta, N., Ronda, A., Berbezier, I.: Early stage of Ge growth on Si(001) vicinal surfaces with an 8°-miscut along [110]. Phys. Rev. B **75**, 033305 (2007)
135. Chen, G., Wintersberger, E., Vastola, G., Groiss, H., Stangl, J., Jantsch, W., Schaffler, F.: Self-assembled Si0.80Ge0.20 nanoripples on Si(1 1 10) substrates. Appl. Phys. Lett. **96**, 103107 (2010)
136. Persichetti, L., Sgarlata, A., Fanfoni, M., Balzarotti, A.: Ripple-to-dome transition: the growth evolution of Ge on vicinal Si(1 1 10) surface. Phys. Rev. B **82**, 121309 (2010)
137. Persichetti, L., Sgarlata, A., Fanfoni, M., Balzarotti, A.: Pair interaction between Ge islands on vicinal Si(001) surfaces. Phys. Rev. B **81**, 113409 (2010)
138. Johnson, H.T., Freund, L.B.: Mechanics of coherent and dislocated island morphologies in strained epitaxial material systems. J. Appl. Phys. **81**, 6081 (1997)
139. Latyshev, A., Aseev, A., Krasilnikov, A., Stenin, S.: Transformations on clean Si(111) stepped surface during sublimation. Surf. Sci. **213**, 157 (1989)
140. Stoyanov, S.: Electromigration induced step bunching on Si surfaces: how does it depend on the temperature and heating current direction? Jpn. J. Appl. Phys. **30**, 1 (1991)
141. Homma, Y., McClelland, R., Hibino, H.: DC-resistive-heating-induced step bunching on vicinal Si (111). Jpn. J. Appl. Phys. **29**, L2254 (1990)
142. Yang, Y.: An STM study of current-induced step bunching on Si(111). Surf. Sci. **356**, 101 (1996)
143. Stoyanov, S., Tonchev, V.: Properties and dynamic interaction of step density waves at a crystal surface during electromigration affected sublimation. Phys. Rev. B **58**, 1590 (1998)

144. Liu, D., Weeks, J.: Quantitative theory of current-induced step bunching on Si(111). Phys. Rev. B **57**, 14891 (1998)
145. Fujita, K., Ichikawa, M., Stoyanov, S.: Size-scaling exponents of current-induced step bunching on silicon surfaces. Phys. Rev. B **60**, 16006 (1999)
146. Homma, Y., Aizawa, N.: Electric-current-induced step bunching on Si(111). Phys. Rev. B **62**, 8323 (2000)
147. Sgarlata, A., Szkutnik, P.D., Balzarotti, A., Motta, N., Rosei, F.: Self-ordering of Ge islands on step-bunched Si(111) surfaces. Appl. Phys. Lett. **83**, 4002 (2003)
148. Pascale, A., Berbezier, I., Ronda, A., Videcoq, A., Pimpinelli, A.: Self-organization of step bunching instability on vicinal substrate. Appl. Phys. Lett. **89**, 104108 (2006)
149. Tripathi, J.K., Garbrecht, M., Sztrum Vartash, C., Rabani, E., Kaplan, W., Goldfarb, I.: Coverage-dependent self-organized ordering of Co- and Ti-silicide nanoislands along step-bunch edges of vicinal Si(111). Phys. Rev. B **83**, 165409 (2011)
150. Omi, H., Ogino, T.: Positioning of self-assembling Ge islands on Si(111) mesas by using atomic steps. Thin Solid Films **369**, 88 (2000)
151. Liang, S., Zhu, H.L., Kong, D.H., Wang, W.: Formation trends of ordered self-assembled nanoislands on stepped substrates. J. Appl. Phys. **108**, 073512 (2010)
152. Abramo, M., Wasielewski, R.: FIB for failure analysis. Semicond. Int. **20**, 133 (1997)
153. Harriot, L.R., Wagner, A., Fritz, F.: Integrated circuit repair using focused ion beam milling. J. Vac. Sci. Technol. B **4**, 181 (1986)
154. Nakamura, H., Komano, H., Ogasawara, M.: Focused ion beam assisted etching of quartz in XeF2 without transmittance reduction for phase shifting mask repair. Jpn. J. Appl. Phys. **31**, 4465 (1992)
155. Patella, F., Sgarlata, A., Arciprete, F., et al.: Self-assembly of InAs and Si/Ge quantum dots on structured surfaces. J. Phys. Condens. Mat. **16**, S1503 (2004)
156. Szkutnik, P., Sgarlata, A., Motta, N., Placidi, E., Berbezier, I., Balzarotti, A.: Influence of patterning on the nucleation of Ge islands on Si and SiO2 surfaces. Surf. Sci. **601**, 2778 (2007)
157. Patella, F., Sgarlata, A., Arciprete, F., Nufris, S., Szkutnik, P.D., Placidi, E., Fanfoni, M., Motta, N., Balzarotti, A.: Self-assembly of InAs and Si/Ge quantum dots on structured surfaces. J. Phys. Condens. Mat. **16**, S1503 (2004)
158. Kammler, M., Hull, R., Reuter, M.C., Ross, F.M.: Lateral control of self-assembled island nucleation by focused-ion-beam micropatterning. Appl. Phys. Lett. **82**, 1093 (2003)
159. Portavoce, A., Hull, R., Reuter, M.C., Ross, F.M.: Nanometer-scale control of single quantum dot nucleation through focused ion-beam implantation. Phys. Rev. B **76**, 235301 (2007)
160. Karmous, A., Cuenat, A., Ronda, A., Berbezier, I., Atha, S., Hull, R.: Ge dot organization on Si substrates patterned by focused ion beam. Appl. Phys. Lett. **85**, 6401 (2004)
161. Ishikawa, T., Kohmoto, S., Asakawa, K.: Site control of self-organized InAs dots on GaAs substrates by in situ electron-beam lithography and molecular-beam epitaxy. Appl. Phys. Lett. **73**, 1712 (1998)
162. Kim, E.S., Usami, N., Shiraki, Y.: Control of Ge dots in dimension and position by selective epitaxial growth and their optical properties. Appl. Phys. Lett. **72**, 1617 (1998)
163. Hu, H., Gao, H.J., Liu, F.: Theory of directed nucleation of strained islands on patterned substrates. Phys. Rev. Lett. **101**, 216102 (2008)
164. Vastola, G., Montalenti, F., Miglio, L.: Understanding the elastic relaxation mechanisms of strain in Ge islands on pit-patterned Si(001) substrates. J. Phys. Condens. Mat. **20**, 454217 (2008)
165. Pascale, A., Berbezier, I., Ronda, A., Kelires, P.C.: Self-assembly and ordering mechanisms of Ge islands on prepatterned Si(001). Phys. Rev. B **77**, 075311 (2008)
166. Srolovitz, D.J.: On the stability of surfaces of stressed solids. Acta Metall. **37**, 621 (1989)
167. Kolobov, A., Shklyaev, A., Oyanagi, H., Fons, P., Yamasaki, S., Ichikawa, M.: Local structure of Ge nanoislands on Si(111) surfaces with a SiO_2 coverage. Appl. Phys. Lett. **78**, 2563 (2001)

168. Shklyaev, A., Shibata, M., Ichikawa, M.: High-density ultrasmall epitaxial Ge islands on Si(111) surfaces with a SiO_2 coverage. Phys. Rev. B **62**, 1540 (2000)
169. Barski, A., Derivaz, M., Rouviere, J.L., Buttard, D.: Epitaxial growth of germanium dots on Si(001) surface covered by a very thin silicon oxide layer. Appl. Phys. Lett. **77**, 3541 (2000)
170. Baron, T., Pelissier, B., Perniola, L., Mazen, F., Hartmann, J.M., Rolland, G.: Chemical vapor deposition of Ge nanocrystals on SiO_2. Appl. Phys. Lett. **83**, 1444 (2003)
171. Karmous, A., Berbezier, I., Ronda, A.: Formation and ordering of Ge nanocrystals on SiO_2. Phys. Rev. B **73**, 075323 (2006)
172. Sutter, E., Sutter, P.: Assembly of Ge nanocrystals on SiO_2 via a stress-induced dewetting process. Nanotechnology **17**, 3724 (2006)
173. Zacharias, M., Fauchet, P.M.: Blue luminescence in films containing Ge and GeO_2 nanocrystals: the role of defects. Appl. Phys. Lett. **71**, 380 (1997)
174. Niquet, Y.M., Allan, G., Delerue, C., Lannoo, M.: Quantum confinement in germanium nanocrystals. Appl. Phys. Lett. **77**, 1182 (2000)
175. Kan, E.W.H., Chim, W.K., Lee, C.H., Choi, W.K., Ng, T.H.: Clarifying the origin of near-infrared electroluminescence peaks for nanocrystalline germanium in metal-insulator-silicon structures. Appl. Phys. Lett. **85**, 2349 (2004)
176. Wilcoxon, J.P., Provencio, P.P., Samara, G.A.: Synthesis and optical properties of colloidal germanium nanocrystals. Phys. Rev. B **64**, 035417 (2001)
177. Nakamura, Y., Watanabe, K., Fukuzawa, Y., Ichikawa, M.: Observation of the quantum-confinement effect in individual Ge nanocrystals on oxidized Si substrates using scanning tunneling spectroscopy. Appl. Phys. Lett. **87**, 133119 (2005)
178. Oliver, W.C., Pharr, G.M.: Measurement of hardness and elastic modulus by instrumented indentation: advances in understanding and refinements to methodology. J. Mater. Res. **19**, 3 (2004)
179. Cheng, Y.-T., Cheng, C.-M.: Relationships between hardness, elastic modulus, and the work of indentation. Appl. Phys. Lett. **73**, 614 (1998)
180. Fisher-Cripps, A.C.: Nanoindentation. Mechanical Engineering Series. Springer, New York (2004)
181. Ruffell, S., Bradby, J.E., Williams, J.S.: High pressure crystalline phase formation during nanoindentation: amorphous versus crystalline silicon. Appl. Phys. Lett. **89**, 091919 (2006)
182. Taylor, C., Marega, E., Stach, E., Salamo, G., Hussey, L., Munoz, M., Malshe, A.: Directed self-assembly of quantum structures by nanomechanical stamping using probe tips. Nanotechnology **19**, 015301 (2008)
183. Rastelli, A., Stoffel, M., Tersoff, J., Kar, G.S., Schmidt, O.G.: Kinetic evolution and equilibrium morphology of strained islands. Phys. Rev. Lett. **95**, 026103 (2005)
184. Kitayama, D., Yoichi, T., Suda, Y.: Artificially positioned multiply-stacked Ge dot array. Thin Solid Films **508**, 203 (2006)
185. Snow, E.S., Campbell, P.M., McMarr, P.J.: Fabrication of silicon nanostructures with a scanning tunneling microscope. Appl. Phys. Lett. **63**(6), 749–751 (1993). doi:10.1063/1.109924
186. Snow, E.S., Campbell, P.M.: Fabrication of Si nanostructures with an atomic force microscope. Appl. Phys. Lett. **64**(15), 1932 (1994)
187. Kohmoto, S., Nakamura, H., Ishikawa, T., Asakawa, K.: Site-controlled self-organization of individual InAs quantum dots by scanning tunneling probe-assisted nanolithography. Appl. Phys. Lett. **75**(22), 3488–3490 (1999)
188. Szkutnik, P.D., Sgarlata, A., Nufris, S., Motta, N., Balzarotti, A.: Real-time scanning tunneling microscopy observation of the evolution of Ge quantum dots on nanopatterned Si (001) surfaces. Phys. Rev. B **69**(20), 201309 (2004)
189. Tersoff, J.: Step energies and roughening of strained layers. Phys. Rev. Lett. **74**(24), 4962 (1995)

190. Jesson, D.E., Kastner, M., Voigtlander, B.: Direct observation of subcritical fluctuations during the formation of strained semiconductor islands. Phys. Rev. Lett. **84**(2), 330–333 (2000)
191. Lyding, J.W., Shen, T.C.: Nanoscale patterning and oxidation of H-passivated Si(100)-2x1 surfaces with an ultrahigh vacuum. Appl. Phys. Lett. **64**(15), 2010 (1994)
192. Ruess, F.J., Oberbeck, L., Simmons, M.Y., Goh, K.E.J., Hamilton, A.R., Hallam, T., Schofield, S.R., Curson, N.J., Clark, R.G.: Toward atomic-scale device fabrication in silicon using scanning probe microscopy. Nano Lett. **4**(10), 1969–1973 (2004). doi:10.1021/nl048808v
193. Simmons, M.Y., Ruess, F.J., Goh, K.E.J., Pok, W., Hallam, T., Butcher, M.J., Reusch, T.C. G., Scappucci, G.: Atomic-scale silicon device fabrication. Int. J. Nanotechnol. **5**(2–3), 352–369 (2008)
194. Hallam, T., Reusch, T.C.G., Oberbeck, L., Curson, N.J., Simmons, M.Y.: Scanning tunneling microscope based fabrication of nano- and atomic scale dopant devices in silicon: the crucial step of hydrogen removal. J. Appl. Phys. **101**(3), 6 (2007). doi:034305 10.1063/1.2433138
195. Shen, T.C., Kline, J.S., Schenkel, T., Robinson, S.J., Ji, J.Y., Yang, C., Du, R.R., Tucker, J.R.: Nanoscale electronics based on two-dimensional dopant patterns in silicon. In J. Vac. Sci. Technol. B **22**, 3182–3185 (2004)
196. Ruess, F.J., Pok, W., Reusch, T.C.G., Butcher, M.J., Goh, K.E.J., Oberbeck, L., Scappucci, G., Hamilton, A.R., Simmons, M.Y.: Realization of atomically controlled dopant devices in silicon. Small **3**(4), 563–567 (2007). doi:10.1002/smll.200600680
197. Scappucci, G., Capellini, G., Simmons, M.Y.: Influence of encapsulation temperature on Ge: P delta-doped layers. Phys. Rev. B **80**(23), 233202 (2009)
198. Scappucci, G., Capellini, G., Lee, W.C.T., Simmons, M.Y.: Atomic-scale patterning of hydrogen terminated Ge(001) by scanning tunneling microscopy. Nanotechnology **20**(49), 495302 (2009)
199. Klesse, W.M., Scappucci, G., Capellini, G., Simmons, M.Y.: Preparation of the Ge(001) surface towards fabrication of atomic-scale germanium devices. Nanotechnology **22**(14), 145604 (2011)
200. Zhang, J.J., Stoffel, M., Rastelli, A., Schmidt, O.G., Jovanovic, V., Nanver, L.K., Bauer, G.: SiGe growth on patterned Si(001) substrates: surface evolution and evidence of modified island coarsening. Appl. Phys. Lett. **91**(17), 173115 (2007)
201. Zhang, J.J., Rastelli, A., Schmidt, O.G., Bauer, G.: Compositional evolution of SiGe islands on patterned Si (001) substrates. Appl. Phys. Lett. **97**(20), 203103 (2010)

Index

S. Bellucci (ed.), *Self-Assembly of Nanostructures: The INFN Lectures, Vol. III*, Lecture Notes in Nanoscale Science and Technology 12,
DOI 10.1007/978-1-4614-0742-3, © Springer Science+Business Media, LLC 2012